ʌn
ate

SOLAR FLARES

GEOPHYSICS AND
ASTROPHYSICS MONOGRAPHS

AN INTERNATIONAL SERIES OF FUNDAMENTAL TEXTBOOKS

VOLUME 8

SOLAR FLARES

by

ZDENĚK ŠVESTKA

American Science and Engineering, Inc., Cambridge, Mass. 02139, U.S.A.

D. REIDEL PUBLISHING COMPANY

DORDRECHT-HOLLAND / BOSTON-U.S.A.

Library of Congress Cataloging in Publication Data
Švestka, Zdeněk.
    Solar flares.

    (Geophysics and astrophysics monographs; v. 8)
    Bibliography: p.
    Includes index.
    1. Solar flares.   I.   Series.
QB526.F6588   523.7′5   75–37939
ISBN 90–277–0662–X
ISBN 90–277–0663–8 pbk.

Published by D. Reidel Publishing Company,
P.O. Box 17, Dordrecht, Holland

Sold and distributed in the U.S.A., Canada, and Mexico
by D. Reidel Publishing Company, Inc.
Lincoln Building, 160 Old Derby Street, Hingham,
Mass. 02043, U.S.A.

Printed in Belgium

To Leen de Feiter

TABLE OF CONTENTS

PREFACE

This book is the first part of the originally planned publication by Z. Švestka and L. D. de Feiter 'Solar High Energy Photon and Particle Emission'. The second part, with the original title, was to be published by de Feiter in about one year from now. However, to the deep sorrow of all of us, Dr de Feiter died suddenly and unexpectedly when the present book was in print. Thus, unfortunately, de Feiter's second part may not appear.

Due to the fact that the originally planned publication was divided into two parts, the present book is mainly descriptive and concerned with the flare morphology. It was expected that theoretical interpretations would be extensively developed in the second part, prepared by de Feiter. In particular, this refers to the theoretical backgrounds of radio emissions, particle acceleration and particle propagation in space. Only in Chapter II, concerning the 'low-temperature' flare, do we go deeper into the theoretical interpretations, anticipating that de Feiter would have been concerned mainly with the 'high-energy' physics. Still, the book includes discussions on all important aspects of flares and thus can present the reader with a complete picture of the complex flare phenomenon.

It is clear that many observed data on flares can be interpreted in different ways. I have tried to mention most of these different opinions, pointing out what is good or bad in any of them according to my subjective point of view. An alternative approach might have been to exclude all the doubtful data and interpretations and thus present a book that would be more concise and easier to read. But, on the other hand, such an approach would also greatly restrict the reader's appreciation of the controversies surrounding the flare problem. Therefore, I have decided against it.

The physics of solar flares develops very fast. If one compares the last book on solar flares, twelve years old (by Henry J. and Elske v. P. Smith), one finds that there are very few sections which might be taken over without any substantial change. Thus it is essentially impossible to present the reader with a book that would be up-to-date for many years. The final manuscript of this book was finished in May 1975 including, as I hope, all important literature of 1974 and a few 1975 publications available to the author.

At the end I would like to express my most sincere thanks to two colleagues, who contributed to this book in an outstanding way: the late Dr L. D. de Feiter of Utrecht, with whom I had the privilege to discuss in detail how the subject of solar flares should be approached in our closely linked books; and to Dr D. M. Rust of American Science and Engineering who not only made from my manuscript a 'book in English', but who also insisted on interpretations different from mine and thus forced me to rewrite some sections in a way that certainly improved the book.

ZDENĚK ŠVESTKA

Cambridge, Massachusetts, 31 October, 1975

ACKNOWLEDGEMENTS

Parts of this manuscript were prepared during author's stay at the Space Science Department of ESTEC in Noordwijk, Netherlands, and at the Fraunhofer Institute, Freiburg, Germany.

The author is greatly obliged to Drs David Rust and Michael Shaw who carefully read the manuscript and contributed to its improvement by correcting the English and by drawing the author's attention to many items which were not properly explained in the original draft. The author's thanks are also due to Drs M. A. Altschuler, A. Bruzek, L. D. de Feiter, A. D. Fokker, S. W. Kahler and J. Vorpahl for important comments on several parts of the manuscript.

It is the author's pleasure to acknowledge generous cooperation of many scientists and institutions who kindly provided him with originals or glossy prints of the illustrations used in this book: Drs J. M. Beckers, G. E. Brueckner, A. Bruzek, J. P. Castelli, E. L. Chupp, H. J. Crawford, D. L. Croom, D. Datlowe, H. L. DeMastus, W. F. Dietrich, M. Dizer, H. W. Dodson, J. W. Evans, L. Golub, D. A. Guidice, R. T. Hansen, E. R. Hedeman, D. M. Horan, C. L. Hyder, T. J. Janssens, S. W. Kahler, S. R. Kane, A. S. Krieger, L. J. Lanzerotti, R. P. Lin, P. McIntosh, R. W. Milkey, R. W. Noyes, R. Petrasso, M. Pick, R. Ramaty, J. R. Roy, D. M. Rust, J. A. Simson, H. Tanaka, G. S. Vaiana, B. Valníček, K. P. White III, G. Wibberenz, K. G. Widing, and J. P. Wild, American Science and Engineering, Cambridge, Massachusetts; Astronomical Institute of the Czechoslovak Academy of Sciences, Ondřejov; CSIRO Radiophysics Division, Sydney; Fraunhofer Institute, Freiburg; High Altitude Observatory, NCAR, Boulder; Sacramento Peak Observatory, New Mexico; and San Fernando Observatory of the Aerospace Corporation, California.

The author also acknowledges the kind permission to reproduce illustrations published in the *Astrophysical Journal*, University of Chicago Press, Copyright by the American Astronomical Society (Figs. 24a, 40, 54a, 54c, 59, 63, 103, 104, 109, 115, 116), *Bulletin of the Astronomical Institutes of Czechoslovakia*, Publishing House of the Czechoslovak Academy of Sciences (Figs. 18, 21, 31, 101b, 122), *Journal of Geophysical Research*, Copyright by the American Geophysical Union (Figs. 58, 65, 90, 101a), as well as in *Solar Physics* and other publications of D. Reidel Publishing Company.

The author also wishes to thank the Drawing Office of the ESTEC Space Science Department headed by Mr A. Schepers, Miss W. Gartemann of the Photolaboratory of the Fraunhofer Institute at Freiburg, and the Technical Publications Office (Mr P. Maggio) and Photolaboratory (Mr R. J. Haggerty) of American Science and Engineering for careful preparation of the figures. The author is especially grateful to Miss M. Grady for final typing of the manuscript.

CHAPTER I

INTRODUCTION

Solar flares emit electromagnetic radiation within a very broad range of wavelengths; in extreme cases from 0.002 Å (2×10^{-11} cm, 6.1 MeV) up to more than 10 km (10^6 cm, 30 kHz). A schematic summary of all these kinds of radiation is given in Figure 1, and examples of the flare development curves at different wavelengths are shown in Figure 2. This radiation may be continuous emission in some parts of the spectrum, line emission in others, or a combination of both.

The different kinds of radiation come, of course, from different heights above the solar photosphere and thus from different parts of the flare, as is indicated schematically in the upper part of Figure 1. The extremely short waves (γ- and hard X-rays) and impulsive enhancements of radiation at longer wavelengths are produced by streams of accelerated particles and thus they come from the depth where these particles interact with the solar atmosphere. In the shortwave region (optical, UV, and soft X-rays) the height of the thermal radiation is determined by the flare's temperature distribution with height and, in strong spectral lines and continua, by the optical thickness of the flaring chromosphere. In the radio range, the height dependence is determined largely by the electron density distribution with height. Because our knowledge of the temperature and electron density height distribution in flare regions is unsatisfactory, the heights indicated in Figure 1 must be considered only approximate.

As in the undisturbed solar atmosphere, temperature in the flare region must rise steeply between its chromospheric and coronal parts. While analysis of the chromospheric lines in flare spectra implies that the electron temperature is close to 10^4 K, X-rays at a height of about 20 000 km give temperatures exceeding 10^7 K. Since the low- and high-temperature parts of a flare differ greatly in their physical conditions as well as in the methods of their observation, it is reasonable to discuss them separately. However, the flare structure is highly inhomogeneous and we meet with a great variety of physical conditions even at the same altitude. The observations clearly show that in some flare regions the 'low-temperature flare' extends up to 15 000 km or more above the photosphere, while in other areas the 'high-temperature' part penetrates into the chromosphere. Therefore, there is no uniquely defined height that would separate these two flare parts, and for example, when observing at the limb, one can see one or the other part of the flare at the same height, depending on the method of observation.

Individual solar flares differ greatly in size and importance. What is common to all of them is a rapid temporary heating of a restricted part of the solar corona and

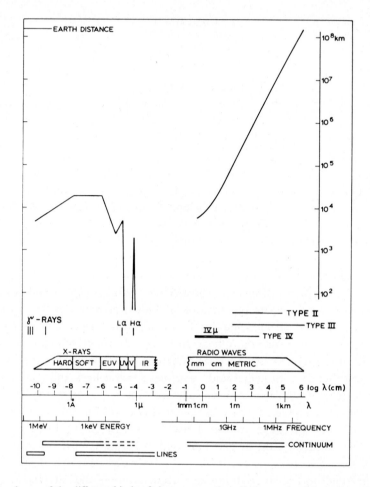

Fig. 1.   A scheme of the different kinds of electromagnetic radiation emitted from flares. The upper part of the figure indicates the estimated heights (above the base of the chromosphere) from which the different kinds of radiation come. Below, Lα and Hα denote the Lyman-α and Balmer-α lines of hydrogen, EUV = extreme ultraviolet (UV), V = visual range, IR = infrared. The wavelength ranges, in which different types of radio bursts are observed, are indicated at the right-hand side of the figure. Scales at the bottom are in wavelengths (log λ) and below it in frequencies for the radio waves and in energies for the X- and γ-rays.

chromosphere (thermal part T in Figure 2). Depending upon the magnetic configuration, non-thermal processes may also occur at some places inside the flaring volume giving rise to accelerated electrons and atomic nuclei. The accelerated electrons manifest themselves through impulsive non-thermal X-ray and radio bursts which occur during the rising phase of the flare development (part NT in Figure 2). In some flares the non-thermal component is dominant. However, there are many flares where only the thermal effects are observed, without any indication that particles in them have been accelerated to suprathermal energies.

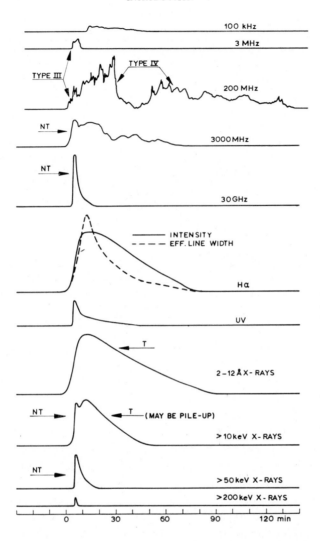

Fig. 2.   Typical curves of the flare time development at different wavelengths: 3 km (100kHz)
100 m (3 MHz), 1.5 m (200 MHz), 10 cm (3000 MHz), 1 cm (30 GHz), 6563 Å (Hα), 300–1500 Å
(UV), 2–12 Å, <1.2 Å (>10 keV), <0.25 Å (>50 keV), and <0.06 Å (>200 keV). *NT* indicates a
non-thermal and *T* a quasi-thermal component of a burst. The lowest frequency is recorded in a
radio type III, and the highest one in a hard X-ray burst. In exceptional cases, a γ-ray burst at still
higher frequencies can be observed, with time development probably similar to the hard X-ray burst,
but slightly delayed (cf. Figure 118).

Figure 3 shows in a schematic way the energy range of protons and electrons that
are produced in flares and either can be observed directly in space or deduced from
effects they produce on the Sun. The curve shows the maxwellian distribution of
thermal particles which exists in each flare at temperatures between several million
degrees and more than 10^7 deg (here $T = 10^7$ K has been chosen as an example).

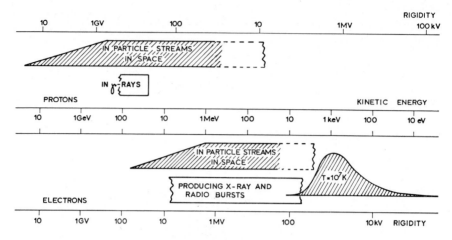

Fig. 3.   A scheme of the energy ranges of the particle emission from flares. The curve represents the Maxwellian distribution of energies for temperature of 10^7 K. The stripes indicate energy ranges in which the flare-associated particles have been directly or indirectly observed. Quite recently the direct measurements of both protons and electrons have been extended to still lower energies as indicated with the dashed lines.

When non-thermal impulsive acceleration occurs, a fraction of the particles of this thermal assembly are accelerated to much higher energies. We observe impulsive X-ray and radio bursts which tell us of the existence of non-relativistic and mildly relativistic electrons in the flare region. Some of the non-relativistic electrons may also be detected in space, but as long as they are not accompanied by protons, we in fact have no evidence that protons too were accelerated in the flare region at the same time. When protons are detected, we obviously meet with a strong acceleration process on the Sun (sometimes the term 'proton flare' is used in such an event) and then usually relativistic electrons accompany them in space. In most proton events only energies of the order of 10 MeV or less are recorded, but occasionally a flare occurs which gives rise to protons with energies exceeding 100 MeV and, in the most outstanding events, protons with energies close to or exceeding 1 GeV produce an increase in the neutron flux at sea-level on the Earth. These very outstanding flare events are some-times called 'cosmic-ray flares'.

Important flares are also sources of wave motions, which are the probable source of some types of radio bursts in flares. The flare waves can propagate through space up to the Earth distance, producing shocks which are observable when encountered in situ by spacecraft or when hitting the Earth's magnetosheath. Blast waves propa-gating along the solar surface manifest themselves by disturbing distant existing plasma configurations in the solar atmosphere (quiet filaments and prominences, the spicule pattern), and on rare occasions they can be observed in the Hα line as dark expanding features on the solar disc. Finally, solar flares also are associated with fast mass motions that can be seen as streams of outflowing or inflowing material (surges), or ejections of matter with velocities exceeding the escape velocity on the Sun (sprays).

In order to make our discussion simpler and more understandable, we shall discuss the low-temperature flare, the high-temperature flare, and the various flare-associated phenomena separately in the different coming chapters. Only after finishing these separate discussions shall we try to combine all the results and present a synthesized picture of the complex flare phenomenon as a whole.

THE LOW-TEMPERATURE FLARE

Flares obviously do not affect the photosphere in any significant way, because apart from a few exceptional cases, they cannot be observed in integrated white light. From about 1000 Å through the visible part of the spectrum the flare brightening

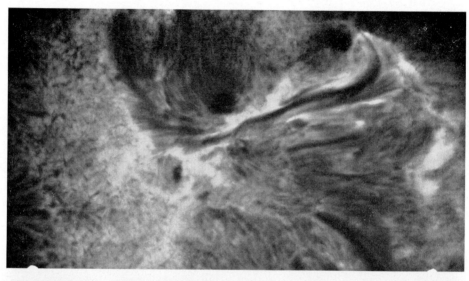

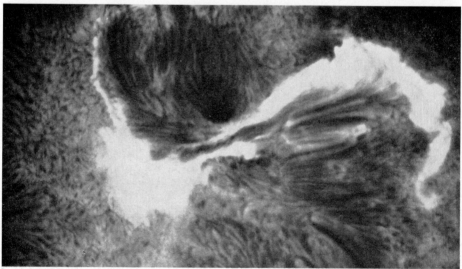

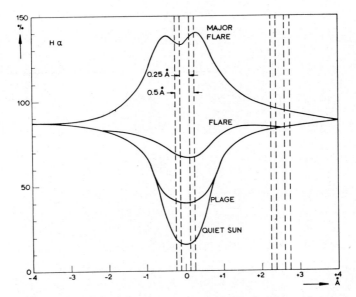

Fig. 5. The Hα line profile in the quiet Sun, a plage, and in small and large flares, respectively. The pass-bands of 0.25 Å and 0.50 Å (the most widely used) are indicated by the dashed vertical lines, in a position centered at the Hα line and in position +2.5 Å.

appears only in individual spectral lines, and it is most outstanding in those formed high in the chromosphere.

In the last few years flare observations have been extended to the EUV spectral range (below ~1500 Å), where the lines are produced throughout the transition region, from the Lyman lines of hydrogen still close to 10 000 K up to highly ionized iron lines excited at temperatures of 10^6 K. This intermediate region as a whole will be discussed in Chapter III, but some of the results will be mentioned in this chapter since they have immediate relevance to 'low-temperature flare' problems.

In the visible part of the spectrum, the hydrogen Hα and singly ionized calcium H and K lines are most strongly affected by the flare. Since the H and K lines are near the limit of the visible spectrum, most optical data on solar flares have been obtained in the light of the Hα line by means of instruments that allow the selection of only a narrow (~0.5 Å) pass-band from the solar spectrum at the Hα wavelength. Therefore, we shall start our discussion with a description of flares as seen in Hα.

2.1. Flare Observations in the Hα Line

When the Sun is observed in the Hα line, active regions appear as bright chromospheric plages surrounding and overlying the sunspots which characterize the active region in the photosphere (Figure 4). A flare appears either as a brightening of a part

Fig. 4. An example of a plage (above) and of a flare (below) which developed in it. Note that most of the bright flare areas were already enhanced in the plage prior to the flare occurrence. (Courtesy of D. Rust, Sacramento Peak Observatory.)

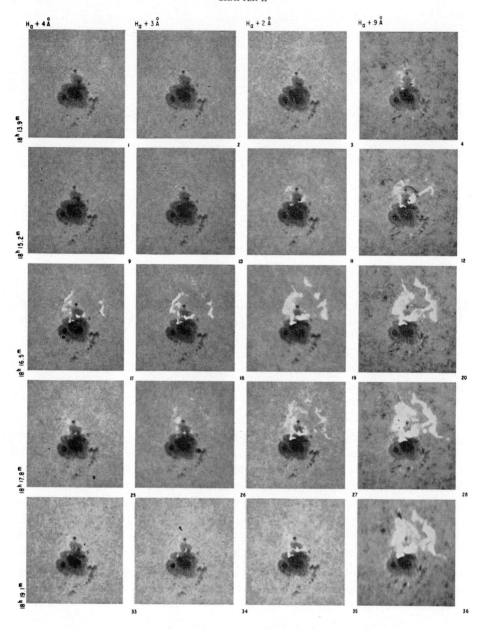

Fig. 6. Set of photographs of the flare of 1968, August 8 (importance 2b) taken at different distances from the center of the Hα line, from +4 to −4 Å. In pictures of this kind one can see the structure of the flare brightenings which is lost in photographs taken in the line center. When comparing the pictures taken in the red (+) and blue (−) wings of the line one should be aware of the time sequence going from left to right (the times of the +4 and −4 Å pictures are given at the ends of each row). Thus the existence of the more pronounced emission on the red wing (a red asymmetry of

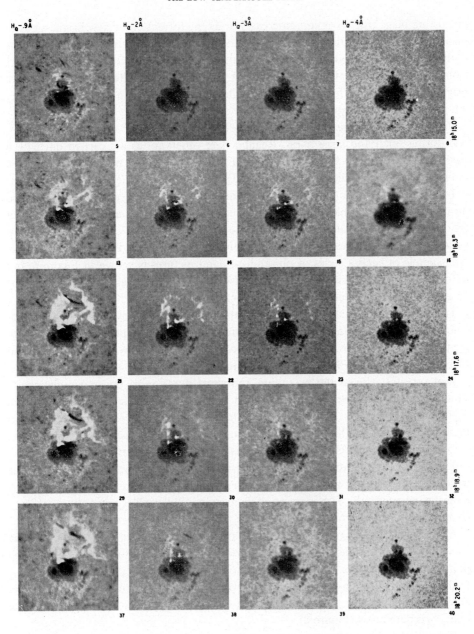

the line) is obviously real, while the blue asymmetry in the second row may be an apparent effect due to this time sequence.

Note how the flare knots avoid the spot umbra.

Descending absorption material is seen in frame 12, while rising absorption material is visible in frames 21, 29 and 37. (After Janssens and White, 1970.)

of existing plage (the most common case) or as the formation of new bright areas in places where the plage did not exist before (see example in Figure 4). In regions close to the peak of their activity, brightness variations in the plage occur almost continually (some observers say that the region 'boils'). The characteristic feature of actual flare brightenings is a very fast increase in brightness followed by a much slower decay (cf. the Hα curves in Figures 2, 7, and 8).

Figure 5 shows the profile of the Hα line as observed on the quiet Sun, in plages, and during small and large flares. Regular flare observations are carried out with a pass-band of 0.25 to 0.5 Å in the center of the Hα line. Thus we observe the most intense and optically thick portion of the flare in the core of the Hα line. This yields only limited information about the structure of the flare and about the associated moving phenomena which, due to the Doppler shift, emit radiation at the wings of the line. More information can be obtained if the pass-band is moved and observations carried out at various distances from the line center in the wings of the line. An example of such a series of on- and off-band pictures is shown in Figure 6, but unfortunately, only a few observatories are able to obtain off-band pictures, and they do not carry out this type of observation on a regular basis.

To some extent, similar measurements can be made visually by means of a spectrohelioscope with a moving line-shifter. Figure 7 shows an example of a flare development curve measured with a spectrohelioscope. The effective widths of the red and blue wings of the flare Hα emission line are plotted here against time. It shows the typical very fast rise to the maximum and short duration of the maximum phase. This part of the flare development is often called the 'flash phase' and as shown in Figure 2 this appears to be the most important time interval of the flare development because essentially all the energetic flare processes are accomplished during this time.

One should not interchange the term 'flash phase' with the term 'explosive phase', often used in the literature. The flash phase means a rapid increase of the Hα line width and intensity in the brightest region of the flare, whereas the explosive phase indicates an unusually fast increase of flare area. Whilst all 'typical' flares show a flash phase, an occurrence of the explosive phase is a relatively rare phenomenon.

As one can see in Figure 8, individual flare knots may develop differently and the curve in Figure 7 is in fact only an envelope obtained by always following the brightest point in the flare region. This is also the reason for some of the secondary brightenings, in the curve: they often represent maxima of secondary flare knots. Therefore, a set of photographs is always better since one can see the detailed development of the individual flare areas (Figure 8).

Figure 7 also clearly illustrates that the Hα line in flares is asymmetrical, with a very pronounced 'red' asymmetry in its maximum phase. The same effect can be seen in Figure 6. We shall say more about this effect when discussing the flare spectra (Section 2.2.9), since this is a common phenomenon observed in other spectral lines as well but which is still puzzling and unexplained.

So far we have been discussing flares observed on the solar disc. As a limb phenomenon, a typical small flare appears as a bright starlike point in the Hα chromosphere

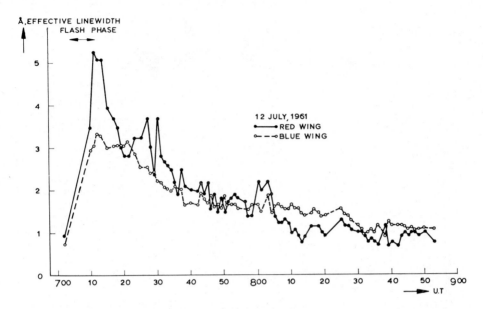

Fig. 7.   The time variation of the Hα effective line width in the flare of 1961, July 12, as measured by the spectrohelioscope at Ondřejov. The effective line width is the distance (in Å) between two positions of the slit in Figure 5, in the red and blue wings, respectively, at which the relative brightness of the flare decreases to some 10% of the adjacent continuum (Fritzová, 1964). Note the very steep rise during the flash phase, and the pronounced red asymmetry at the maximum. (Courtesy of the Astronomical Institute of the Czechoslovak Academy of Sciences, Ondřejov.)

and this point changes rapidly into a brilliant hill, which may show temporary extensions, pulsations, and ejections. A resulting conical shape is very common. A good description of such events has been given by Severny (1964a).

Another fairly frequent type of limb flare is a loop-like form (for a more detailed description see, e.g. Smith and Smith, (1963)). Other flare shapes have been observed on the limb also, particularly in association with larger flares, and quite often it is fairly difficult to decide what is the flare proper and what belongs to the flare-associated prominences. Spectral analysis can resolve this (cf. Section 2.2.10), but spectra are rarely available. Whilst on the disc flare-associated prominences mostly appear as dark filaments, above the limb they are bright and therefore easily mistaken for the flare proper. Two examples of limb flares are reproduced in Figure 9.

Observations at the limb may be used to obtain an estimate of the height of the low-temperature flare in the solar atmosphere. When one summarizes the studies made by Giovanelli and McCabe (1958) and Warwick and Wood (1959), the average height of flares on the limb is found close to 7000 km above the base of the chromosphere. Some 30% of them are so low that one does not see them on the limb at all (since the chromosphere is essentially opaque on the limb below some 3000 km), but on the other hand about 25% of flares have heights exceeding 10 000 km, and a few low-temperature flares going up to 30 000 to 50 000 km have been observed as well. The associated prominences go still higher.

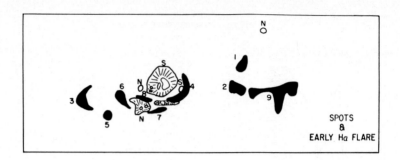

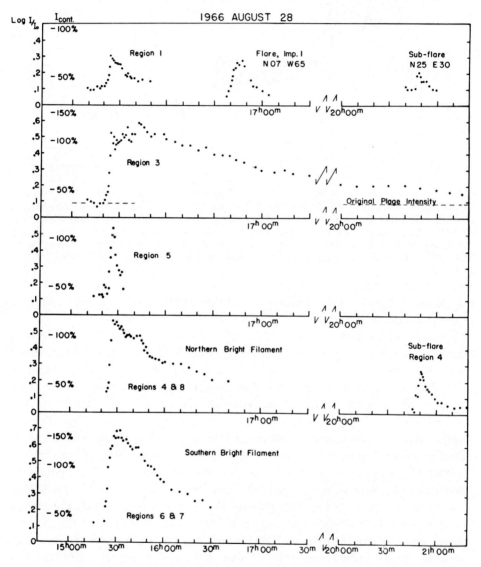

Fig. 8. Photometric light curves for specific regions (as numbered on the top) of the flare of 1966, August 28. I_{cont} gives the intensity of the Hα flare in percentage of the intensity of the continuous spectrum near Hα at the center of the solar disc. (After Dodson and Hedeman, 1968a.)

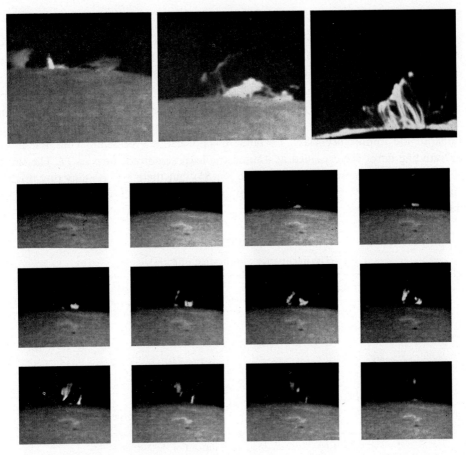

Fig. 9. Examples of two limb flares. The upper three photographs show three aspects of the cosmic-ray limb flare of 1961, July 20, the last one showing the loop-prominence system late in the flare development. The other photos illustrate the development of an importance 2 limb flare of 1951, May 8. UT in the first row: $14^h 33^m 29^s$, $15^h 00^m 10^s$, $03^m 30^s$, $05^m 11^s$; in the second row: $15^h 05^m 43^s$, $06^m 18^s$, $06^m 52^s$, $07^m 25^s$; in the last row: $15^h 09^m 58^s$, $11^m 54^s$, $13^m 02^s$, and $15^m 26^s$. (Courtesy of H. W. Dodson, McMath-Hulbert Observatory, Michigan.)

The net $H\alpha$ emission in the maximum phase of a large solar flare (area of 25 square degrees) is of the order of 10^{26} erg s^{-1}. For comparison, the energy emitted by the same area in the whole visible spectral region is about 4×10^{29} erg s^{-1}. Thus, as the $H\alpha$ emission is the strongest in the visible part of the flare spectrum, a comparison of these two quantities clearly shows that even the largest flares cannot be observed in integral white light unless they emit in the continuum, which is an extremely exceptional event. As we shall see later, this is not the case in the EUV, X-ray, and radio spectral regions, where the flare emission exceeds or is comparable to the amount of energy emitted by the whole solar disc.

2.1.1. THE FLARE IMPORTANCE

According to the appearance that they have in the light of the Hα line all flares are classified in four groups by their 'importance', 1, 2, 3, and 4. Their areas at the time of maximum brightness are the primary criteria for this importance classification.

A transient brightening is called a flare if its area in the chromosphere exceeds 3×10^8 km². Smaller events are called subflares and their importance is denoted by an S (Table I). The maximum brightness of the flare in the Hα line is added as a suffix to the importance class: f = faint, n = normal, b = brilliant. Thus the most outstanding flares are classified as 4*b* and the least significant ones as 1*f*. The same subdivision is also used for subflares (*Sf*, *Sn*, *Sb*), but there are so many tiny brightenings in some active regions that it is impossible to record all of them in the lists of flares regularly issued.

TABLE I

The importance classification of flares

Area at center of solar disc		Importance (Class)[a]	
in millionths of hemisphere	in square degrees[b]	before	after
		January 1st, 1966[c]	
<100	<2.06	1−	S
100–250	2.06–5.15	1	1
250–600	5.15–12.4	2	2
600–1200	12.4–24.7	3	3
>1200	>24.7	3+	4

[a] The term 'Class' is sometimes used instead of 'Importance'.
[b] 1 square degree = 1.476×10^8 km² of solar surface.
[c] Before 1st January, 1966, there were only three importance classes based upon the flare area without any direct reference to the flare brightness. Intermediate flares with regard to area were denoted as 1+ and 2+ and the same notation also applied to unusually bright flares of importance 1 or 2, respectively. Extraordinarily large flares were classified as 3+ and the subflares as 1−. Some observatories also used notations 2− and 3− without making clear the meaning. Probably the new classification provides better information on both the size and brightness of flares, since these two characteristics were hopelessly mixed in the classification used before.

This fact was illustrated quite well during the period of the CINOF project in June 1972 (cf. de Jager (1975) and references therein). The CINOF project was aimed at observing small flares and subflares and therefore the observing stations were asked to report all the tiny flare brightenings they could see. As a consequence of it the number of reported subflares increased enormously; for example, in the active region McMath number 11 911 the reports gave 124 flare brightenings, 120 of them being subflares (Švestka *et al.*, 1974).

The lower limit in size for subflares is obviously far below the resolving power of solar patrol telescopes. Hyder *et al.* (1973) have reported an Hα flare point brightening of about 1 arc sec (~ 700 km) in diameter, with a lifetime of 20 s, still apparently associated with a soft X-ray and possibly even a type III radio burst. Petrasso *et al.* (1975) report a soft X-ray brightening (observed on Skylab, cf. Figure 11) with typical X-ray flare characteristics which the Hα patrol did not classify even as a subflare. It is also known that some 15% of the coronal X-ray bright points (Golub *et al.*, 1975), which probably can be identified with the ephemeral active regions (Harvey and Martin 1973; Harvey *et al.*, 1975), produce flare-like brightenings (Figure 10) that are not reported (and almost never noticed) by the Hα observers. About 30% of EUV bursts (Section 3.2.2) and more than 70% of type III bursts (Section 3.2.6.A) have no reported Hα counterpart.

Thus all lists of subflares are incomplete and also greatly inhomogeneous, because different solar observatories set different lower limits to their reports. The most complete lists of flares can be found in the *Quarterly Bulletin on Solar Activity* (published by the Eidgenössische Sternwarte Zürich) and in *Solar and Geophysical Data* (published by NOAA Boulder).

The duration of flares in Hα light varies from a few minutes for reported subflares up to more than 7 h for the most powerful flares of importance 4. According to H. J. Smith (1962a) the mean duration is 25 min for flares of importance 1, 55 min for flares of importance 2, and slightly more than 2 h for flares of importance 3 or 4.

However, as one can see from Figure 2, the life time of the flare in Hα light may be quite different from the flare duration at other wavelengths. Thermal X-ray and gradual microwave bursts have durations similar to the Hα flare. Impulsive bursts on the other hand, in the hard X-ray, EUV and millimeter radio regions last for only tens of seconds in small flares, and for a few minutes in more important events. Type III radio bursts at metric wavelengths have similar durations. However, long-lived continuum bursts at metric wavelengths can last for many hours after disappearance of the chromospheric flare in Hα light.

This demonstrates that the Hα emission is only one aspect of the complex flare phenomenon and it does not necessarily reflect the flare behaviour at other wavelengths. It is a fairly good flare characteristic when considered statistically because generally flares of large Hα importance also produce important X-ray, radio, and particle events. But this need not be true if individual cases are considered. There are many Hα subflares, for example, which are associated with quite intense X-ray and radio phenomena, and even with ejection of non-relativistic electrons into space. On the other hand, we know of many flares of importance 2 and even higher which in fact have been quite insignificant events when other kinds of radiation or the particles are considered. It is obviously not the Hα flare size, but probably rather the location of the flaring patches with respect to the magnetic field distribution in the active region which determines the actual power of the energy conversion processes.

Efforts therefore have been made to establish other flare classifications which would take more aspects of the flare phenomenon into account. As an example we will mention here the proposal made by Dodson and Hedeman (1971). They introduce a Comprehensive Flare Index (CFI) = $A + B + C + D + E$, where

A = importance of ionizing radiation as indicated by time-associated SID, on the scale 1–3;

B = importance of Hα flare, on the scale 1–3 (3 and 4 being classified as 3, subflare as 0);

C = log of 10 cm radio flux, in units of 10^{-22} Wm^{-2} Hz^{-1};

D = effects in dynamic radio spectrum: Type II = 1, continuum = 2, Type IV = 3;

E = log of 200 MHz flux in the same units as C.

The values of CFI then range from 0 for subflares without any significant ionizing or radio emission to 15 to 17 for flares that were the most outstanding in all aspects of electromagnetic flare radiation. Flares with CFI > 10 can be considered as verified major flares.

During the years 1956–1969, 20 flares had CFI ≥ 15. All but two of them were associated with strong proton events (polar cap absorption) and 4 of them were cosmic-ray flares. Also all but two were of Hα importance 3. This shows that the most important flares are also important in Hα light. But the opposite is not true. Out of a total of 164 flares of importance 3 during 1955–1969, 18 had CFI ≥ 15, but 86 of them had CFI < 10. Out of the 210 flares that had CFI ≥ 10, and hence can be consid-

TABLE II

Annual numbers of major flares in the years 1955–1969

Year	Annual number of						Yearly relative sunspot number[b]
	Cosmic-ray flares	Flares associated with a PCA[a]	Flares with a PCA >3 db	Flares with CFI ≥ 15	Flares with CFI ≥ 10	Flares with importance ≥ 3	
1955	0	1 (4)	0	0	1	5	38
1956	2	6 (10)	4	0	6	23	142
1957	0	22 (39)	9	3	29	30	190
1958	0	15 (21)	7	5	28	21	185
1959	1	13 (14)	4	2	29	29	159
1960	5	32 (49)	11	4	36	19	112
1961	2	11 (20)	5	2	12	9	54
1962	0	1 (5)	0	0	4	2	38
1963	0	8 (14)	2	1	6	3	28
1964	0	1 (1)	0	0	1	0	10
1965	0	3 (5)	0	0	0	0	15
1966	1	4 (5)	2	2	16	6	47
1967	1	6 (24)	2	2	8	8	94
1968	2	14 (21)	4	1	13	4	106
1969	2	13 (21)	0	0	22	5	106

[a] The number in brackets gives the total number of PCA's recorded during the year. Some of them were produced by flares on the invisible solar hemisphere; some might have been delayed events associated with a ssc, and some may represent misinterpretation of the ionospheric data.
[b] Yearly means taken from Waldmeier (1956–1970).

ered as major events, 78 were of importance 3, 106 of importance 2, 23 of importance 1, and 3 were subflares.

In order to show the frequency of occurrence of large solar flares, Table II summarizes the annual numbers of flares that can be considered, from some point of view, as outstanding events. One can see from it that the very large 'catastrophic' events which produce a cosmic-ray increase at the Earth, or are characterized by $CFI \geqslant 15$, are very rare phenomena. There are many years when no such event occurs at all, and even in the very active years the annual number of such flares is only 2 to 3 on average. The number of major flares, characterized by $CFI \geqslant 10$ or by an associated polar cap absorption (PCA) event was about 20 to 30 yr^{-1} in the years of the maximum of cycle 19, and about 10 to 15 during the maximum of the last cycle No. 20. On the other hand, the total number of all flares reported during that period was many thousands, and thousands of other events were classified as subflares. This clearly shows that whilst solar flares are very common on the Sun, a really large flare is a fairly rare phenomenon. (These statistics have been based on papers by Švestka and Olmr (1966), Švestka (1966b), Fritzová and Švestka (1966), Hultqvist (1969), Dodson and Hedeman (1971), and the *Catalog of Solar Particle Events* edited by Švestka and Simon (1975).)

Flare occurrence rate roughly follows the relative sunspot number; nevertheless there are some bursts of activity which deviate from this simple rule, as one can see in Table II. In the preceding solar cycle, the most flare-abundant year was 1960, three years after the maximum, and a burst of flare activity occurred even in 1963, one year

TABLE III

Regions of outstanding activity formed late in the declining phase of the cycles 17–20

Year	CMP date	Latitude	Years after cycle maximum	Number of PCA's	Number of GLE's	Remark
1972	Aug. 04.5	13 N	3.7 yr	3	1	(1)
1963	Sept. 20.5	13 N	5.9 yr	5	0	(5)
1961	July 14.1	06 S	3.9 yr	5	2	(5)
1960	Nov. 12.7	27 N	3.1 yr	7	3	(5)
1951	June, 2 regions		4.0 yr	2	0	(2), (3)
1950	Feb. 20.0	10 N	2.7 yr	2	0	(2)
1942	Feb. 28.8	07 N	4.8 yr	2	2	(2), (4)
1940	March 26.2	12 N	2.9 yr	2	?	(2)

Remarks:

(1) Lincoln and Leighton (1972).
(2) Indirect PCA observations, hence only the strongest events are known (Švestka, 1966b; Levitsky, 1970).
(3) One very strong, or two PCA events, but the source of the particles on the Sun is ambiguous.
(4) The first two GLE's and the first solar radio noise ever recorded were produced in this region.
(5) *Catalog of Solar Particle Events* (Švestka and Simon, 1975).

before the solar minimum. Similarly, the famous flare activity in August 1972 (Lincoln and Leighton, 1972; Coffey, 1973) occurred 3.7 yr after the maximum of the last solar cycle.

Such late activity bursts appear to be a common characteristic of the Sun (Dodson and Hedeman, 1973; Fritzová and Švestka, 1973), and Table III gives their occurrence in the last four solar cycles. It seems that the solar cycle is composed of several such activity bursts and even the last of them may produce very powerful flares.

2.1.2. Flares in relation to magnetic fields

Although the mapping of solar magnetic fields invented by Babcock in 1953 has developed rapidly during the past two decades, our knowledge of the magnetic field configuration and its time variations is still unsatisfactory from the point of view of studies of flares. In principle we meet with two main difficulties: (1) the details of the magnetic field structure are most probably much smaller than the resolving power of the solar magnetographs (see, e.g., Stenflo, 1973); and (2) the magnetic measurements so far are reliable only in photospheric lines so that we know the distribution of the magnetic field fairly well in the photosphere, but poorly in the chromosphere and corona, where the flares occur.

From the measurements of the longitudinal magnetic field in the photosphere one can compute theoretical three-dimensional maps of the magnetic field in the corona under the assumption that the field is potential (Schmidt, 1964; Altschuler and Newkirk, 1969). Whilst these computations may represent pretty well the space configuration of the large-scale coronal magnetic field and even some post-flare situations in active regions (cf. Section 4.4), they do not give correct information on the magnetic configuration in developing active regions where the existence of strong electric currents makes the computation of the space configuration unreliable. A modification of these computations for force-free fields, where the electric current and the magnetic field are everywhere parallel and have a constant ratio of magnitudes (Nakagawa and Raadu, 1972), can be an improvement in some special cases (e.g. Rust *et al.*, 1974), but generally it can be again quite far from the actual situation in a developing and magnetically complex active region.

Qualitative information on the coronal magnetic field structure can be obtained from the Skylab observations in soft X-rays and in EUV lines (provided that the coronal loops follow the magnetic field lines). However, in relation to flares, the information so far published is scarce and it only confirms, more or less, what already has been known about flare relations to the photospheric magnetic fields. Thus most of our conclusions on the relationship of flares to solar magnetic fields are still based upon poorly resolved photospheric magnetic data.

X-ray photographs on Skylab (Figure 10) have shown that flare-like brightenings (flaring bright points) occur as soon as tiny bipolar magnetic structures emerge in the solar atmosphere (Golub *et al.*, 1974). Petrasso *et al.* (1975) have observed an X-ray flare (Figure 11) which occurred in a bipolar active region only 1 day old. Also Hα flares have been observed many-a-time in newly born active regions prior to the

appearance of the first recognizable sunspots. All this indicates that flares begin
to occur as soon as a bipolar magnetic configuration is formed in the solar atmosphere.

However, the frequency and importance of flares generally increase with the rate
of change of the magnetic field and with its increasing complexity (Giovanelli, 1939;
Kleczek, 1953; Bell and Glaser, 1959; Künzel, 1960). There are many small or short-
lived active regions which may not produce any significant flare or any flare at all

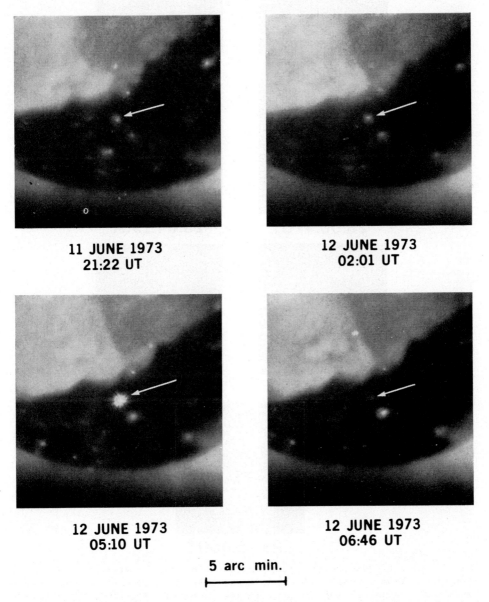

Fig. 10.   An example of flare-like brightening in an X-ray bright point (arrow). Also note how fast
are the changes in the bright points pattern. (Courtesy of L. Golub, AS & E.)

SEPT. 1, 1973

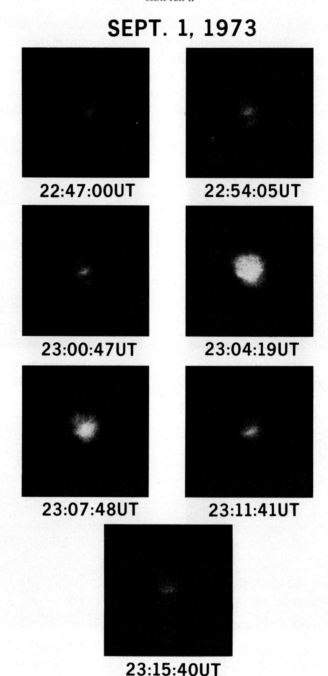

Fig. 11. A flare-like brightening in a young (about one-day-old) active region as observed in soft X-rays by the AS & E telescope on Skylab. Note that the loop in which the flare occurred was pre-heated some 10 minutes before the flare onset. The brightened point is not in the center of the loop, indicating that the flare began near the western foot and not at the top of the loop. (Courtesy of R. Petrasso, AS & E.)

during their whole lifetime; on the other hand, the number of flares in very large and highly developed active regions may exceed 100 events, several of them being flares of outstanding importance. Generally, frequency of flare formation is highest shortly before maximum development of the sunspot group.

The problem of whether flares can occur in a simple bipolar magnetic configuration or whether they necessarily need a magnetically complex structure is of essential importance for their theoretical interpretation. As we shall see in the coming sections, there is no evidence on deviations from bipolar configurations for many flaring regions, in particular if the flares in them do not show any non-thermal impulsive components. The fact that the occurrence of non-thermal acceleration processes (impulsive bursts) greatly increases in magnetically complex active regions (Švestka and Simon, 1969; Krüger, 1969b; Švestka, 1971a; de Jager, 1975) supports the conclusion that the occurrence of quasi-thermal flares in simple bipolar fields is real. Nevertheless, one has to expect that complexities of the magnetic field may be so small that they are below the resolving power of our magnetographs. Thus certainly at least some 'bipolar' fields are actually complex. It is also necessary to say that 'bipolar' need not resemble a simple dipole field. The bipolar fields can be twisted or sheared.

A. *Flare Positions Relative to Sunspots and Plages*

Until 1958, when the pioneer work on the relationship between flares and magnetic fields was started by Severny's group at the Crimean Astrophysical Observatory, our knowledge was limited to the position of flares with respect to the sunspots and plages. In more recent years magnetograms have given us additional knowledge and physical insight into the relationship between flares and magnetic fields. Nevertheless, even today much information is still based on sunspot and plage observations which represent a more complete set of data for statistical analysis. This information can be summarized as follows:

(a) All solar flares occur in active regions characterized by sunspots and plages, or (in old regions) by quiescent filaments.

(b) By far the most flares occur in active regions with developed sunspots. Flares which appear in an active region without spots are exceptional events. According to a study made by Dodson and Hedeman (1970), in the years 1956–1968 only 7% of all flares of importance $\geqslant 2$ occurred in plages with only small or no spots. Most of these flares appeared during the late, decaying phase of an active region which had been highly developed some time before. These 'spotless' flares are usually fairly isolated in the dying active region, and are characterized by a slow rise to the maximum, i.e. by a poorly developed flash phase.

(c) Flares, even when concentrated in sunspot groups, avoid big spots (cf. Figure 6). The Hα flare patches always form outside spots and even when they expand to cover a large area they prefer the spot-free space in the active region. Quite often, flare expansion is stopped at the boundary of a large spot and the Hα emission increases at this place (the brightest flare parts are often found at the edges of sunspot penum-

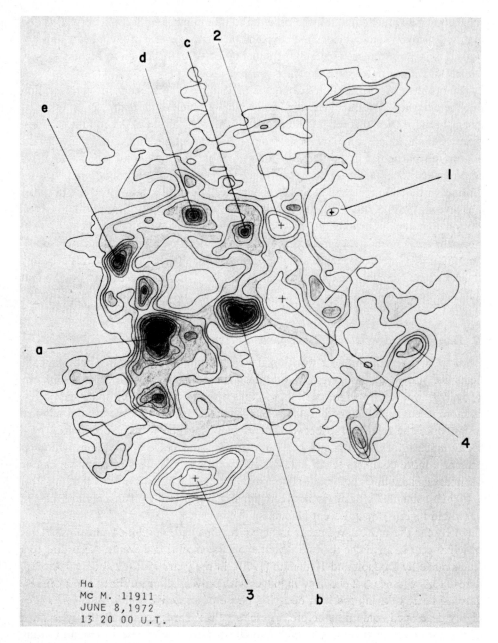

Ha
Mc M. 11911
JUNE 8,1972
13 20 00 U.T.

Fig. 12.   Hα-isophotes of the plage Mt. Wilson No. 18849 at $13^h 20^m$ UT in 1972, June 8. 1, 2, 3, and 4 are spots. *a* and *b* are two knots of a subflare that flared at that time. However, one can also clearly distinguish three other 'live points' above *a* and *b* (marked *c*, *d* and *e*) that formed another subflare about a half-hour later. These five 'live points' outline the border of one supergranular cell and mark the points where three supergranuli were in contact. (Courtesy of M. Dizer, Istanbul.)

brae (Švestka *et al.*, 1961). If a flare penetrates onto a spot, which does happen with large flares and with exceptional flares of smaller size, the flare, as a rule, is very productive in X-rays and radio microwave radiation (Dodson and Hedeman, 1960; Martres and Pick, 1962; Malville and Smith, 1963; Hagen and Neidig, 1971).

(d) Flares are most frequent in magnetically complex groups (γ or δ types of the Mt. Wilson classification)*. According to Bell and Glaser (1959) the average number of flares during the transit of a magnetically complex γ group over the solar disc is 19.1 (22.7 if the strength of the magnetic field on the Mt. Wilson visual patrol exceeds 2000 gauss), and 11.2 for complex $\beta\gamma$ groups*; whilst for magnetically simple β and α groups the average numbers of flares are only 1.9 and 0.7, respectively. The complex groups also produce the most important flares, and flares from them are the most productive in impulsive X-rays and radio bursts (Švestka and Simon, 1969; Krüger, 1969b; de Jager, 1975).

(e) Smaller flares usually consist of small, separated bright areas, which, for the most part, are brightenings of parts of the chromospheric plage. Using Hα observations Dizer (1969) has shown that subflares very often occur at places that were brighter than the surroundings prior to the subflare onset and he calls these bright pre-subflare patches the 'live points'. During the CINOF period Švestka *et al.* (1974; de Jager, 1975) confirmed Dizer's observation for two subflares they studied in detail (Figure 12). In addition they showed that the five knots which formed the two subflares were distributed along the border of one supergranule cell, at, or very close to the places where three supergranules met.

In larger flares the originally small nodules link up and often form ribbon-like structures, which, particularly in very large flares, resemble twisted hanks of yarn (Smith and Smith, 1963). These ribbons tend to lie along the lines of demarcation between regions of opposite magnetic field polarity. As we shall see in the following chapters, a well-developed ribbon-like structure with separating ribbons embedded among large spots is characteristic for large proton flares and cosmic-ray flares (Ellison *et al.*, 1961; Avignon *et al.*, 1963, 1965).

However, there is a great variety in flare shapes, and it is not clear whether we meet in all cases with exactly the same phenomenon. Some brightenings, often classified as subflares or even small flares are possibly feet of surges (Section 4.2.1), parts of sheared filaments, or products of falling prominence material (Hyder, 1967a, b).

(f) High-resolution and in particular off-band Hα photographs show intense short-lived brightenings in limited parts of the flare area, which coincide in time with the impulsive microwave and hard X-ray bursts (de Jager, 1967; Vorpahl and Zirin, 1970; Zirin *et al.*, 1971). They have been called *flare kernels*. (One should keep this term for this particular, short-lived phenomenon, obviously produced by streams of high energy electrons penetrating into the chromosphere (cf. Section 2.3), and not to interchange it for bright *flare knots* of apparently thermal origin.)

---

* $\beta\gamma$ is a bipolar group in which one or more spots are out of place relative to a typical bipolar configuration; γ is a group in which the polarities are completely mixed; δ is a γ-group in which spots of opposite polarity are embedded in the same penumbra.

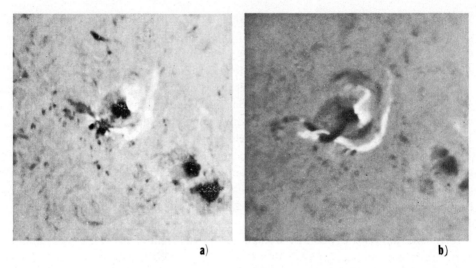

a)   **b)**

Fig. 13.   An example of homologous flares. Both pictures were taken 0.8 Å to the red in the Hα line:
(a) at 18^h18^m UT in 1968, July 9, about 11 min. after the flare onset; (b) at 00^h08^m UT on July 12
(i.e., about 54 h. later), 9 min. after the onset of the homologous flare.
(After White and Janssens, 1970.)

Vorpahl (1972) has summarized properties of these flare kernels: They have
luminosity several times greater than the surrounding flare and diameter of 3000 to
6000 km, which is only a small fraction of the total flare area. A typical Hα kernel
begins to brighten 20 to 30 s before the recognizable onset of the hard X-ray burst,
peaks 20 to 25 s after it and lasts about twice as long as the hard X-ray spike (cf.
Section 3.2.1.A). Similar to the white-light flares (Section 2.2.11) the kernels often
appear as two separate areas in regions of opposite magnetic polarity.

(g) In some active regions especially, and in particular parts of the active region,
flares show a striking tendency to recur in nearly identical form. This fact was first
noted by Waldmeier (1938), and Ellison *et al.* (1960) have proposed for such events
the term 'homologous flares'. A very detailed study of this homology was carried out
for three active regions by Dodson and Hedeman (1949) and for the interesting sunspot
group of September 1963 by Zirin and Werner (1967). Homologous radio bursts have
been studied by Fokker (1967). An example of two large homologous flares, separated
by 54 h in time, is shown in Figure 13, after White and Janssens (1970).

B. *Flare Positions in the Magnetic Field*

The first direct comparisons of solar magnetograms (maps of photospheric magnetic
fields) with flare positions were made by Bumba (1958) and Severny (1958b, 1960).
Later on, similar studies were carried out by Howard and Babcock (1960), Gopasyuk
(1961), Severny (1962, 1963, 1964c, 1969b, 1971), Howard and Severny (1963),
Howard and Harvey (1964), Martres *et al.* (1966, 1971), Moreton and Severny (1968),
Smith and Ramsey (1967), Rust (1968b), Zvereva and Severny (1970), Michard (1971),

and Vorpahl (1972). The most reliable results from these studies can be summarized as follows:

(1) The flare usually starts with two or more bright points or areas. When there are two of them, they appear on opposite sides of the zero line ($H_\parallel = 0$) of the longitudinal magnetic field (Martres et al., 1966). Moreton and Severny studied 27 flares and 25 of them followed this rule. The same is true for the bright flare kernels which coincide in time with the hard X-ray bursts (Vorpahl).

(2) Usually flares avoid the exact position of the zero line, but on the other hand, at least one of the initial knots tends to lie immediately adjacent to the zero line, in a region of the greatest longitudinal field gradient. Out of 80 flare nodules studied by Moreton and Severny, 60 were situated within 10″ of the zero line, but only four of them overlapped it. Also, positions of the bright kernels are within 6000 km of the zero line (Vorpahl).

(3) Measurements of the transverse magnetic fields have shown that in many cases the first Hα brightenings of flares coincide with the places where the transverse field bifurcates or becomes greatly disordered, even to the extent that transverse field lines

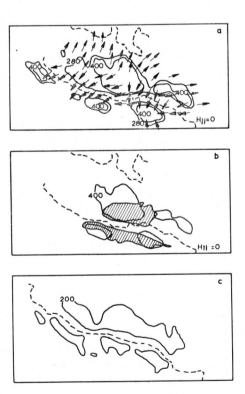

Fig. 14.   (a) A composite map of the longitudinal (280 and 400 G contours) and transverse (arrows) fields in the region which produced the major flare of 1966, July 7. The map shows the situation about 18 hours before the flare. (b) Position of the flare in relation to the 400 G contours and the zero line of the longitudinal field $H_\parallel = 0$. (c) The 'simplified' longitudinal field (200 G contours) about 5 h after the flare onset. (After Zvereva and Severny, 1970.)

of different directions apparently cross. Such a disordered situation can be seen frequently near the zero line of the longitudinal field where, instead of simply crossing the line in the direction of the gradient of $H_{\parallel}$, the transverse field is distorted in various directions, sometimes even being nearly parallel to the $H_{\parallel} = 0$ line (Severny). The strength of the transverse field in regions of flares can reach high values of 200 to 1000 G (Zvereva and Severny). Figure 14 shows an example of the position of a large (cosmic-ray) flare on maps of the longitudinal and transverse magnetic fields in the active region.

Observations carried out during the strong flare activity in August 1972 confirmed the sheared structure of the magnetic field near the zero line in regions where the big flares occurred (Zirin and Tanaka, 1973; Tanaka and Nakagawa, 1973). The penumbral fibrils, which probably trace the transverse field, ran almost parallel to the zero line; also the initial loops connecting the two bright flare ribbons in the large white-light flare of August 7 (Figure 15) crossed the zero line at a sharp angle. As time proceeded, however, the new loops, formed at larger altitudes, made progressively greater angles to the zero line. Thus either the low magnetic loops were more sheared than the higher ones, or the shear relaxed with time.

(4) When all components of the magnetic field are measured, there is also the possibility to compute the electric currents (cf. Section 6.4.3.B.) Having done this, Severny and his co-workers have found a coincidence between the bright knots which appear during the initial phase of flares and areas with an increased vertical current density.

Rust and Bar (1973) used photospheric magnetic maps for a computation of the coronal magnetic field configuration during the large flares of 1972, August 2, 7 and 11, under the assumption that no currents flow in the corona above the active region. This assumption proved wrong: the orientation of the current-free lines differed by up to 45° from the orientation of the loops actually observed, and this disagreement improved only in a later phase of the flare development. This can be considered as evidence that strong currents indeed play a significant role in the initial phase of flares, as Severny's photospheric observations have indicated.

(5) Severny (1960) and Martres *et al.* (1971) also find that flares tend to occur in places where the line-of-sight component of the photospheric velocity $V_{\parallel}$ is zero; according to them crossings of the $H_{\parallel} = 0$ and $V_{\parallel} = 0$ lines are the most likely positions of the initial flare brightenings. But generally the relation of flare occurrence to the velocity fields in the photosphere is still unclear, in particular because measurements of velocity fields are very scarce.

Yoshimura *et al.* (1971) observed curved absorption lines under a subflare region, which indicated large-scale mass motions in the photosphere, decaying during the flare development. However, the subflare tentatively associated with these motions was so insignificant that the association is open to some doubt.

More convincing is an observation associated with the great flare of 1972, August 7, reported by Rust (1973a) and Zirin and Tanaka (1973). They observed unusual blue shifts in the photospheric lines close to the $H_{\parallel} = 0$ line below the flare site prior to the

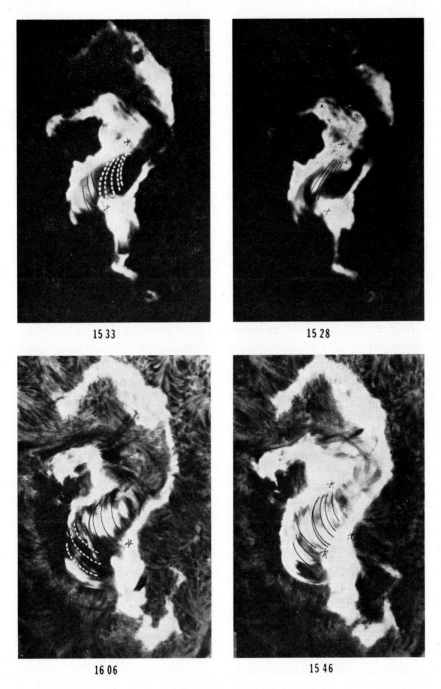

Fig. 15.   The comparison of the computed force-free magnetic field lines and the observed loops in Hα off-band ($+0.5$ Å) filtergrams of the major flare of 1972, August 7. The first picture (at 15^h28^m, upper right-hand corner) shows the greatest shear. The shear relaxes as the time proceeds. (After Tanaka and Nakagawa, 1973.)

flare, but the same region became red-shifted after the flare onset. However, the interpretation of these velocity changes is ambiguous. Zirin and Tanaka speak about a shearing (i.e. lateral) motion with velocity of 4 to 5 km s^{-1}, whereas Rust interprets the same observations as upward and subsequently downward motions in a sunspot penumbra. It appears that more frequent and sophisticated velocity field measurements are needed in order to know more about photospheric motions related to flares.

(6) When the initial flare areas extend, this extension tends to proceed parallel to the zero line of the longitudinal magnetic field. Thus the shapes of flares follow the shape of the zero line and cover areas of steep longitudinal field gradient, since this is a characteristic of areas closely adjacent to the zero line (Smith and Ramsey). The expansion of individual flare parts however, always occurs within the region of one magnetic polarity (Gopasyuk).

(7) Flares generally cover areas with strong fields, but some bright patches may also form in, or extend into, areas with fairly weak fields on the periphery of the active region; this happens predominantly in their later development. We have also already mentioned in item (b) in the preceding section that some flares of large Hα importance appear in active regions without spots where the magnetic field cannot be very strong.

Thus the strength of the magnetic field does not seem to be the most decisive characteristic of the flare locations. More important seems to be the gradient of the magnetic field. Gopasyuk et al. (1962) have shown that the energies of particles accelerated in flares (cf. Table IV) increase with the increasing magnetic field gradient. Extraordinarily high horizontal gradients in the longitudinal field also were observed in association with the large 1972 August flares (Rust, 1973a) and 1966 July 7, cosmic-ray flare (Severny, 1969a).

C. *Flare-Associated Changes in the Magnetic Field*

These observational results demonstrate a strong coupling between the magnetic field and the flare phenomenon. Also most theoreticians believe that flares derive their energy from the magnetic field (cf. Section 6.4). The activation of quiescent filaments preceding flares (Section 4.1) shows that at least some rearrangement of magnetic field occurs prior to the flare onset. Therefore, the obvious question arises as to whether there appear any observable changes in the magnetic field strength and configuration during the occurrence of a large flare on the Sun.

C.1. *Old-type measurements*

Many studies of this problem were made in the past, investigating sunspot areas, sunspot magnetic fields, or magnetic maps of active regions (Severny, 1958a, 1960, 1962, 1964b, 1969a, b; Howard et al., 1959; Chistyakov, 1959, Howard and Babcock, 1960; Gopasyuk, 1961, 1962; Michard et al., 1961; Gopasyuk et al., 1962; Leroy, 1962; Howard and Severny, 1963; Howard, 1963; Sawyer, 1968; Rust, 1968; Sivaraman, 1969; Zvereva and Severny, 1970; Mayfield, 1971; Wiehr, 1972; Rust, 1972, 1973a). The results are conflicting, since both positive and negative results were obtained by these different authors.

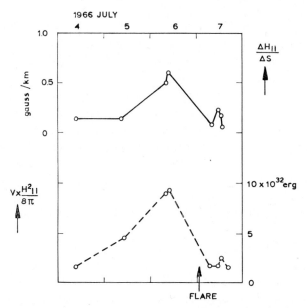

Fig. 16a.   The magnetic field gradient (full curve) and magnetic energy (dashed curve) variations before and after the major flare of 1966, July 7. The measurements have been carried out in the λ 5250 Å line. The gradient refers to the two 400 G 'hills' on the right-hand side of Figure 14 b. (After Severny, 1969 a.)

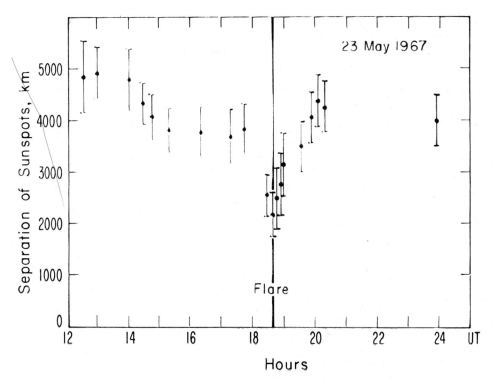

Fig. 16 b.   Time variation in the separation of the big spots between which the white-light flare of 1967, May 23 (Figure 36) occurred. (After McIntosh, 1969).

Positive results come mainly from detailed studies made at the Crimean Astrophysical Observatory, and one can summarize them in the following way (it is to be understood that only large flares have been considered):

The most characteristic phenomenon is a simplification of the magnetic field configuration after a flare occurrence. Before the flare, the magnetic fields become more and more complicated, the longitudinal field gradients for certain specific directions, the magnetic energy and the total magnetic flux all increase and the magnetic peaks draw near to each other. The occurrence of a flare results in the decrease of the magnetic gradients in the vicinity of the zero line, either because of decrease of the field strength or because of a separation of the nearby magnetic peaks (Severny). Figure 16a demonstrates this behaviour as observed in the case of the flare of 1966, July 7 (see also Figure 14).

Whilst Severny's results are based upon magnetic maps, Gopasyuk (1962) has also found similar changes in the positions of sunspots adjacent to the flare regions. The spots first seem to approach the location of the flare and then are pushed away soon after the flare. These observations are not as convincing as the measurements of the magnetic field; nevertheless in one case McIntosh (1969) obtained direct photographs of a similar approach, i.e. a merging or darkening of penumbrae of two spots of opposite polarity before a major flare (Figure 16b). Therefore, slight sunspot motions prior to large flares cannot be excluded. Zirin and Lazareff (1975) claim to observe many flares in front of a rapidly moving spot (1000 km h^{-1}) whereas flares ceased to occur when the motion stopped.

All the effects so far mentioned have been observed in a few individual cases only. However, support for the reality of some of them comes also from a statistical study, which leads to similar results. Gopasyuk et al. (1962) studied the changes of magnetic field gradient for 51 flares of importance $\geqslant 2$. Their results are summarized in Table IV. The most striking change of gradient has been found in connection with flares which produce strong fluxes of energetic particles in space (cosmic ray flares and proton flares). These are the flares for which the changes in the magnetic field strength and structure have also been detected individually. In other flares the gradient and its variation are smaller, and so far no changes in the configuration of the magnetic field could be found for flares of importance smaller than 2. This of course, might be explained by the low resolving power of the magnetic records.

TABLE IV

Change of the magnetic field gradients during flare events, after Gopasyuk et al. (1962)

Flare importance	Flares associated with	Number of events	Gradient (G km^{-1})	
			before the flare	after the flare
3+	cosmic rays (GLE)	13	0.73	0.25
3 and 3+	PCA	11	0.46	0.27
3	no geophysical effect	12	0.18	0.13
2 and 2+	no geophysical effect	15	0.054	0.038

Chistyakov (1959), Michard *et al.* (1961) and Zvereva and Severny (1970) have noted that the field pattern recovers to its pre-flare state fairly soon after the flare occurrence. This observation is very important in relation to homologous flares, which appear twice or several times in very much the same positions and possess quite similar shapes and properties. The homology has always been one of the basic arguments of those theoreticians who do not believe in the magnetic field as the source of energy for flares.

On the other hand, by using similar data, many other authors did not find any magnetic field changes associated with flares. Some of them obviously considered flares so small (e.g. Mayfield, 1971 and Wiehr, 1972), that any expected changes were probably below the resolving power of their equipment. But the main reason for the contradictory results seems to be, as Rust (1975b) points out, that quite generally most of the observations in the 1960's were made at a level of precision comparable with the maximum field changes expected. In such situations, of course, one hardly can expect any decisive and fully convincing results to be obtained.

Several authors (Bappu and Punetha, 1962; Zirin and Werner, 1967) have also tried to detect changes in the calcium and hydrogen plages before and after a flare occurs in them, since their shapes closely follow the magnetic field structure in the active regions. However, no definite changes could be detected. On the other hand, Ellison *et al.* (1960) have observed that the fine fibril structure which surrounds sunspot groups in the light of the Hα line, is liable to sudden changes during great flares and is restored again after the flare. Since this fibril structure reflects the structure of the transverse field (Tsap, 1965), this effect might indicate that the magnetic field changes. However, high-resolution pictures of later years have never shown this Ellison's effect again.

C.2. *High-resolution measurements*

Starting in about 1968, with magnetographs and Hα solar telescopes of improved resolving power, the reality of flare-associated changes of magnetic field has become more convincing. However, another question has arisen: which magnetic field changes lead to the flare event and which are the consequence of it?

The Meudon group (Martres *et al.*, 1968a, b; Ribes, 1969; Michard, 1971) has divided each complex pattern of longitudinal field into a number of what they call 'evolving magnetic features', each containing one polarity. They have shown that in many cases flares involve at least two adjacent features of this kind, and these evolve differently from each other in the period of flare occurrence.

These are clearly flare-associated changes in the magnetic field, but significantly different from those we described earlier. They characterize the evolutionary variations typical of a flaring region rather than changes caused by the flare process itself. The French authors are of the opinion that one should be very careful when ascribing changes in the magnetic field (the decrease or disappearance of magnetic peaks, for example) to the flare process. Due to usually very long time-intervals between individual successive measurements of the magnetic field (cf. Figure 16a) we do not know exactly when the change occurred: it could be before or after as well as during the flare.

In more recent years Rust (1972, 1973 b) and his co-workers (Rust and Roy, 1974; Rust *et al.*, 1974; Rust and Bridges, 1975) have confirmed that flares occur close to evolving magnetic features. The likelihood of occurrence of a flare in a large bipolar region increases if, in the vicinity of the zero line, there is rapid development of small, reversed-polarity magnetic features (Rust, 1973 b). As a matter of fact this only confirms, by using greatly improved resolving power, what we already mentioned earlier: that magnetically complex active regions are more capable of producing flares.

Almost all the flares which Rust and his co-workers studied apparently were triggered by emergence of new magnetic flux through the photosphere. This also agrees with Vorpahl's (1973 a) Hα observation that most flares are triggered when and where emerging flux regions appear. These can be recognized in the Hα line as a bright plage crossed by a dark arch-filament system (after Bruzek, 1967 and Zirin, 1972).

Emergence of new flux in an active region is certainly not a sufficient condition for producing flares, since many such variations are also observed without any associated flarings. Schoolman (1973) has found that new flux regions emerge at a rate of one or two per hour in a developing active region. According to Rust (1973 b) even reversed polarity in the emerging flux need not necessarily produce a flare. Thus many associations of this kind still may be chance coincidences, in particular when only single and isolated events are described.

Rust (1972) based his analysis on several observed events and arrived at the following conclusions: The apparent magnetic flux in Martres *et al.*'s 'evolving magnetic features' increases or decreases by about a factor of two in the half-hour before flare onset, but the changes are confined to very small areas; mainly islands of opposite polarity with $< 10''$ diameter. One needs very good seeing to resolve these regions, but if the seeing is good, the field in these features appears to strengthen for about half-an-hour preceding the flare. During the flare the field drops and then returns to its previous level.

C.3. *Mathematical models of coronal magnetic fields*

Another approach to this problem can be made by means of theoretical computations of the magnetic field in the corona, which extrapolate the measured photospheric magnetic field upwards in a current-free (Altschuler and Newkirk, 1969) or force-free (Nakagawa and Raadu, 1972) approximation.

This method was first applied by Valdes and Altschuler (1970) to proton flares. They were able to present some evidence that after proton flares the surrounding coronal magnetic field decreases in flux and changes from a close-loop structure to an open or diverging field. Later on Altschuler (1974) repeated this study by using the *Atlas of the Coronal Field* prepared by Newkirk *et al.* (1972). He has confirmed the previous result and found that as a rule, the coronal magnetic field changes drastically in strength and geometry around the flare region. Low-lying loops disappear, or at least decrease significantly in field strength. Since the low magnetic loops representing strong magnetic field are coupled with strong electric currents flowing in the underlying photosphere, their disappearance should mean that photospheric electric currents disappear or disintegrate at the time of large flares.

These methods have also been applied by several authors to the large August 1972 flares. Rust and Bar (1973) applied the current-free approximation to the loop systems associated with the flares on August 2 and 7. As we mentioned in the preceding Section (B), items (3) and (4), the disagreement found between the computed field lines and observed low loops decreased significantly for higher loops recorded in later phases of development. This might mean that strong currents flow in the flaring corona at the flare onset, whereas the fields become practically current-free in the late phase of the flare development.

In a plasma where the magnetic energy density exceeds the kinetic energy density by many orders of magnitude the currents flow parallel to the lines of force, i.e. we have a force-free field. This approximation has been applied to the 1972 August flares by Tanaka and Nakagawa (1973). In the current-free (potential) configuration the field lines (and the visible loops marking them) should be perpendicular ($\gamma = \pi/2$) to the $H_\parallel = 0$ line, representing the minimum energy state. On the other hand they may be sheared in a force-free configuration (so that $\gamma < \pi/2$) and the surplus energy stored in the non-potential field is then proportional to $\sin^{-1} \gamma$, i.e. to the amount of shearing. Thus the stored energy can be computed from the observations made by Zirin and Tanaka (1973) for any time the observations are available.

According to Tanaka and Nakagawa's computations the stored surplus energy increased from $\sim 3 \times 10^{30}$ erg on July 31, through 8×10^{31} erg on August 2, to values exceeding 10^{32} erg on all the following days. During the flares of August 4 and 7 the shear became smaller, thus indicating relaxation towards a potential magnetic field configuration. Similar results have been reported by Rust et al. (1975) for another flare. On the other hand, some Crimean maps from other periods (e.g. Zvereva and Severny, 1970) show sheared fields at the photospheric level even after flare occurrences. Thus either the relaxation is much more pronounced at higher altitudes than in the photosphere, or we interpret altitude-dependent shear erroneously as a time-dependent phenomenon (cf. item (3) in the preceding section).

C.4. Other methods

Several other methods have been proposed on how to attack this difficult problem. Schröter (1973) has suggested that more attention should be paid to the old spotless active regions that occasionally produce fairly large flares (cf. Section 2.1.2.A(b)). Equation (104) in Section 6.4.3 shows that energy of 10^{32} erg can be gained through magnetic field annihilation in a volume of 10^{29} cm^3 if the original field strength of 500 G is decreased by only 26 G. Such a change would hardly be discovered. On the other hand, if the field strength is only 160 G, as it may be in an old region, the whole field must be completely annihilated in order to get the same amount of energy. Therefore, the flare-associated magnetic field variations in old spotless regions should be much more pronounced than in developed regions with strong fields.

As a matter of fact to some extent this has been checked. A flare of this type (shown in Figure 20) was actually observed by Michalitsanos and Kupferman (1974) in 1973, July 29 and the Kitt Peak magnetograms did not show any obvious changes in the morphology of the magnetic field after the flare. However, no information could be obtained on magnetic field strength variations in the active region.

Janssens (1972) discussed changes in the coronal magnetic field configurations that should follow reconnection of the magnetic field lines, annihilation of the magnetic field, or spreading and motion of flux bundles. The quantity which should show a change in any case is the magnetic energy, whereas the flux changes only for annihilation and reconnection. In the latter case some magnetic field lines also 'change their feet' in the photosphere. Janssens applied his topological considerations to 5 subflares, but did not discover any indication of any kind of magnetic field change.

Kundu and McCullough (1972) suggested that magnetic field changes can be also deduced from changes in the degree of polarization on millimetre wavelengths. They report a change in the degree of polarization at 9.5 mm before and after the start of a bright subflare. Their result can be interpreted as a decrease of longitudinal magnetic field by about 200 G at the chromospheric levels where the millimetre radiation originates.

C.5. *Summary*

It is certainly difficult to draw any definite conclusions from these inhomogeneous and contradictory results. When studying the photospheric fields one obviously needs magnetic field measurements of better space and time resolution. A magnetograph in space might be the best solution because the 'seeing' variations on the ground contribute significantly to the uncertainty of the results obtained. When considering the mathematical models of the coronal fields, one has to be well aware of all the dangers hidden in the approximations applied. As Levine and Altschuler (1974) have demonstrated, agreement with the current-free approximation, $\nabla \times \mathbf{H} = 0$, does not preclude the possibility that electric currents are still present in the flare region, since quite large coronal currents are required before significant topological deviations from the potential magnetic field configuration can be noticed. Neither can one be sure that the scale factor α in the force-free approximation, $\nabla \times \mathbf{H} = \alpha \mathbf{H}$, is constant throughout the flaring region, as the approximation assumes.

Nevertheless, in spite of all the uncertainties involved, some conclusions can be drawn from the observations described above. It is true that each observation is open to doubt. But several points seem to be common to many of them, to the extent that a chance coincidence is improbable.

First of all, it seems to be well established that flares preferably occur in active regions which show fast changes in magnetic field structure, mainly due to emergence of new magnetic flux. Particularly favourable for flare occurrence is an emergence of flux with reversed polarities, which makes the active region magnetically complex. As Altschuler (1973) has pointed out, one should naturally expect that the regions of great magnetic complexity are the most likely places for flares to occur, because there the antiparallel magnetic fields and metastable current sheets are most likely to form.

Since flares originate close to the $H_{\parallel} = 0$ line, magnetic changes close to this zero line appear to be of greatest importance. Changes there may be due to emergence of new flux, as well as to deformations of the field caused by motions and rotation of sunspots (Tanaka and Nakagawa, 1973; Piddington, 1974). Shears of the magnetic field lines near the zero line seem to be of great importance in association with the occurrence of major flares.

All this concerns the situation prior to flare appearance. The question of whether magnetic field changes take place during a flare is still unanswered. If the field does not change, the flare energy obviously cannot be the transformed energy of the magnetic field. Even when we cannot fully exclude this possibility, let us assume that this is not the case and that the flare energy of 10^{28} to 10^{32} erg is taken from the magnetic energy stored in the active region. It is important to realize that even then it may be rather difficult to discover the magnetic energy deficit with the means we presently have at our disposal.

Usually what is measured in the photosphere is only the longitudinal component of the magnetic field. Based on these measurements, the potential field in higher atmospheric layers can be computed. However, potential field is the configuration with minimum energy and therefore, energy of this field is not freely available for release in a flare. What we need is surplus energy in the form of non-potential fields which may return to the lower-energy state through relaxation at the time of a flare. The trouble is that we possess very few means to detect the non-potential fields and the currents associated with them. Most magnetographs measure one component of the field only; the current-free approximation for computing the coronal fields a priori excludes any deviations from the potential configuration; the force-free approximation makes only one special type of current-induced deviations detectable.

Nevertheless, the evidence for currents in the flaring region, based on Severny's (1965) computations in the photosphere or Rust and Bar's (1973) and Tanaka and Nakagawa's (1973) extrapolations in the corona, makes it easy to see how magnetic energy could play an active role for the flare build-up; namely, the electric current regions mark exactly those regions in the solar atmosphere where magnetic energy is available for conversion into various kinds of kinetic energy (for a good discussion of these problems see Altschuler (1974)). One can expect that photographs of flare regions in X-rays and EUV lines might contribute to the resolution of this problem by showing flare-associated reconnection of magnetic field lines and improved agreement between observed coronal structures and potential fields as a flare proceeds. So far no such results have been presented by the Skylab study groups; this need not mean that such changes do not happen in the flare regions; they may be below the resolving power of the Skylab instruments which only on rare occasions had a resolution better than 5″.

2.1.3. MOTIONS IN FLARES

While flare-associated phenomena (discussed in Chapter IV) show a great variety of
motions with appreciable velocities, observations relating to mass motions in the
flaring areas themselves are both subtle and contradictory. Bright Hα-flare elements
observed on the solar disc often show expansion, slow shifts and lengthening in some
particular direction, but it is difficult to decide whether we meet with an actual transport of solar gases there or with a travelling excitation in an essentially stationary
chromospheric material. The absence of any significant Doppler shifts indicates that
in most cases the second explanation is the correct one. On the other hand, on the
limb, flares often ascend and descend and sometimes there are lateral shifts which in
a few cases have been confirmed by Doppler shifts (Dodson and Hedeman, 1968b)
or by splitting of lines (Severny, 1968). Since we observe different parts of the flare
on the disc and on the limb (low flare parts are obscured by the chromosphere at the
limb; higher parts, observed on the limb, are transparent when projected on the disc),
these observations indicate that motions of the flaring material, very small (if present
at all) in the chromosphere, may become significant in the low-temperature flare parts
at coronal altitudes.

As we shall see in Section 2.2.3, similar behavior also appears when the internal
motions of flares are investigated. Whilst the hydrogen lines in spectra of disc flares
appear to be intensely broadened by the Stark effect, the metal emission lines are quite
narrow; this indicates an absence of any significant random non-thermal motions in
the flare parts situated in the low chromosphere. The situation changes, however,
if one goes higher in the solar atmosphere and investigates spectra of limb flares.
There the Doppler effect of macroscopic motions seems to play a significant part in the
line broadening, and in some flares, if one exceeds a height of about 10 000 km above
the photosphere, the metal lines appear essentially as broad as the hydrogen lines
(Jefferies and Orrall, 1961a; Švestka, 1965).

Nevertheless, some systematic motions must be present also in the low chromospheric flare parts, since the chromospheric emission lines of flares are asymmetrical
(cf. Section 2.2.9). The nature of this asymmetry is not yet well understood. Increased
random non-thermal motions above the chromospheric level (Brueckner, 1974;
Brueckner *et al.*, 1975) as well as systematic downward streaming (Widing and Cheng,
1974) have been found also in EUV flare spectra observed aboard Skylab (Section
3.2.2).

In very rare cases motions have been reported also in the photosphere underlying
a flare, either in the form of large-scale mass motions (Yoshimura *et al.*, 1971),
up- and downward motions in a sunspot penumbra (Rust, 1973a), or a velocity shear
(Zirin and Tanaka (1973). These motions, however, need not be related directly to
flares. In Yoshimura *et al.*'s case the association of the flaring region with the photospheric motions is doubtful; in the case of Rust, and Zirin and Tanaka, the motions
preceded the flare (Rust) and lasted significantly longer than the flare (Zirin and
Tanaka).

The very subtle problem of the mass motions in flares was discussed several years ago at a special symposium by Švestka (1968 b), Dodson and Hedeman (1968 b), Bruzek (1968), Severny (1968), Öhman *et al.* (1968) and E.v.P. Smith (1968). Summarizing the results, one can suspect that the following forms of mass motions occur in the low-temperature parts of flares of different types (in fact, even at the limb, one never can be perfectly sure that the effect is not due to a travelling excitation, as long as spectral measurements are not available).

A. *Slow Ascent*

On some occasions, a slow continuous ascent with mean velocity of several kilometers per second is observed in flares at the limb or close to the limb (Dodson and Cornwall, 1939; L. d'Azambuja, 1942, 1949; Waldmeier, 1945; Švestka, 1962). All of these observations involve large flares, and one is strongly inclined to believe that in most (or all) of these cases the observers saw, not the flare itself, but the brightest part of the associated loop prominences that formed in the flare region and expanded upwards. An illustrative example of such a 'loop-flare' has been presented by Elliott *et al.* (1960).

B. *Rise and Fall*

Another type of motion observed in limb flares has been described by Severny and Shaposhnikova (1960) and was later confirmed by Guseynov and Gasanalizade (1960) and Švestka (1962). Some flares expand upwards in the starting phase of their development, with velocities mostly in the range between 50 and 150 km s^{-1}, i.e. with velocities one order higher than observed in loop prominence systems, and reach heights of 10 000 to 50 000 km above the base of the chromosphere. This fast ascent is then followed by a descent with lower velocities, i.e. several tens of kilometers per second. An example of such an event is shown in Figure 17.

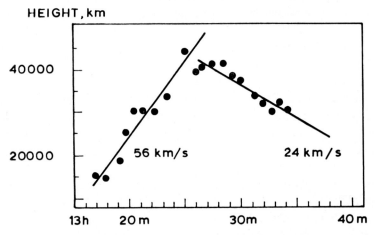

Fig. 17.   Measured height and velocities of the rise and descent in the limb flare of 1956, May 8.
(After Severny and Shaposhnikova, 1960.)

If these observations correspond to actual mass motions, we should have to verify their existence in the spectra of disc flares, where the expansion would proceed along the line of sight. With a velocity of 50 km s^{-1} such motions would cause a Doppler shift of about 1 Å in the Hα line, and it is well known that no such shifts have ever been observed. Thus, either the 'rise' merely reflects an extension of the flare exciting agent into higher layers of the solar atmosphere, or only the higher layers, which are seen clearly above the limb, expand upwards, whilst the motion tends to zero velocity in the chromospheric part of the flare, which is observed on the disc. In such a case, an asymmetrical line profile might be expected in the Hα line, first extended towards the blue and later on towards the red side of the spectrum. However, as we mentioned before (Figure 7), in the vast majority of flares the red wing of the line is stronger essentially throughout the whole life of the flare (Švestka *et al.*, 1962). The opposite asymmetry only occurs in some flares during the first two or three minutes of their development and sometimes in their decaying phase (Section 2.2.9), and this does not correspond to the observed scheme of velocities. Therefore, spectrographic observations do not yield any support for this type of motion in flares, but it might be real in the case that it affects only the uppermost layers of the Hα flare which are completely transparent in projection on the disc.

C. *High-Speed Ejections*

Another type of motion associated with limb flares is a high-speed ejection of flare matter. One can mention several examples here:

In 1951, May 8, Dodson and McMath (1952) observed a limb flare that was associated with a sudden ejection of matter, reaching a height of 50 000 km in 90 s, with a velocity of about 700 km s^{-1}. In 1958, May 23, Severny and Shaposhnikova (1960) also observed a limb flare which in a short time reached a height of 50 000 km. The initial velocity of 180 km s^{-1} increased to 560 km s^{-1} before the ejection became invisible. In 1958, August 7, Reid (1959) observed a flare on the disc accompanied by a fast ejection (bright in projection on the disc) that extended beyond the limb and travelled with a mean velocity of 300 km s^{-1}. Lateral motions in limb flares up to a distance of 50 000 km with velocities of several hundred km s^{-1} have also been reported by Dodson and Hedeman (1968 b). The most striking event, however, was observed by Valníček (1962) in 1961, September 16, when a flare close to the limb ejected a portion of its matter to a height of 250 000 km. The initial velocity of 400 km s^{-1} increased to 1400 km s^{-1} and remnants of the flare dissipated high in the corona (Figure 18). All these events differed from other flare-associated moving phenomena by their great, flare-like brightness, but, depending on the definition, they certainly also might be considered to be sprays (Section 4.2.2).

It is clear that any observation of these phenomena on the disc is difficult, and one can expect that only laterally moving ejections of this type might be detected. As a matter of fact, similar phenomena have been observed several times on the disc, in the form of emission or absorption clouds originating in the flare and travelling along paths from 60 000 to 350 000 km with lateral velocities of 250 to 2500 km s^{-1}

Fig. 18. Ejection of a portion of the flare of 1961, September 16, located close to the limb. The picture is composed of eight subsequent photographs, the lowest one showing the flare in the chromosphere and the others the ejected cloud at different heights. The time difference between the last two pictures corresponds to the cloud velocity of 1400 km s^{-1}. (After Valníček, 1962.)

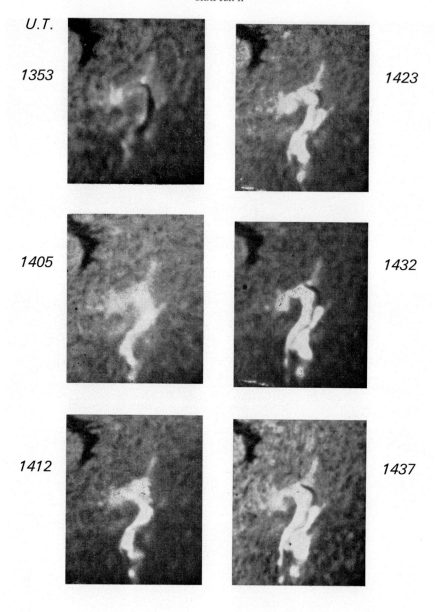

Fig. 19.   Development of the two-ribbon flare in 1959, July 14 which formed in a spotless region at the location of a disappeared dark filament (the transformation from absorption to emission can be seen between 1353 and 1405 UT). The velocity of expansion of the bright ribbons is shown in Figure 21. (Courtesy of the Fraunhofer Institute, Freiburg.)

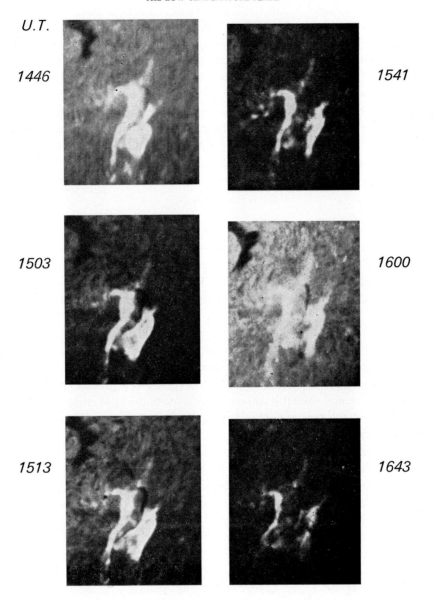

(Gopasyuk, 1960; Athay and Moreton, 1961). Many of these fast-moving disc events are moving skirts in the chromosphere of coronal flare-waves (cf. Section 4.3.1), but some of them may be identical with the high-speed ejections observed on the limb.

Generally, the associations among different types of high-speed phenomena observed on the disc and limb (ejections, sprays (4.2.2), blown-off prominences (4.1), blast waves (4.3), radio type II bursts (3.2.5), and moving type IV bursts (3.2.4.A)) are not yet clear enough. Munro (1975) reports a good correlation between high-speed flare-associated Hα phenomena and mass ejections through the white-light corona observed during the Skylab mission. Six coronal transients of this type were associated with flares, and all these flares were accompanied by high-velocity ejections, sprays or eruptive prominences. However, 24 other coronal transients were associated with other dynamic phenomena in the chromosphere which occurred without flares, and 36 had no chromospheric counterpart at all (probably due mostly to the fact that the source was on the invisible solar hemisphere). Thus the only conclusion one can make is that in some flares some parts can be thrown off with very high speed in various directions, these flare remnants can be followed in the Hα line as far as some 300 000 km and still much farther in the white-light corona, where however they represent only a small fraction of the coronal disturbances observed.

D. *Rotation Effects*

Öhman (1968a, b, 1972) and his co-workers (Öhman *et al.*, 1968a, b) have drawn attention to tilted emission lines often observed in the spectra of prominences and moustaches and sometimes in flares as well. This might well be evidence for some type of rotational mass motion in flares, but its occurrence is very infrequent.

E. *Expansion of Flare Ribbons*

Probably the best established, most general, and most important type of motion in flares is the expanding motion of the two flare ribbons. This expansion, first observed by Servajean and Olivieri (1946) and Dodson (1949), is a general characteristic of two-ribbon flares, and since this category also involves the most energetic proton and cosmic-ray flares, this type of motion certainly is of great importance from the theoretical point of view.

We have seen in Section 2.1.2.B that the flare nodules first form on both sides of the zero line of the longitudinal magnetic field and then extend along the zero line. Thus a two-ribbon flare shape is formed quite naturally in any flare large enough to complete the fully developed two-ribbon configuration. The two-ribbon shape is most striking to the observer when the zero line is straight, as in Figure 20. But two-ribbon flares can also have very curvilinear shapes as in Figure 22.

The condition of a straight zero line is primarily fulfilled in old active regions, where the zero line manifests itself through the position of a quiescent dark filament. Flares in such regions always appear in some association with the existing filament (since both the flare and the filament are associated with the zero line) and the most typical case is a flare that forms at the place where a filament disappeared immediately before (cf. Section 4.1). A bright ribbon forms on each side of the pre-existing filament and both ribbons move away continuously from the zero line, thus slowly separating one from the other (Figure 19). Such a transformation of a dark filament into a two-ribbon flare was first noted by Waldmeier (1938), discussed by Bruzek (1951, 1957), and analyzed in detail by Hyder (1967a). According to Dodson and Hedeman (1970) essentially all large flares in old active regions develop in this way (another example showing the detailed structure of the ribbons is in Figure 20a).

The two bright flare ribbons separate with velocities of the order of 1 to 10 km s^{-1}, the velocity being either constant or slightly decreasing in the later phase of the flare development (Dodson, 1948; Dodson and Hedeman, 1960; Valníček, 1961; Švestka, 1962; Malville and Moreton, 1963). At the end of the visible chromospheric flare the motion still continues, and the flare ribbons fade to invisibility whilst their remnants are still moving. X-ray photographs reveal (Figure 20b) that the Hα ribbons are connected by hot and bright coronal loops which are still visible in X-rays after the Hα flare disappears. Figure 21 shows the velocities of the ribbon separation for five different flares observed in old active regions having either small spots or none at all. The velocities always refer to one ribbon only, hence the total velocity of separation is approximately twice as high as given in the figure.

A similar phenomenon is also observed in young and well-developed sunspot groups, but with three important distinctions: many-a-time there is no quiescent dark filament to be dissolved prior to the flare; the separation of the ribbons is usually stopped shortly after its onset; and sometimes the shape of the zero line and the brightness distribution in the ribbons is so complex that it is difficult to recognize the resulting two-ribbon shape of the flare at all.

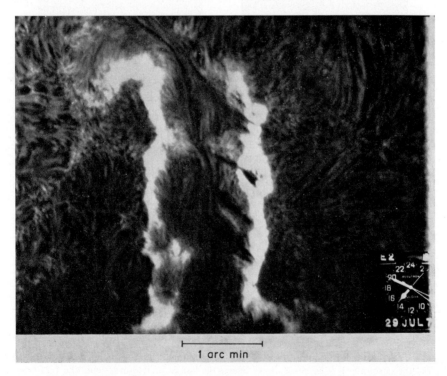

1 arc min

Fig. 20a.   On-band Hα filtergram of the 'spotless' two-ribbon flare of 1973, July 29. The two ribbons were formed at both sides of a disappeared dark filament and the filament channel (along the $H_{\parallel} = 0$ line) is still clearly indicated in the filtergram. Dark material flowing toward the east (right) ribbon indicates existence of a loop prominence system even in this spotless flare. (After Michalitsanos and Kupferman, 1974.)

Fig. 20b.   Same flare as in Figure 20a 2 h later (the upper and right two pictures). The X-ray photo-graph at the bottom (AS & E, 1975) shows clearly the hot coronal loop system. The high-resolution off-band Hα photograph on the top (Michalitsanos and Kupferman, 1974) indicates existence of a few cool loops that are apparently co-spatial with the hot ones. Note that the sharp left rim in the X-ray photograph is the left flare ribbon. The brightest region is obviously the top of the loops. The left two photographs show the pre-flare dark filament in Hα with essentially no response in X-rays.

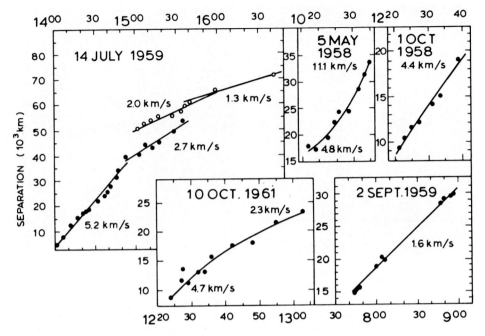

Fig. 21. Velocities of the separation of the flare bright ribbons in five different flares, all observed away from any large sunspot group. Distance between the ribbons is plotted on the vertical axis and the velocities refer to the motion of one ribbon. The flare of 1959, July 14, is shown in Figure 19. (After Švestka, 1962.)

When a large flare appears in such a group, it starts as an elongated bright filament along the zero line, which soon splits into two clearly visible bright ribbons. These ribbons move away from the zero line, but in this case their motion appears to be strongly influenced by the magnetic field in the sunspot group. As soon as they approach the umbrae of large spots or strong 'hills' of the longitudinal magnetic field, the motion is usually slowed down significantly or even stopped. Thus it may eventually happen that the flare remains motionless, anchored inside the strong sunspot group, or that only one ribbon is still moving while the other has been stopped (cf. Figure 22).

Examples of the velocity development in these cases are shown in Figure 23. In the major flare of 1972, August 7, Křivský et al. (1973) found a separation velocity of the bright ribbons equal to 50 km s^{-1} in the initial phase of the flare, but it decreased sharply to 1.3 km s^{-1} when the separation slowed down. During the fast initial expansion the ribbons were visible in white light (Rust and Hegwer, 1975). This initial expansion is probably identical with what Moreton (1964) has called the 'explosive phase'.

The dependence of the rate of lateral shift on the strength of the magnetic field was demonstrated by Smith and Ramsey (1967) on a smaller flare, with a moving ribbon formed outside the sunspot group. They found that the rate of lateral displacement

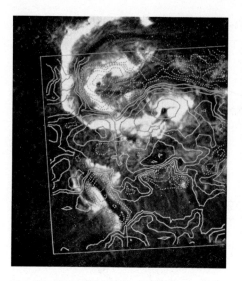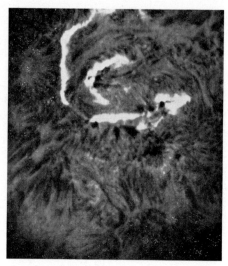

Fig. 22.   Two-ribbon flare of 1971, September 17. The right-hand picture shows the position of the two bright ribbons relative to the chromospheric structures and sunspots. In the left-hand picture a magnetic map is superposed on the flare. One can see how the two ribbons run along the $H_{\parallel} = 0$ line dividing the polarities; remnants of the dark filament are still visible along the zero line. The figure also shows how the bright ribbons rest on the 'hills' of the magnetic field, not propagating behind them. The right-hand picture shows that some of these hills are spots that the bright ribbons do not cross. (Courtesy of D. Rust, Sacramento Peak Observatory.)

increases with decreasing field strength and a deceleration occurs as the flare segments are displaced laterally in the direction of increasing field strength.

In some cases flares in developed sunspot groups also form at the location of a dark filament, e.g. the large flare of 1966, August 28 (Hyder, 1966b; Dodson and Hedeman, 1968b), but this is more an exception than the rule, simply for the reason that groups of this type usually are young and the filaments do not yet exist in them.

As was first recognized by Ellison *et al.* (1961) and later confirmed in a more general way by Avignon *et al.* (1963, 1965), the two bright ribbons embedded in a large sunspot group are a characteristic feature of the majority of proton flares. Therefore, this flare configuration is obviously favourable for the origin of acceleration processes which give rise to high energy nuclei in space. We shall speak more about it in Section 5.1.1.

In many cases the development of a two-ribbon flare in an active region is associated with the occurrence of a system of loop prominences visible in Hα (cf. Figures 88, 89 and Section 4.4). A very intuitive drawing of the development of a two-ribbon flare and the associated loops is shown in Figure 24a, after Bruzek (1964b). The observations indicate that throughout the flare, from its first appearance to its decay many hours later, the flare consists of an arcade of loops marking the magnetic field lines and spanning the zero line. With time these loops expand, as manifested initially by a separation of the two bright ribbons, which are in fact the visible feet of the row of loops in the chromosphere. Later, the expanding arcade becomes visible as a set of

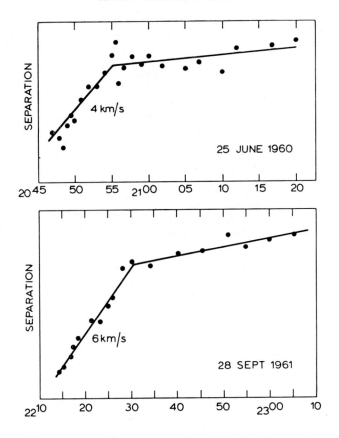

Fig. 23. Same as Figure 17 for flares which occurred in well-developed sunspot groups. The velocity decreases occur when the moving ribbons encounter strong magnetic fields. (After Malville and Moreton, 1963.)

rising loop prominences, probably fed by the accelerated particles in the flare region. It is to be emphasized that loops are rarely observed in association with the two-ribbon flares outside sunspots, probably due to the fact that particles are not usually accelerated in them (Dodson and Hedeman, 1970). However, as Figure 20 demonstrates, there are events when X-ray loops are very pronounced even in spotless flares; at least traces of the loops can be seen also in the Hα photographs.

An alternative explanation is that even in this case there is no actual mass motion, but only an agent propagating through a stationary magnetic field. Let us take, as an example, field between two sunspots, schematically drawn in Figure 24b. As the exciting agent propagates upwards along the arrow, progressively higher loops become affected and visible and feet farther from the zero line emit in Hα. As time proceeds, dz/dt corresponds to progressively smaller dx/dt. This can explain the slowing down of the Hα ribbons near sunspots. Observations of loop prominences on the limb support this latter point of view, since they grow through appearance of progressively higher loops and not through expansion (Section 4.4).

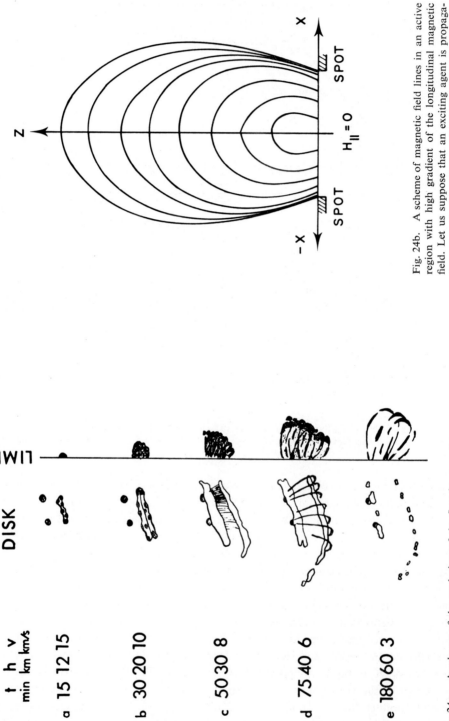

Fig. 24b. A scheme of magnetic field lines in an active region with high gradient of the longitudinal magnetic field. Let us suppose that an exciting agent is propagating upward with constant speed along the z-axis. Then the height of newly excited loops will increase with a constant speed, but the shift of their foot-points along the x-axis will be progressively slower as the field lines approach the sunspots.

Fig. 24a. A scheme of the evolution of the flare-loop system as seen projected against the disc (two expanding bright ribbons) and on the limb (loop prominences). (After Bruzek, 1964b, *Astrophysical Journal*, University of Chicago Press, Copyright by the American Astronomical Society.)

F. *Fast Expansion*

In some events expansion with much higher velocities has also been reported. In 1956, August 31, Gopasyuk (1958) observed a flare consisting of two regions, of which one was fixed, but the other moved away with a velocity of almost 100 km s^{-1}. In 1957, July 3, Bruzek (1958) found expansion velocities of the bright ribbons as high as 140 km s^{-1} in the first phase of the flare development. As we mentioned in the preceding section, the initial velocity of the ribbons in the white-light flare of 1972, August 7, was also quite high, about 50 km s^{-1} (Křivský *et al.*, 1973). In this case the fast expanding ribbons were visible in white light (Rust and Hegwer, 1975). Thus in spite of the fact that such high expansion velocities seem to occur very rarely, they might be worth more detailed study. They certainly represent the most powerful examples of Moreton's 'explosive flares'.

G. *In-Fall Motions*

Hyder (1967a, b) suggested that some flares, or at least parts of them, can be produced by material falling down into the chromosphere. This is probably true at weak chromospheric brightenings associated with the disparitions brusques (cf. Section 4.3.2), and Kubota *et al.* (1974) also invoke this mechanism to explain temporal umbral brightenings not associated with flares. But it is unlikely that this process can be generally applicable in flares as Hyder (1973) himself has realized. Even in a flare which was closely associated with the destabilization phase of a quiescent filament (Figure 20a) the visible falling material did not show spatial correspondence with bright emission knots in the Hα ribbons (Michalitsanos and Kupferman, 1974). Also in other flares where downstreaming material and Hα brightenings could be compared, these two phenomena were not cospatial (Rust *et al.*, 1975; de Jager, 1975). Nakagawa *et al.* (1973) tried to identify the optical flare with the response of the chromosphere to a shock propagating downwards, instead of with infalling material. However, Canfield and Athay (1974) have shown that the shock would give rise to flare emission line profiles which do not fit the observed flare spectra (Section 2.3.1.D).

2.2. Flare Spectra

Observations of flares in Hα light give information on the occurrence, importance, shape, development and lateral motions in flares; but information on the physical properties can only be obtained from an analysis of flare spectra. The first qualitative descriptions of flare spectra were given by Richardson and Minkowski (1939) and by Allen (1940) for several flares observed in the years 1937 and 1938. Quantitative discussion of flare spectra started only after the second world war, when the first Hα line profile was obtained by Ellison (1946) and when Ellison and Hoyle (1947) suggested that the Stark effect is the broadening mechanism for hydrogen lines in flares. Other lines in the flare spectra also were photographed in later years, but high-dispersion photographs at widely separated wavelengths could not be obtained simultaneously; because flares develop very rapidly a simultaneous record of the

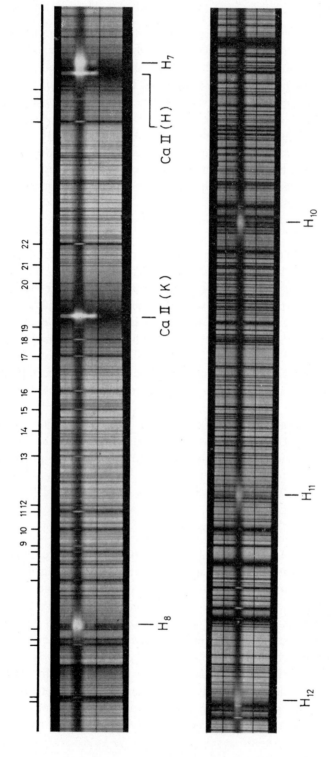

Fig. 25. Two sections of the spectrum of the major flare of 1961, July 12. The upper part extends from λ 3873 to λ 3978 Å, including the outstanding emissions in the Hε and H8 Balmer lines and H and K Ca II lines. It also includes the whole spectral region listed in Table VI (the numbers on the top refer to this table). The lower part extends from λ 3747 to λ 3819 Å (with higher dispersion) and shows three higher lines of the Balmer series of hydrogen. (Courtesy of the Astronomical Institute of the Czechoslovak Academy of sciences, Ondřejov.)

whole spectrum is necessary for obtaining correct information on the physical parameters in the flaring region.

Ten more years elapsed until special flare spectrographs fulfilling this demand were
built for the period of the International Geophysical Year, and complete flare spectra
(Figure 25) were obtained at several solar observatories (H. J. Smith, 1957; Michard
et al. 1959; Valníček et al., 1959; Suemoto and Hiei, 1959; Severny et al., 1960;
Gurtovenko and Didytchenko, 1960). Analyses of them have given us information
on the physical conditions in the low-temperature chromospheric part of the flare
phenomenon.

2.2.1. Characteristic features of the flare spectrum

Flare radiation in the optical region and in the UV up to the Lyman limit of hydrogen
is emitted in spectral lines; only very rarely does continuum emission also occur.
Figure 25 shows an example of the spectrum of a major solar flare and Table V gives
a list of the most prominent lines in the visible part of the flare spectrum. In order to
give the reader an idea of how many spectral lines are excited in a large flare, Table VI
gives the complete list of all observed lines in a short selected wavelength interval of
Table V (reproduced in the upper spectrum in Figure 25). These tables are based upon
the lists of flare lines prepared by Jefferies et al. (1959), Švestka et al. (1961), Blaha
et al. (1962), Stepanyan (1963), Koval and Steshenko (1963), Banin (1965a), Švestka
(1966), Polupan (1968), Grossi Gallegos et al. (1971) and Machado and Rust (1974).
We have no knowledge about the flare lines excited below 3400 Å in the UV, since
experimental data on flare spectra in this region are completely missing. Further
along, we know spectra of flares in the EUV region, but these already concern the
transition region between the low- and high-temperature flare and we shall discuss
them in Section 3.2.2.

At the very beginning of a flare, emission becomes visible in the lowest lines of the
Balmer series and in the H and K lines of singly ionized calcium, and slight absorption
appears in the D_3 line of the He I. Very soon, broad wings begin to appear in the
Balmer lines and as soon as the wings are formed, higher lines of the Balmer series
also become visible (Švestka et al., 1961). As soon as the Balmer series excitation is
sufficiently strong for lines higher than H_{10} to be seen in the spectrum, the emission
in metal lines also appears, mostly in quite a regular sequence of appearance (Blaha
et al., 1962; Banin et al., 1967). Absorption in the D_3 line weakens.

The next development of the flare is characterized by increasing intensity in the
cores and further extension of the wings of the Balmer lines. This manifests itself in
the Hα time-development curves by the steep rise in the flash phase (cf. Figures 2, 7,
and 8), and it is accompanied by increasing intensity of the metal lines. In contrast
to the greatly broadened hydrogen lines, however, the metal emission lines remain
sharp and narrow (cf. Figure 25). The absorption D_3 line disappears, and is replaced
by emission in the flash phase of the flare (Waldmeier, 1955; Švestka et al., 1962).
As soon as the D_3 emission appears, other He lines, too, start to be visible in the flare
spectrum (Švestka et al., 1962). In the IR, however, the λ 10 830 Å helium line seems

TABLE V

The most prominent emission lines in the visible part of the flare spectrum

Wavelength[a] (Å)	Identification[b]	Intensity[c]	Wavelength[a] (Å)	Identification[b]	Intensity[c]
3440.6	Fe I (6)	3	3829.4	Mg I (3)	3
3441.0	Fe I (6)	3	3832.3	Mg I (3)	3
3442.0	Mn II (3)	3	3834.2	Fe I (20)	3
3444.3	Ti II (6)	3	3835.9	H9	4
3461.5	Ti II (6)	4	3838.3	Mg I (3)	4
4365.9	Fe I (6)	3	3856.4	Fe I (4)	3
3474.1	Mn II (3)	4	3859.9	Fe I (4)	3
3475.5	Fe I (6)	3	3886.3	Fe I (4)	3 ⎤
3476.7	Fe I (6)	3	3889.1	H8	4 ⎥
3477.2	Ti II (6)	3	⎡3905.5	Si I (3)	4 ⎥
3482.9	Mn II (3)	3	⎢3922.9	Fe I (4)	3 ⎥
3489.7	Mn II (3)	3	e ⎨3927.9	Fe I (4)	4 ⎥
3490.6	Fe I (6)	3	⎢3930.3	Fe I (4)	4 ⎬ d
3495.8	Mn II (3)	3	⎢3933.7	Ca II (1)	5 ⎥
3497.8	Fe I (6)	3	⎣3944.0	Al I (1)	3 ⎥
3570.1	Fe I (24)	3	3961.5	Al I (1)	3 ⎥
3581.2	Fe I (23)	3	3968.5	Ca II (1)	5 ⎥
3631.5	Fe I (23)	3	3970.1	H7	5 ⎦
3647.8	Fe I (23)	3	4045.8	Fe I (43)	3
3679.9	Fe I (5)	3	4063.6	Fe I (43)	3
3705.6	Fe I (5)	3	4101.7	H6	5
3706.0	Ca II (3)	4	4226.7	Ca I (2)	3
3719.9	Fe I (5)	3	4233.2	Fe II (27)	3
3721.6	Ti II (13)	3	4307.9	Fe I (42)⎫	
3722.6	Fe I (5)	3		Ti II (41)⎭	3
3733.3	Fe I (5)	3	4340.5	H5	5
3736.9	Ca II (3)	4	4351.8	Fe II (27)	3
3737.2	Fe I (5)	4	4383.5	Fe I (41)	3
3745.6	Fe I (5)	4	4471.7	He I (14)	4
3745.9	Fe I (5)	3	4861.3	H4	5
3748.3 ⎤	Fe I (5)	3	4923.9	Fe II (42)	4
3758.2 ⎥	Fe I (21)	3	5015.7	He I (4)	3
3759.3 ⎥	Ti II (13)	3	5018.2	Fe II (42)	4
3761.3 ⎬ d	Ti II (13)	3	5167.3	Mg I (2)	3
3770.6 ⎥	H11	3	5169.0	Fe II (42)	4
3797.9 ⎥	H10	3	5172.7	Mg I (2)	3
3798.5 ⎥	Fe I (21)	3	5183.6	Mg I (2)	3
3799.5 ⎦	Fe I (21)	3	5234.6	Fe II (49)	3
3820.4	Fe I (20)	3	5316.6	Fe II (49)	3
3824.4	Fe I (4)	3	5875.6	He I (11)	5
3825.9	Fe I (20)	3	5890.0	Na I (1)	4
3827.8	Fe I (45)	3	6562.8	H3	5

[a]   The data below 3500 Å are less reliable since they are based on measurements of only one author (Polupan, 1968).

[b]   The multiplet number is given in brackets. Hn means the nth line of the Balmer series of hydrogen.

[c]   The following scale has been used (identical to that one in Table V): 3 – moderately strong, 4 – strong, and 5 – very strong emission lines.

[d]   These two spectral regions correspond to the spectra reproduced in Figure 25.

[e]   This spectral region coincides with Table VI.

TABLE VI

A section of Table V between 3900 and 3950 Å comprising all emission lines observed in the flare spectra in this spectral region. This part of the spectrum is also illustrated in Figure 25

[c]	Wavelength (Å)	Identification[a]	Intensity[b]	[c]	Wavelength (Å)	Identification[b]	Intensity[b]
(9)	3900.5	Ti II (34)	2	(18)	3930.3	Fe I (4)	4
(10)	3903.0	Fe I (45)	2		3931.1	Fe I (565)	0
	3903.9	Fe I (429)	1	(19)	3932.0	Ti II (34)	2
(11)	3905.5	Si I (3)	4		3932.6	Fe I (280)	0
(12)	3906.5	Fe I (4)	2		3933.7	Ca II (1)	5
	3907.5	Fe I (284)	0		3934.8	Zr II (43)	0
	3907.9	Fe I (280)	1		3935.3	Fe I (282)⎫	
	3909.8	Fe I (364)	0			Fe II (562)⎭	0
	3910.8	Fe I (284)	0		3935.8	Fe I (362)	0
(13)	3913.5	Ti II (34)	2		3937.3	Fe I (278)	1
	3913.6	Fe I (120)	0	(20)	3938.3	Fe II (3)	2
	3914.6	Fe I (120)	1	(21)	3940.9	Fe I (20)	1
(14)	3917.2	Fe I (20)	2		3942.5	Fe I (364)	1
	3918.4	Fe I (364)	0		3943.3	Fe I (72)	1
	3918.6	Fe I (430)	1	(22)	3944.0	Al I (1)	3
	3919.1	Fe I (430)	1		3945.1	Fe I (280)	0
(15)	3920.3	Fe I (4)	2		3945.2	Fe II (3)	1
	3921.8	Zr I (8)	0		3947.0	Fe I (561)	0
(16)	3922.9	Fe I (4)	3		3947.5	Fe I (361)⎫	
	3924.5	Ti I (13)	0			Fe I (426)⎭	0
	3925.2	Fe I (567)	1		3948.1	Fe I (562)	1
	3926.0	Fe I (364)	1		3948.8	Fe I (604)	1
	3926.7	Fe I (364)	0		3950.0	Fe I (72)	1
(17)	3927.9	Fe I (4)	4				
	3929.1	Fe I (280)	0				

[a] Same as in Table V.
[b] The following scale has been used:
   0 – weak, and found by one author only,
   1 – weak, but confirmed independently by two or more authors,
   2 – moderate,
   3 – moderately strong,
   4 – strong,
   5 – very strong emission lines.
   The scale refers to the maximum phase of large flares.
[c] The numbers refer to the line numbering in the upper part of Figure 25.

to be in emission from the flare beginning (Rust and Bridges, 1975). In limb flares, and occasionally in the brightest disk flares, the λ 4686 line of He II can also be seen in the flash phase. On the other hand, in the EUV, as we shall see in Section 3.2.2, the Lyman series of He II, and the He II Lyman-α line at λ 304 Å in particular, show strong emission in the flare spectrum (Tousey *et al.*, 1973).

In the maximum of large flares we finally see the very rich emission spectrum, a part of which is shown in Table VI. H15 or H16 are usually the highest Balmer lines clearly visible in spectra of large solar flares on the disc. Above the limb, Balmer lines up to H30 can be observed on rare occasions (Jefferies and Orrall, 1961a). In smaller flares the development stops before it reaches this fully developed phase. Thus, even in the maximum phase of small flares only a few metal lines are reversed, the D_3 line remains in absorption, wings of hydrogen lines are only slightly developed or do not exist at all, and the visibility of the Balmer series ends at H12 or lower.

Generally, spectra of different flares observed on the disc vary within a wide intensity range, but the main characteristic features of all spectra are fairly uniform. Attempts to find distinctive characteristics in flares of some particular type (proton flares, e.g.) have failed (Blaha et al., 1962; Koval and Steshenko, 1963; Banin et al., 1967; Steshenko, 1971b). The sequence of appearance for emission lines appear to be quite regular, and the intensity ratios of various lines are similar in all flare regions. Exceptions to this rule are very rare, and they seem to be mostly or entirely due to absorption effects and to differences in positions of flares on the disc (Blaha et al., 1962; Švestka, 1963); some differences appear in flare spectra above sunspots (Machado and Seibold, 1973), but these are probably due partly to the increased contrast against the background spectrum and partly to smaller optical thickness of the flare parts penetrating the chromosphere over sunspots.

This uniformity of flare spectra is in apparent disagreement with the great variety of flare shapes as observed above the limb, with the suggestions of some authors (Slonim, 1962; Dolginova and Odincova, 1967; McKenna-Lawlor, 1968) that the so-called flare includes several different types of phenomena, and with the fact that some flares do, and others do not, produce the impulsive flash phase, radio bursts, X-rays, and particle streams in space. For example, as Donnelly (1973) has pointed out, in the spectra of impulsive kernels one would expect an enhancement of lines which originate in the low chromosphere and near the temperature minimum region.

This discrepancy between what we expect and what we observe might be explained in three different ways:

(a) Flares differ greatly in the coronal parts of the atmosphere, but their effect on the chromosphere, where the spectra are obtained, is essentially the same in all the events. This may be the case particularly if flares originate in the corona and the Hα emission is only a secondary product of the coronal energy release, as most authors now believe (cf. Švestka (1973) and Section 2.3).

(b) Differences in the spectrum exist only in the very initial phase of the flare development, for which very few spectra have been obtained (it takes some time to start operation of a spectrograph after a flare has been sighted).

(c) Since flares possess a fine structure (cf. Section 2.2.7), differences might exist in the individual flare elements and we are unable to distinguish them when integrating a large area on the solar disc through the spectrograph slit.

Spectra of limb flares differ from disc-flare spectra in six main aspects (Jefferies and Orrall, 1961a, b; Zirin, 1964; Švestka, 1965; Zirin and Acton, 1967; Dodson and Hedeman, 1968; Severny, 1968):

(1) Very high Balmer lines, in some cases as high as H30 can be distinguished in the spectrum;

(2) metal lines lose the sharp and narrow character and, in some cases, their width is comparable to that of high Balmer lines;

(3) in some limb flares Doppler shifts appear;

(4) He II emission at λ 4686 Å is observed;

(5) enhanced coronal line emission becomes visible at high altitudes;

(6) sometimes the Balmer continuum appears.

These differences are due partly to the increased contrast of the flare spectrum, which is no longer projected on the bright background of the solar disc. Items (1), (4), (5), and (6) can be ascribed, at least partly, to this effect. Apart from this however, (1) indicates lower density, (4) and (5) higher temperature, and (2) and (3) the existence of non-thermal motions in the higher parts of the flares observed above the disc.

The continuous spectrum observed in some limb flares (item (6)), due to free-bound transitions beyond the Balmer limit of hydrogen, is too weak to be observed in projection on the solar disc. It may perhaps contribute to the continuum in some exceptional flare events (Machado and Rust, 1974), but in general the continua of flares on the disc must have another source for their origin. We shall discuss this problem in more detail in Section 2.2.11.

2.2.2. METHOD OF ANALYSIS OF HYDROGEN LINES

When discussing the hydrogen lines in flare spectra, the equation

$$I(\lambda) = I_0(\lambda) \exp[-\tau_0(\lambda)] + S\{1 - \exp[-\tau_0(\lambda)]\} \tag{1}$$

is usually used, where $I(\lambda)$ denotes the emerging intensity at a wavelength λ, I_0 is the intensity entering the flare from below, and τ_0 is the optical thickness of the flare, all of these quantities being considered along the line-of-sight. S is the source function in the flare region. In limb flares the first term on the right-hand side disappears.

This equation, however, involves several simplifying assumptions (Švestka, 1965, 1972a). First, the source function S is not constant. It certainly varies with optical depth, τ, as evidenced by the central reversal often observed in lower Balmer lines. Thus S in Equation (1) should represent the mean value of the source function in the flare. However, its actual value is

$$S = \int_0^{\tau_0(\lambda)} S(\tau) \exp[-\tau(\lambda)] \, d\tau \Big/ \int_0^{\tau_0(\lambda)} \exp[-\tau(\lambda)] \, d\tau, \tag{2}$$

which equals the mean value of the source function $S(\tau)$ in the interval $(0, \tau_0(\lambda))$ only if $\tau_0(\lambda) \ll 1$. If this condition is not fulfilled throughout the whole line, S is no longer a constant, but varies with the distance from the line center, depending on the $S(\tau)$ variation with depth. Therefore, Equation (1) can be applied safely only to lines, or line parts (wings), where $\tau_0(\lambda) \ll 1$, or at east < 1. We also neglect the change of S with λ due to frequency redistribution (cf., e.g., Athay, 1972), which is another reason why the central parts of the lines should be avoided in the analysis.

Another very serious simplification is the assumption of a homogeneous flare.

As we shall see later, analysis shows that flares most probably are not homogeneous radiating regions, but may be composed in fact of small flaring elements, whose dimensions are well below the instrumental resolution. Thus, with Δx and Δy being the dimensions of the region the light of which we integrate (i.e. the widths of the spectrograph and microphotometer slits), we simplify:

$$\int_{\Delta x} \int_{\Delta y} I(\lambda, x, y) \, dx dy = I(\lambda). \tag{3}$$

This simplification may have serious consequences. If the flare is inhomogeneous, then wherever the flare optical thickness is small (high Balmer lines, extreme wings of strong lines, helium lines other than D_3 and λ 10 830, and most metal lines) only the flare elements contribute to the observed emission, hence $I(\lambda)$ represents an integration of the flare emission proper. But the radiation may be diluted if the elements do not completely cover the integrated area. In such a case the analysis gives correct results only if *all* flare elements are optically thin (Švestka, 1972c). Relative measurements are still reliable, since the dilution affects all wavelengths in the same way, but one should avoid drawing any conclusions based upon the absolute intensities. Also, the flare elements obviously excite the surrounding 'interflare' matter and produce, thereby, some additional radiation in the central parts of strong lines. This is another reason why the optically thick regions should be avoided.

The individual elements of flares must differ greatly as to the physical conditions in them (or alternatively, the physical conditions differ greatly within one single element), since temperatures from about 7000 K, corresponding to metal lines, up to above 25 000 K for He II can be deduced from a single optical spectrum of a flare. In fact the chromospheric flare resembles a very heterogeneous medium. Therefore, when discussing a particular line, we have to assume that its emission is produced in elements of approximately identical physical properties and that the contribution of elements at different temperatures and densities to the radiation in this line can be neglected. This certainly is not fully true, but there is no other way to handle the problem. Some satisfaction may perhaps derive from the fact that stellar astronomers work with integrated light of what are certainly very complicated objects, and they do not care (and cannot care) for the fine structure. Yet they still get reasonable results on the physical parameters in stars.

Summarizing, use of the simplified Equation (1) is permitted if (a) the optical thickness $\tau_0(\lambda)$ of the flare in the spectral region considered is less than unity, and (b) only lines originating under similar conditions are compared. In such a case relative measurements can be considered reliable, provided that the dilution of radiation is not too pronounced.

2.2.3. LINE BROADENING

In a major flare the Balmer lines are very broad (cf. Figure 25) and the first problem to be solved is the mechanism of this broadening. The two competitive broadening mechanisms are the Doppler effect due to non-thermal motions (micro- or macro-turbulence) and the Stark effect. For many years, contradictory results were obtained for different flares and the correct broadening mechanism could not be definitely established (for references to some 30 papers discussing this problem see Švestka (1965)). Looking back with our present knowledge, one can find four main reasons for it:

(a) Often the restricting assumptions upon which Equation (1) is based were forgotten. The value of S was taken as being constant throughout the line, apparently with misleading results.

(b) In the wings, bad seeing favours the Doppler-like profile over the Stark profile (Švestka, 1965).

(c) The Stark profile also resembles a Doppler profile when the flare radiation is diluted (Švestka, 1972b).

(d) There is probably a real difference in the broadening of flare lines above the limb and on the disc, and this fact contributed to the confusion in the results.

It is now completely clear that the Balmer lines of all disc flares are broadened by the Stark effect, and there are several convincing independent arguments for it:

(1) While Balmer lines are very broad, metal lines remain extremely narrow (see

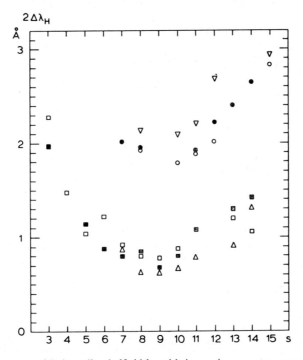

Fig. 26a.   The increase of Balmer line halfwidths with increasing quantum number s for $s \geq 9$ in small (Suemoto and Hiei, 1959) and large (Švestka, 1965) flares, due to Stark effect. Different marks refer to different flares. The halfwidth increase from $s \simeq 9$ toward lower s values is due to increasing self-absorption.

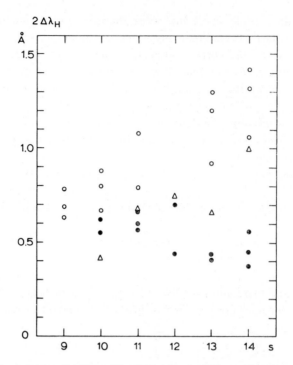

Fig. 26b.   Same dependence as in Figure 27a plotted for small flares (circles). In addition, we have plotted halfwidths in a limb flare (triangles, after Hiayama, 1961) and in an active prominence (dots, after Švestka, 1964b). Due to low electron density no halfwidth increase appears in the prominence lines.

Figure 25). This cannot be the case with Doppler broadening due to non-thermal motions, and the ratio of Doppler widths for hydrogen and metals (Fe, Ti) greatly exceeds the value of about 7 expected for thermal broadening.

(2) The halfwidth of high Balmer lines increases with series number s, starting at about $s = 10$ (Figure 26a). Only high electron density can produce this rise. For a pure Doppler broadening the halfwidth should steadily decrease with increasing s. Since the increase of halfwidths of high Balmer lines in flare spectra has been confirmed for large numbers of flares (Suemoto and Hiei, 1959; Hirayama, 1961; Švestka, 1963, 1965; de Feiter and Švestka, 1964; Banin, 1965b; Stepanyan, 1969; Kurochka, 1970), it represents evidence for the Stark broadening that can hardly be doubted.

(3) The wings of low Balmer lines extend over photospheric absorption lines, many of which are not enhanced in the flare spectrum. The visibility of these background lines depends on the transparency of the Balmer wings, i.e. on the optical thickness of the flare at the particular wavelength (cf. Section 2.2.5(a)). Thus, several values of τ_0 in the wings of low Balmer lines can be determined from the visibility of the Fraunhofer lines and compared with theoretical values resulting for different modes of broadening. This has been done by Švestka (1961) and E. v. P. Smith (1963) with the result that the optical thickness in higher Balmer lines, when compared with

that in the Hα line, is too large (by a factor of ~ 100 for H8) if Doppler broadening is assumed. On the other hand this factor is less than 2 if the Stark effect with $n_e = 3 \times 10^{13}$ cm^{-3} is considered.

Quite similar results have also been obtained by Semel and Soru-Escaut (1971). Even when their assumption that the maximum intensity in each line approximates the value of S may be doubted, it is essentially impossible to change the source functions to such an extent that the observed profiles agree with the Doppler ratio of $\tau_0(\mathrm{H}\alpha)/\tau_0(\mathrm{H10}) = 280$ for the same $\Delta\lambda/\lambda$ ($\Delta\lambda$ being the distance from the line center). For the Stark effect this ratio (for the same $\Delta\lambda$) is about 17, in good agreement with observation.

In spectra of limb flares the difference in width of hydrogen and metal lines is much less striking; in some limb flares the ratio of the halfwidths of high Balmer lines to metal lines is below the value for thermal Doppler broadening, thus indicating an increase of non-thermal motions with height in the flare region (Jefferies et al., 1961 a; Švestka, 1965).

2.2.4. ELECTRON DENSITY

The broadening mechanism being known, we can determine the electron density from the halfwidths of those high Balmer lines for which the condition $\tau_0 < 1$ is fulfilled. It is important to realize that the halfwidth of an optically thin line is completely independent of the simplifying assumptions we specified in Section 2.2.2. Even with diluted radiation the halfwidth remains the same.

A. *Electron Density in Disc Flares*

Redman and Suemoto (1954) were the first to show that the profiles of emission Balmer lines in the chromosphere can be approximated by the Voigt function, i.e. one can put

$$\tau_0(a, v) = \tau_0(0) \, H(a, v), \tag{4}$$

where a is the damping width and $v = \Delta\lambda/\Delta\lambda_t$ with $\Delta\lambda_t$ denoting the Doppler width. $\tau_0(0)$ is the optical thickness in the line center,

$$\tau_0(0) = 5.0 \times 10^{-13} \frac{\lambda^2 f}{\Delta\lambda_t} n_2 z. \tag{5}$$

Here n_2 is the average number of two-quantum hydrogen atoms cm^{-3}, z the linear thickness of the flare along the line of sight and f the oscillator strength of the line considered. The Voigt function $H(a, v)$ can be taken from appropriate tables (e.g. Posener, 1959). Then, according to Equations (1) and (4), as long as $\tau_0(0) \ll 1$, the net emission is given as

$$I(v) - I_0(v) = [S - I_0(v)] \, \tau_0(a, v). \tag{6}$$

These relationships are exact only for a combination of Doppler effect and damping. However, as has been shown by several authors (Suemoto and Hiei, 1959; Švestka,

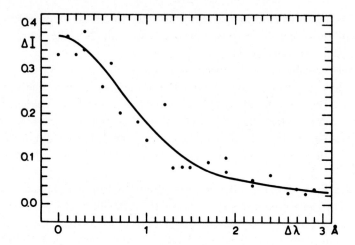

Fig. 27.   The measured profile of the H11 line in the spectrum of the flare of 1960, August 7 (dots) compared with a Voigt profile computed for $a = 1.20$. (After Švestka, 1965.)

1963, 1965; de Feiter, 1964, 1966), it also represents a good approximation to the actual line profiles in flares when the Stark effect is added, with deviations affecting only the far wings of the line (Figure 27). Then

$$\Delta\lambda_t = [(\Delta\lambda_D)^2 + (\Delta\lambda_F)^2]^{1/2} \qquad (7)$$

and

$$a = (\gamma_i + \gamma_e)/\Delta\lambda_t, \qquad (8)$$

where $\Delta\lambda_D$ is the actual Doppler width, $\Delta\lambda_F$ is the fictitious width of the line broadened by the Stark effect only, while γ_i and γ_e are the ion and electron damping constants, respectively.

For Holtzmark (pure ion) broadening

$$\gamma_i = 0.4\,\Delta\lambda_F, \qquad (9)$$

and the halfwidth $\Delta\lambda_F$ can be approximated according to Griem (1960) for a Balmer line with the upper quantum number s as

$$\Delta\lambda_F = 3.88\,K(s)\,F_0, \qquad (10)$$

where

$$K(s) = 5.5 \times 10^{-5}(2s)^4/(s^2 - 4) \qquad (11)$$

and F_0 is the normal field strength for an ion concentration (in the hydrogen atmosphere it is equal to the electron concentration) n_e,

$$F_0 = 1.25 \times 10^{-9}\,n_e^{2/3}. \qquad (12)$$

However, the most difficult problem remains. It is the determination of γ_e in Equation (8). Its value has been discussed by Griem (1960, 1962, 1967), Minaeva *et al.* (1967), Minaeva and Sobelman (1968), and Minaeva (1968), and with respect to flare spectra by de Feiter (1966) and Kurochka (1969, 1970). Several expressions for γ_e have been obtained, all approximations, and at present it is difficult to judge which is the best. As an illustration, Table VII shows the differences which result when the various approximations are used.

If one of the approximations is adopted (throughout this book we adopt that of Griem (1962; Švestka, 1965)) we get for each Balmer line a dependence of the half-width, $\Delta\lambda_H$, on the electron density n_e for a given temperature T_e. Commonly $T_e = 10\,000$ K is adopted, but the resulting value of n_e is not very sensitive to the variation of T_e, changing only by a factor 1.4 within the limits of $5000\,\text{K} < T_e < 20\,000$ K. A detailed analysis of the errors entering into the described method of n_e determination has been given by Švestka and Fritzová (1967). The greatest uncertainty certainly is in the value of γ_e. However, as Table VII and other comparisons show, even this cannot lead to an error in n_e exceeding a factor 2.

Since the increase of $\Delta\lambda_H$ with s starts with $s = 10$ (Figure 26), one can suppose that H10 is the lowest line for which self-absorption is unimportant. Therefore, the method can be applied to the H10 and higher Balmer lines. The results (Švestka, 1965; de Feiter, 1966; Fritzová and Švestka, 1967; Kurochka, 1970; Švestka, 1972a; and other references in these summarizing papers) show that in the maximum phase of all disc flares of importance 1 + or higher the electron density is always greater than 10^{13} cm^{-3}. In the later development of these flares n_e can be as low as 6×10^{12} cm^{-3}.

TABLE VII

A comparison of electron density values obtained with different approximations of the electron damping contribution [a] (for the flare of 1960, August 7)

Approximation	Author	n_e (cm^{-3})
Pure ion Holtzmark broadening (electron damping neglected)	Švestka (1963)	6.4×10^{13}
Holtzmark broadening with the same contribution by ions and electrons		3.2×10^{13}
computed according to Griem (1960) without the Lewis (1961) correction (the effect of γ_e overestimated)	Švestka (1963)	2.1×10^{13}
computed according to Griem (1962) with the Lewis correction included	Švestka (1965)[b] de Feiter (1966)[b]	2.9×10^{13} 2.1×10^{13}
computed according to Minaeva et al. (1967)	Kurochka (1970)	2.0×10^{13}

[a] Based on Švestka and Fritzová (1967) and Kurochka (1970).
[b] The difference between these two results is due to the fact that γ_e in Griem's (1962) approximation varies with λ within the line. Therefore, γ_e for some representative wavelength must be selected as entering into Equation (8) and thus into the Voigt profile. The two authors differed slightly in their choice of this representative wavelength.

However, we have essentially no information on n_e in smaller flares, nor in the very early or very late phases of large flares, simply because of the absence or weakness of high Balmer lines in the spectrum.

The highest electron density ever found was deduced from the spectrum of the large proton flare of 12 July 1961, $n_e = 4.4 \times 10^{13}$ cm^{-3} (de Feiter and Švestka, 1964). This value decreased to 2×10^{13} cm^{-3} after 17 min. The average time variation of the electron density in large flares (based on measurements of five events) is shown in Figure 28.

Fritzová and Švestka (1967) have found the electron density in different bright nodules of the same flare to be approximately identical. Of course, these measurements always referred to the brightest flare regions. Stepanyan (1969), who studied one rather unimportant knot above a penumbra in the proton flare of 2 September, 1966, found $n_e \leqslant 7 \times 10^{12}$ cm^{-3}, which indicates that the electron density in some parts of major flares (and consequently in small flares, too) might be lower than 10^{13} cm^{-3}. Stepanyan's result has not been confirmed by Machado and Seibold (1973) who have found in 9 spectra of 6 flares above spots n_e always in excess of 10^{13} cm^{-3}. However, this value might have been overestimated, since Machado and Seibold used the Hε line for the n_e determination. Though seemingly optically thin above spots, the line still may be optically thick in the individual (small) emitting elements. The resulting self-absorption then would make the n_e higher (cf. Švestka, 1972c).

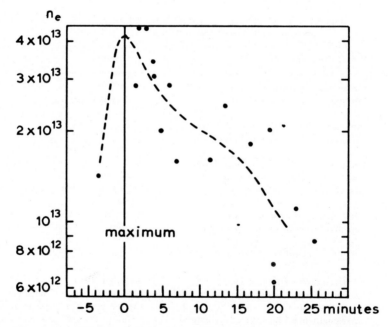

Fig. 28.   The electron density variation during the flare development as composed from measurements of five flares of importance $\geqq 2$. (After Fritzová and Švestka, 1967.)

Recently, Spicer and Davis (1975) suggested that Stark broadening might be increased (for the same n_e) in plasma which is driven unstable by shock waves or energetic particle beams which generate turbulent fields. This effect might influence n_e determinations in the flare kernels during the impulsive phase. The highest n_e values, however, have been found in the maximum of Hα flares that occurs after the impulsive phase, and almost all other n_e values have been determined during the flare decay.

Semel and Soru-Escaut (1971) have pointed out that n_e values in individual flare elements may vary and that suitable line profiles can also be obtained when combining widely different electron densities in different flare elements. This certainly may be true, but of course it is hardly possible to deduce the real n_e variations from the observed line profiles and we must be content with a constant n_e expressing some representative (but maybe not the exact 'mean') value of the electron density in the flare elements.

Electron density determinations based on lines of elements other than hydrogen do not give us any more information. According to Athay and Johnson (1960) and Krat and Sobolev (1960) n_e must exceed 10^{12} cm^{-3} in all flare regions where the D$_3$ line of neutral helium is observed. Estimates of the electron density from the He II lines are very unreliable, leading to values between 10^{10} and 10^{14} cm^{-3} (Jefferies et al., 1959; Sobolev, 1962; Goldberg-Rogozinskaya, 1962). An analysis of metal lines has led to the limits $10^{13} < n_e < 10^{14}$ cm^{-3} (Letfus, 1964).

B. Electron Density in Limb Flares

In the case of limb flares the halfwidth method could be applied only in two events marked 2 and 5 in Figure 29 (Fritzová and Švestka, 1967; Kurochka, 1970). In several other cases, however, n_e could be determined (or estimated) by other methods. All the results of these electron density measurements are summarized in Figure 29 (Švestka, 1972a), where an attempt has been made to present a rough picture of the decrease of n_e with height. 'Table' shows the n_e-range in disc flares.

The other methods used can be summarized as follows:

(a) Electron density can be deduced from the intensity of the Balmer continuum when present in the spectrum above the limb (Jefferies and Orrall, 1961 a, b; Orrall in Zirin and Acton, 1967). One can suppose that the continuum before the Balmer limit (after eliminating scattered light, which is no easy problem) is formed by four processes: emission of negative hydrogen ions, with

$$E_{\text{H}^-}\, d\lambda = 4.0 \times 10^{-42}\, n_e^2\, n_i\, T_e^{-1/2}\, \exp[14.4 \times 10^4 / T_e]\, d\lambda \qquad (13)$$

(thermodynamic equilibrium being assumed; in fact this contribution is negligible if the electron temperature T_e is high enough); free-free transitions,

$$E_{ff}\, d\lambda = 1.2 \times 10^{-19}\, n_i\, n_e\, T_e^{-1/2}\, \exp[-3.94 \times 10^4 / T_e]\, d\lambda\,; \qquad (14)$$

scattering of photospheric light by free electrons,

$$E_s\, d\lambda = 2.8 \times 10^{-11}\, n_e\, d\lambda\,; \qquad (15)$$

and by free-bound emission due to captures to hydrogen quantum levels higher than 3,

$\sum_{s=3}^{\infty} E_{s, \infty}$, where

$$E_{s, \infty} \, d\lambda = 3.9 \times 10^{-14} n_i n_e T_e^{-3/2} s^{-3} \exp[1.579 \times 10^4/s^2 T_e - $$

$$- 3.964 \times 10^4/T_e] \, d\lambda. \tag{16}$$

Behind the Balmer limit, the free-bound emission to the second quantum level is to be added to these emission sources,

$$E_{2, \infty} \, d\lambda = 4.9 \times 10^{-15} n_i n_e T_e^{-3/2} \, d\lambda. \tag{17}$$

All the emissions are in units of erg cm^{-3} s^{-1} sr^{-1}.

Thus, if one measures the intensity before and behind the Balmer limit, one can deduce the quantity

$$R_B = \frac{E(<3646) - E(>3646)}{E(>3646)} = \frac{E_{2, \infty}}{E_{H^-} + E_{ff} + E_s + \sum_{3}^{\infty} E_{s, \infty}}. \tag{18}$$

After substituting in Equation (18) the expressions (13) through (17) and assuming $n_i = n_e$, one gets a relation between T_e and $E_{2, \infty}$, with n_e being eliminated, from which T_e can be determined. Then, knowing T_e, the electron density can be found from Equation (17).

The n_e values denoted by the numbers 1', 4', and 9' in Figure 29 have been found in this way. In fact they are based on absolute photometry, i.e. on the measured value of $E_{2, \infty} = E(<3646) - E(>3646)$. Therefore, if the flare structure is filamentary, the electron density in the flare filaments may greatly exceed the values of n_e found by this method.

(b) The reliability of the preceding method can be checked by applying the corrected Inglis-Teller formula (taking electron damping into account) to the same limb events. This formula can be written (Kurochka and Maslennikova, 1970) as

$$\log n_e = 22.0 - 7.0 \log s_M, \tag{19}$$

where s_M denotes the series number of the highest line still resolved in the spectrum (its halfwidth being equal to half the distance to the line $s_M + 1$). Thus we obtain the n_e values marked as 1 and 4 in Figure 29. Since Equation (19) is based on the line halfwidth, it should not be influenced by the inhomogeneous flare structure. One can see that whilst 1 does not differ significantly from 1', the value 4 exceeds by a factor greater than 2 the value 4' determined by the previous method. Thus one can suppose that the inhomogeneous structure did not essentially influence the deduced values of n_e in the first limb flare (1, 1'), but that it caused an underestimate of n_e when the second flare (4, 4') was studied. The measurement 5 also refers to the flare (4, 4'), but value 5 has been deduced from the observed halfwidths of high Balmer lines (Kurochka, 1970). As one should expect, it is in good agreement with the values 4, but it disagrees with 4'.

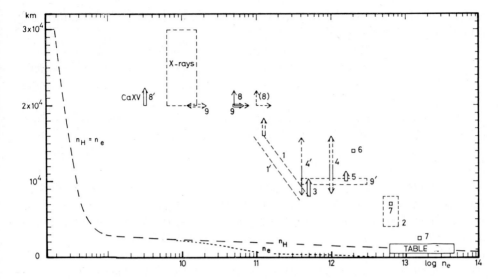

Fig. 29.   A summary of the n_e values measured in the limb flares. See the text for explanation of the numbers 1 to 9 and 'Table'. A comparison with the decrease of hydrogen density (n_H) and electron density (n_e) with height in the quiet solar atmosphere shows that the density in flares is generally by about two orders of magnitude higher. (After Švestka, 1972, and de Feiter, 1974).

The Inglis-Teller formula was also applied by Stepanyan (1963) to two regions of the limb flare of 1959, July 25 (notation 7 in Figure 29, the original values have been corrected for electron damping).

(c) Polupan and Yakovkin (1966) determined the n_e value in a limb flare (number 6 in Figure 29) by solving the steady state equations for the first six levels of the hydrogen atoms. This method, of course, meets with the serious difficulty that we have no data on the population of the ground level of hydrogen. Populan and Yakovkin assumed that the kinetic temperature coincides with electron temperature and is equal to the temperature of Lα radiation.

(d) In very high regions of flares one can suppose that the observed continuum emission is due solely to Thomson scattering which, after using van de Hulst's (1950) formulae, immediately leads to the value of the electron density provided that one knows the linear thickness z of the flare along the line-of-sight. For a flare right at the limb (Zirin, 1964)

$$n_e = 6.3 \times 10^{24} \, (I/I_\odot z),$$ (20)

where I is the intensity observed and $I_\odot$ is the central intensity of the Sun.* z can be approximated by assuming axial symmetry in the limb event observed. Of course, we again only get an average density, whilst some condensations of higher density may exist along the line of sight. This method has been applied by Zirin (1964) and Zirin

* Equation (20) would lead to a constant of 4.0×10^{-11} in Equation (15). The difference is due to the fact that Jefferies and Orrall (1961a) considered Zanstra's (1960) expression which takes the limb darkening into account in a somewhat different way; and of course, the limb darkening is stronger at the Balmer limit than close to the wavelength of 5000 Å considered by Zirin.

and Acton (1967) to two limb events at over 20 000 km above the photosphere (numbers 8 and 9 in Figure 29).

(e) At these great altitudes coronal lines may also be observed. These lines, of course, no longer belong to the low-temperature flare. They correspond to flare regions of very high temperature that are mixed at the same altitude with the low-temperature flare elements.

From the ratio of intensities of the coronal Ca xv lines at λ 5694/λ 5445 Å Zirin (1964) found $n_e = 5 \times 10^9$ cm^{-3} (number 8' in Figure 29) in the same limb event where Thomson scattering gave $n_e = 1 \times 10^{11}$ cm^{-3} (number (8)). Of course, intensity measurements in the coronal lines are difficult and the excitation cross-sections easily can involve an error of one order in magnitude. It is also somewhat doubtful, whether Zirin actually measured what is generally understood to be a flare. It seems more probable that these two measurements – (8) and 8' in Figure 29 – referred to the flare-associated loop prominence in a very late phase of the flare development. Nevertheless, the result indicates that the electron density in the low-temperature elements exceeds the electron density in the high-temperature region at the same level in the atmosphere. This is to be compared with some measurements in soft X-rays (Gabriel and Jordan, 1970; Neupert, 1971; cf. Section 3.1.5), which give opposite results.

(f) Finally, as we shall see in Sections 3.1.5 and 3.1.6, observations of X-ray and microwave bursts lead to an electron density close to 10^{10} cm^{-3}, and this seems to correspond to heights between 20 000 and 30 000 km above the base of the chromosphere. This n_e value is not far from that found by Zirin above 20 000 km from the Ca xv lines.

Summarizing, one can say that the n_e determinations in limb flares, though much less accurate than in flares on the disc, clearly show that the electron density decreases with height. The structure of individual flares, however, may differ greatly in higher atmospheric layers. In the spray-type flare of 7 March, 1959, n_e was found to decrease continuously with height, and at 16 000 km it was a factor of 4 lower than at 7700 km (Jefferies and Orrall, 1961a; Fritzová and Švestka, 1967 – number 1 in Figure 29). In the loop-type flare of 9 June, 1959, however, n_e remained constant over a height difference of 14 000 km (Jefferies and Orrall, 1961b; Fritzová and Švestka, 1967 – number 4).

De Feiter (1974) projected into Figure 29 the n_H and n_e dependence on height for the quiet solar atmosphere (the curves in Figure 29). While n_H is close to the deduced n_e values in the chromospheric flare region (marked 'Table'), n_e in flares is one or two orders greater than $n_H = n_e$ at greater altitudes. Since the flare plasma appears to be only slightly ionized in the chromosphere (cf. Section 2.2.8), this comparison indicates a compression of gas in the flare region throughout the solar atmosphere by an amount up to the order of 10^2.

2.2.5. THE OPTICAL THICKNESS

Most metal and helium lines in the flare spectrum are optically thin. In the strongest metal lines, however, at least in some flares, the optical thickness must exceed 1, because they show self-absorption (Švestka, 1963) and some of them are even self-reversed. The double-peak shape of such lines was first explained as due to Zeeman splitting. If correct, this explanation would imply very high magnetic field strength in flares (Blaha et al., 1960), but later studies have shown that the double peak is a normal self-absorption effect (Kopecký et al., 1963; Banin et al., 1967). However, there also are flares that do not show any traces at all of self-absorption in the metal lines (Banin, 1965a). In a large flare studied by Letfus (1964) most Fe I lines between 4290 and 4400 Å were found to be optically thick whilst Ti II and Fe II line showed an intermediate thickness.

The H and K lines of Ca II are optically thick in flares. In a flare of importance 2 Kazachevskaya (1958) found the column density in the ground state of Ca II equal to 10^{17} cm^{-2} which explains the broadening of H and K line wings by radiation damping. However, analysis of the H and K lines is difficult, because in the H and K absorption lines not only the line cores, but the far wings, too, are formed at a high level in the solar atmosphere. Therefore, we do not know the value of I_0 in Equation (1) and the net emission line profile cannot be obtained with confidence.

The cores of H and K lines appear narrower above sunspots (Švestka et al., 1961; Tandberg-Hanssen, 1967a). The fact that the Hα width and the number of resolvable Balmer lines are also suppressed above umbrae (Švestka et al., 1961) indicates that flare knots may be optically thinner above umbrae. This has also been suggested by Athay and Skumanich (1968). It has been actually confirmed by Machado and Seibold (1973) who have determined the optical thickness from the ratio of central intensities in the H and K lines, I_K/I_H, in several flare nodules above sunspots. The optical thickness of flares above spots is found to be one or two orders of magnitude lower than outside spot penumbrae. In the D_3 line the flare appears to be about six times less thick over an umbra than in the surroundings (Fontenla and Seibold, 1973). An acceptable explanation is that only narrow 'tongues' of the flare emission penetrate above the spots; thus the flare radiation above spots is greatly diluted (cf. Švestka, 1972c) and its analysis then leads to a very small 'average' optical thickness.

Apart from the most prominent lines of neutral helium in the flare spectrum (D_3, λ 6678, λ 10 830 Å) the helium lines in flares, and the singlet lines in particular (Banin, 1965b), are optically thin. The D_3 line is optically thick (Jefferies et al., 1959; Steshenko and Khokhlova, 1962; Severny, 1959), but its thickness cannot be very high; probably between 1 and 10 (Kubeš, 1964; Fontenla and Seibold, 1973).

We can deduce from Figure 26a that the optical thickness of a flare in lower Balmer lines must be fairly high, and $\tau_0(0)$ must be >1 up to about H9, because the halfwidth increase from H9 to Hα can be explained only by the effect of self-absorption. There are several methods by which to estimate the flare optical thickness in the Balmer lines:

(a) The first method has already been mentioned briefly in Chapter 2.2.3, item (3): the determination of $\tau_0(\Delta\lambda)$ from the transparency of the wings of low Balmer lines (Švestka, 1965). If $I-I'$ is the depth of a metal absorption line in the flare spectrum at wavelength distance $\Delta\lambda$ from the line center, and I_0-I_0' is the depth of the same line in the spectrum of the quiet chromosphere, the optical thickness of the flare at this wavelength is given by the equation

$$\exp[-\tau_0(\Delta\lambda)] = (I-I')/(I_0-I_0')$$ (21)

provided that the metal line is not excited in the flare spectrum. Far on the wings, a Stark-broadened line profile can be approximated by the asymptotic formula

$$\tau_0(\Delta\lambda) = [C_D(\Delta\lambda)^{-2} + C_S\, G(\Delta\lambda)\, n_e(\Delta\lambda)^{-5/2}]\, n_2 z,$$ (22)

where $\Delta\lambda$ is the wavelength distance from the line center, C_D and C_S are the appropriate constants for the damping and Stark broadening, respectively (tabulated, e.g. in Švestka, 1959), n_e is the electron density assumed equal to the density of hydrogen ions, and $n_2 z$ is the column density of two-quantum hydrogen atoms, i.e. their number above 1 cm^2 of the base of the flare, along the line of sight. $G(\Delta\lambda)$ is Griem's (1960) correction factor which takes electron damping into account,

$$G(\Delta\lambda) = 1 + [(\Delta\lambda_w)^{-1/2} + R(n_e, T_e)\, g(\Delta\lambda)]\, (\Delta\lambda)^{1/2},$$ (23)

where

$$R(n_e, T_e) = 1.15 \times T_e^{-1/2}\, [(s^5+32)/s^2(s^2-4)^{1/2}] \times$$
$$\times [\log(4\times10^6\, T_e/s^2\, n_e^{1/2}) - 0.125],$$ (24)

$$g(\Delta\lambda) = \log(\Delta\lambda_w/\Delta\lambda)/\log(\Delta\lambda_w/\Delta\lambda_p),$$ (25)

$$\Delta\lambda_w = \lambda^2 kT_e/hcs^2,$$ (26)

$$\Delta\lambda_p = \lambda^2(n_e e^2/\pi mc^2)^{1/2}.$$ (27)

Thus, after determining n_e from the halfwidths of higher Balmer lines, and several $\tau_0(\Delta\lambda)$ values from the filling up of underlying metal lines, $n_2 z$ can be found from Equations (22) through (27) provided that T_e is properly approximated. For $T_e = 10^4$ K this method yields $n_2 z$ from 3×10^{15} to $3\times10^{16}\, cm^{-2}$ in different (major) flares (Švestka, 1965). These in fact are the maximum values of $n_2 z$, since any emission of the metal lines themselves would make the optical thickness seemingly higher.

(b) Equation (6), from which the n_e value is determined, yields the correct value of n_e only if the condition $\tau_0(0) \ll 1$ is fulfilled in the line considered. With decreasing s the optical thickness increases and, due to self-absorption, the halfwidth of the observed $[I(v)-I_0(v)]$ profile exceeds the halfwidth of the profile of $\tau_0(a, v)$. Therefore, the deduced n_e increases, and the optical thickness can be estimated from this increase of n_e with decreasing s. In this way, $2.5\times10^{15} < n_2 z < 4\times10^{15}\, cm^{-2}$ was found from series of spectra of three flares (Švestka, 1965; Švestka and Fritzová, 1967; Stepanyan, 1969). From less reliable single measurements of several other flares Banin (1965b) and de Feiter (1966) give $n_2 z$ values from 1×10^{15} to $2\times10^{16}\, cm^{-2}$.

(c) In limb flares, with their high contrasts against the background sky, the highest Balmer lines merge due to Stark broadening, which defines the limit of the discrete Balmer spectrum. In disc flares however, where the contrast against the background disc is much lower, this is not the case. There, as one can easily prove, the absence of higher lines is simply due to the decrease of the optical thickness of the flare with increasing s. As soon as the product $[S - I_0(0)] \tau_0(a, 0)$ in Equation (6) becomes lower than 5 to 10% of I_0, the line ceases to be visible in the flare spectrum.

With only two exceptions extremely close to the limb (Banin, 1965b; Kurochka, 1970), we have never observed lines higher than H15 in flares on the disc. With H15 being the highest visible line, one obtains $2 \times 10^{15} < n_2 z < 8 \times 10^{15}$ cm^2 for excitation temperatures in the H15 line between 6500 and 7400 K (Švestka, 1972a).

Thus we arrive at the conclusion that the most common value of $n_2 z$ in large flares is close to 3×10^{15} cm^{-2}, this value varying by factors of about 0.5 to 2 in individual flare events. Only exceptionally have values up to one order of magnitude higher been found, but some of these exceptions might easily be due to errors in interpretation (e.g. by considering flare excited lines in method (a), or by underestimating the errors of individual measurements in method (b)). With $n_2 z = 3 \times 10^{15}$ cm^{-2}, $n_e = 3 \times 10^{13}$ cm^{-3}, and $T_e = 10^4$ K one finds $\tau_0(0) = 1.3 \times 10^3$ in the Hα line, 1.6×10^2 in the Hβ line, and the following values for high Balmer lines: 2.64 in H8, 1.28 in H9, 0.66 in H10, 0.35 in H11, 0.07 in H14, and 0.02 in H17. This shows that in a typical flare $\tau_0(0)$ actually becomes less than unity in H10, where the rise of half-widths in Figure 26a begins to be apparent, and that one can hardly expect lines like H17 to be visible against the bright background unless the source function in Equation (6) is extremely high (according to Equation (19) H17 is the highest resolvable line for $n_e = 3 \times 10^{13}$ cm^{-3}).

As we mentioned above, the flare optical thickness is found to be significantly smaller in those flare parts which penetrate over sunspots. Machado and Seibold (1973) give in this case $n_2 z \simeq 2 \times 10^{13}$ cm^{-3}, which is most probably a fictitious value originating through dilution of the flare radiation above sunspots (see the discussion earlier in this section, and Švestka, 1972c).

2.2.6. ELECTRON TEMPERATURE

Letfus (1964) constructed the emission curve of growth for Fe I and Fe II lines in the large flare of 1958 July 20, and found an ionization temperature of 6100 K $< T_i <$ 7100 K for $10^{13} < n_e < 10^{14}$ cm^{-3}. The electron temperature must be higher than T_i, since an assumption of $T_e = T_i$ leads to an electron density exceeding 10^{14} cm^{-3}.

In the flare of 1958, August 7, Cowley and Marlborough (1969) found a strengthening of the Fe I λ 4063 line, which they explained as being due to a selective excitation of the λ 3969.3 line of the same multiplet by the strong Ca II H-line emission (centered at λ 3968.5 Å). If this view is correct, then T_e in the flare part, where these lines were produced, could not much exceed 7000 K for $n_e = 10^{13}$ cm^{-3}. This limiting condition becomes still more strict for higher electron densities. An increase in T_e to 10 000 K would bring such an increase in the rate of collisional excitation that the selective excitation mechanism could not be efficient.

In the hydrogen lines, the electron temperature can be estimated from the population of different quantum states (Švestka, 1964a, 1965). When n_e and $n_2 z$ are known, one can find the source function S from the wings of Balmer lines using Equations (22) and (1). If this analysis is restricted to the far wings, with $\tau_0(\Delta\lambda) < 1$, S can be considered to be the mean value of the source function in the flare, according to Equation (2).

The source function S in a Balmer line of wavelength λ is related to the ratio of the density of hydrogen atoms in the sth and 2nd quantum states,

$$S(\lambda) = [2hc^2/\lambda^5 P(\lambda)] (4n_s/s^2 n_2), \tag{28}$$

where $P(\lambda)$ is the intensity of the adjacent continuum, relative to which the S value and all the intensities are expressed. Therefore, by determining S in the lines $H\alpha - Hs$ we also find values of the ratios n_j/n_i $(j > i)$ for all values $3 \leq j \leq s$ and $2 \leq i \leq s - 1$. These ratios are related to electron temperature T_e through the combined Boltzmann-Saha equation

$$n_j/n_i = (b_j/b_i) (j/i)^2 \exp[-hc/k\lambda_{ji} T_e], \tag{29}$$

where the b factors denote deviations in the level population from LTE at temperature T_e, and λ_{ji} is the wavelength corresponding to the (j, i) transition.

Alternatively, one can define

$$(b_j/b_i) \exp[-hc/k\lambda_{ji} T_e] = \exp[-hc/k\lambda_{ji} T_\alpha(j, i)], \tag{30}$$

where $T_\alpha(j, i)$ denotes the excitation temperature in the (j, i) line considered. Since high-energy levels are predominantly populated by collisions and low levels by radiation, and because the radiation temperature in the flare is lower than T_e, then $b_j > 1$ for all j's, and $b_j = 1|_{j \to \infty}$. Also therefore, from Equation (30), $T_\alpha(j, i) < T_e$ for all values of j and i, and $T_\alpha(j, i) = T_e|_{i \to \infty}$. When applying this method to the lines $H\alpha - H8$, one finds that the electron temperature in disc flares is always lower than 10 000 K, and in some flare regions even as low as 7000 K (Švestka, 1965).

On the other hand, assuming knowledge of T_e, one can determine the ratios b_s/b_2 from Equation (29) and identify $b_s = b_s/b_8$, since b_8 in flares cannot differ much from the value 1.0. The b_s values, however, can also be computed theoretically. This has been done by de Feiter (1966) and a detailed comparison has been made of de Feiter's theoretical values with those deduced from the spectrum of the flare of 1960 August 7. The best agreement has been found when the theoretical b_s's were computed with the assumption of detailed radiative balancing in the Lyman spectrum, a radiation temperature of 6000 K, and $n_e = 3 \times 10^{13}$ cm^{-3} (as found in the flare compared). The electron temperature then came to 7100 to 7800 K for the $b_3 - b_5$ factors, and to 9500 K for the b_2 (de Feiter, 1966; Švestka, 1972a). Thus all these results show that T_e in the hydrogen elements in the chromosphere is always lower than 10 000 K.

As far as electron temperature in the He I regions is concerned, it must be lower than 23 000 K, because this temperature would correspond to the thermal Doppler width of the singlet He I lines when neglecting any other broadening mechanism (Banin, 1965b). Its most probable value is between 15 000 and 20 000 K (Jefferies et al., 1959; Banin, 1965b; Steshenko and Khokhlova, 1962; Fontenla and Seibold, 1973), but we struggle here with our incomplete theoretical knowledge of the population of the helium quantum states for various combinations of n_e and T_e.

From the Balmer continuum, using Equation (18), Jefferies and Orrall (1961 a, b) determined the electron temperature in two limb flares. They found $9500 \text{ K} < T_e < 11\,500$ K and $14\,500 \text{ K} < T_e < 18\,400$ K, respectively. Even with a highly inhomogeneous flare and with errors in the estimate of the scattered light, T_e in the second flare had to be lower than 24 000 K at heights below 20 000 km. In the limb flare of 1961, July 27, Polupan and Yakovkin (1966) solved the steady-state equations for the first six levels of the hydrogen atom and found T_e to be much lower still: equal to 7500 K at a height of 14 000 km.

All of these measurements refer to the low-temperature flare. In the high-temperature flare region temperatures of the order of millions of degrees are obtained from the measurements of flare-associated coronal lines (Zirin, 1964), and even tens of millions of degrees from the X-ray radiation. The only radiation in the visible spectrum which to some extent forms a bridge between the low and high-temperature flare regions, and about which some information is available, is the emission in the lines of ionized helium. Events in which the λ 4686 line of He II was distinguished on the disc have been extremely rare, and their analysis leads to a minimum value of $T_e = 25\,000$ K (Jefferies et al., 1959; Zirin, 1964; Steshenko and Khokhlova, 1960; Sobolev, 1962; Goldberg-Rogozinskaya, 1962). High above the limb, Zirin (1964) found from the width of the He II λ 5411 line a wide range of values of T_e, from 30 000 K up to 300 000 K, the higher values prevailing at higher altitudes.

In the EUV, Linsky et al. (1975) have found that the Lyman continuum of He II in flares corresponds to colour temperature between 24 000 and 41 000 K. Lines of other elements in the transition layer observed in the EUV region show that temperature rises steeply from the 10^4 value in the chromosphere to the 10^6 K values in the coronal flare (Figure 60). We shall speak more about it in Section 3.2.2. Table VIII summarizes the existing measurements of T_e for disc flares and limb flares, respectively.

TABLE VIII

A summary of electron temperature measurements in flares

Method	Section	T_e(K) in flares on the disk	T_e(K) in flares on the limb
metals	2.2.6	7 000	
H	2.2.6	7 000–10 000	10 000–20 000
He I	2.2.6	15 000–23 000	
He II	2.2.6	25 000–40 000	30 000–(3×10^5)
EUV lines	3.2.2	20 000–3×10^6	
thermal microwave bursts	3.1.5	2×10^6–6×10^6	
Ca xv	2.2.6	4×10^6	
thermal X-rays	3.1.4	5×10^6–4×10^7	
hard X-rays	3.2.1	$(10^8$–$2 \times 10^9)$ [a]	

[a] Fictitious temperature, non-thermal effect.

2.2.7. The filamentary structure of flares

When one knows n_e and b_s, one can compute the density of hydrogen atoms in different quantum states, n_s, for properly chosen values of T_e, from the combined Boltzmann-Saha equation,

$$n_s = 4.15 \times 10^{-16} b_s s^2 n_e^2 T_e^{-3/2} \exp[\chi_s/kT_e], \tag{31}$$

where χ_s is the ionization potential of the sth quantum level. As an example, the resulting n_ss for the flare of 1960, August 7, are shown in Table IX (Švestka, 1965, 1972a).

The n_2 values obtained in this way can be compared with the $n_2 z$ values determined before. This comparison leads to astonishingly small values of z, as shown in the last but one line of Table IX. Similar results have been obtained for other flares as well (Švestka, 1965).

In order to verify this result, let us consider the flare emission behind the Balmer limit. The emission, E_B, is given by Equation (17) (transformed to frequency units, for $n_e = n_i$),

$$E_B = 2.1 \times 10^{-34} n_e^2 T_e^{-3/2} \tag{32}$$

and the absorption coefficient is

$$k_B = 1.4 \times 10^{-17} n_2. \tag{33}$$

Thus, since obviously $\tau_{0B} = k_B z \ll 1$, we may use Equation (6), which with Equations (32) and (33) takes the form

$$(I_B/I_{0B}) - 1 = 2.1 \times 10^{-34} n_e^2 z/T_e^{3/2} I_{0B} - 1.4 \times 10^{-17} n_2 z. \tag{34}$$

As soon as I_B is known, the only unknown quantity in this equation is the linear thickness of the flare z.

So far, however, (perhaps with the exception of the large white-light flare of 1972, August 7, studied by Machado and Rust (1974)) no enhancement behind the Balmer

TABLE IX

Physical conditions in the low chromospheric part of the flare of 1960, August 7

T_e	7500 K	8000 K [a]	9000 K	10 000 K
n_2 (cm^{-3})	1.03×10^9	9.00×10^8	7.47×10^8	6.10×10^8
n_3 (cm^{-3})	5.78×10^7	5.10×10^7	4.22×10^7	3.46×10^7
n_5 (cm^{-3})	3.18×10^7	2.83×10^7	2.34×10^7	1.91×10^7
n_8 (cm^{-3})	4.18×10^7	3.71×10^7	3.04×10^7	2.50×10^7
n_1 min. (cm^{-3})	1.8×10^{15}	5.8×10^{14}	9.1×10^{13}	2.1×10^{13}
max. ionization	1.6%	4.9%	25%	59%
$z\ (=n_2 z/n_2)$	33 km	38 km	46 km	56 km
max. z (from absence of Balmer continuum)	15 km			23 km

[a] The most probable values.

limit in the spectra of disc flares has ever been observed. Therefore, one can determine the upper limit of z from Equation (34), utilizing the fact that $I_B/I_{B0} < 1.1$ in all flare events. In this way we get maximum z within 10 and 250 km in different flares for T_e between 7000 and 10 000 K. The resulting maximum values of z in the flare of 1960, August 7, are shown in the last line of Table IX and they are clearly of the same order of magnitude as the values found from the $n_2 z/n_2$ ratio.

This astonishingly small geometrical thickness of flares was first discovered by Suemoto and Hiei (1959) and later confirmed by many other authors (Hirayama, 1961; Švestka, 1963, 1965; De Feiter, 1966; Polupan and Yakovkin, 1966; Stepanyan, 1969; Kurochka, 1970; Semel and Soru-Escaut, 1971; Machado and Seibold, 1973). Some of these confirmations are based on quite different considerations than those described above. Thus Suemoto and Hiei determined the value of $n_2 z$ from the self-absorption that causes the increase of halfwidths in Figure 26a from H9 to the Hα line. Even though they possibly underestimated the value of $n_2 z$ (neglecting the S variation with height) and overestimated n_e (by neglecting electron damping), the disproportion of these two quantities was so striking that it could hardly have been ascribed only to these errors. Hirayama then showed that this disproportion also existed for limb flares. Polupan and Yakovkin found $z = 120$ km in another limb flare by solving the steady state equations for the lowest six levels of hydrogen. Stepanyan found z between 160 and 800 km from the Ca II λ 3737 line; and above spots, Machado and Seibold find z as low as ~ 3 km.

There are two ways these extremely small z values can be interpreted:

One can assume that this is the actual thickness of a shell-like emitting flare layer in the chromosphere; if the primary source of the flare is in the corona or transition layer, flare energy must be brought downwards into the chromosphere (cf. Section 2.3.1) and in that case it could be deposited in a very thin layer of the dimensions we have deduced from the flare spectra. Shmeleva and Syrovatsky (1973) have shown that thermal conductivity should be the dominant mode of energy transport from the corona to the chromosphere. In that case a considerable part of the deposited energy is radiated by a thin layer with a very large temperature gradient. Canfield and Athay (1974) expressed a similar view. Even in the case that energy is transferred to the chromosphere by high-energy electrons, the layer of energy deposition can be of the order of only 100 km (Canfield, 1974); the steeper the energy spectrum of the electrons, the thinner the layer in which most of the energy is released (Syrovatsky and Shmeleva, 1972). This 'shell model' of the flare has been also advocated by Machado and Rust (1974).

Thus one cannot exclude the possibility that the small values of the geometrical thickness z, deduced from the chromospheric flare spectra, are real. However, there are several objections to it:

(a) In contradistinction to Shmeleva and Syrovatsky (1973), Brown (1974) has demonstrated that a power-law stream of energetic electrons in a flare (when present) dominates the energy input over conduction in the optical flare region, and at least in this case the energy deposition can occur over a wide range of altitudes, depending on the shape of the electron energy spectrum (see Section 2.3.1.B). In the case of white-light emission (Section 3.2.11) the energy deposit occurs at the bottom of the chromosphere. It appears reasonable to assume that there is a continuous increase in the depth penetration (and thus thickness) of the layer of energy deposition, from small subflares to the largest flares that are visible in white light. Therefore, at least in the impulsive kernels the thickness of the chromospheric flare is certainly larger than the small values of z we have obtained. This implies an inhomogeneous structure in all flares that show an impulsive phase.

(b) Machado and Seibold (1973) have found z above sunspot umbrae equal to or less than ~ 3 km. This certainly cannot be interpreted as the actual thickness of the layer where energy is deposited above sunspots. On the other hand, as we mentioned earlier, filamentary structure in the region of flare penetration over sunspot umbrae and the dilution of radiation associated with it can reasonably explain the absurdly small (and obviously fictitious) geometrical thickness.

(3) The shell model represents an attempt to interpret the initial phase of the flare, that is, when energy is first deposited from the corona to the chromosphere. But the fully developed flare obviously consists of a system of loops (cf. Section 2.1.3.E) which rise and bring the chromospheric material high into the corona. In one case, in order to remove a strong discrepancy between the temperature determined from

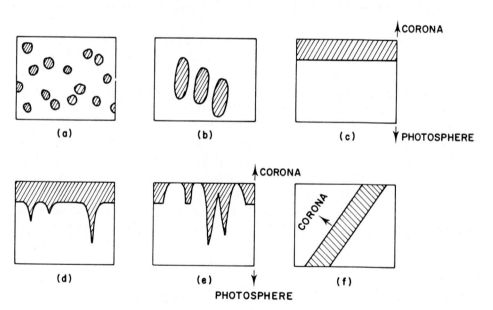

Fig. 30.   Schematic drawing of vertical sections through the solar chromosphere in different flare models: (a) Flare with a filamentary structure that does not lead to dilution of radiation. (b) Flare with a filamentary structure that leads to dilution of radiation. (c) Homogeneous flare with a shell structure. (d) Flare with a shell structure with 'wells' produced by streams of accelerated particles. (e) A combination of the models (b), (c) and (d): a filamentary shell structure, leading to dilution of radiation, with a 'well' produced by a stream of accelerated particles. (f) Even the shell structure (c) may produce dilution of radiation when observed along the solar surface near the limb.

the He II linewidths and the observed ratio of He II/continuum intensities, Zirin (1964) had to assume that less than 10% of the flare material emits helium and hydrogen lines at an altitude of $\gtrsim 20\,000$ km. Thus the flare structure obviously becomes inhomogeneous and filamentary in the corona, and even the lowest parts of the flare should show the inhomogeneities, with, for example, physical conditions in the loop feet different from those between them. One is inclined to believe that the shell (if formed at all) is inhomogeneous from its beginning, because energy is transported downwards along the magnetic field lines, and the magnetic field strength certainly varies throughout the flare region.

Thus we come to the other interpretation of the small values of z, namely, an inhomogeneous and filamentary structure of flares. This is the original explanation of the small z values, first proposed by Suemoto and Hiei (1959) and later adopted by many other authors. Figure 30 summarizes in a schematic way all the interpretations that have been proposed:

(a) 'Dense' filamentary structure (Suemoto and Hiei, 1959; Švestka, 1965). Flares consist of many elements, maybe cospatial with concentrations of magnetic field; if all the elements were contiguous they would form a fictitious layer of the observed thickness z.

(b) 'Diluted' filamentary structure (Švestka, 1972a, c). There are only a few elements (e.g. feet of isolated loops) so that in all observations we always integrate radiation of the flare filaments and the 'interflare' matter. The flare radiation is thus diluted.

(c) Shell structure (Shmeleva and Syrovatsky, 1973; Athay and Canfield, 1974; Machado and Rust, 1974). Energy is deposited in a thin layer of effective thickness z.

(d) 'Well' structure (this discussion; after Brown, 1974). Shell structure (c), produced through heat conduction, is accompanied with deeper 'wells' at places where the chromosphere is heated by energetic particle streams.

(e) Inhomogeneous shell structure (this discussion). Shell structure (c) becomes inhomogeneous, since the energy transport varies from place to place (e.g. in dependence on the magnetic field strength). In addition to it a 'well' can exist where a stream of electrons penetrates into the chromosphere.

In fact, the fully homogeneous shell structure (c) may lead also to a 'filamentary' structure close to the limb since then we observe the shell from its side. With a thickness of 100 km it cannot cover the whole field of view of the spectrograph (case (f) in Figure 30) and the radiation will be diluted.

We have emphasized before that any inhomogeneous structure of flares may invalidate the deduction of many physical quantities in flares (Section 2.2.2). A detailed analysis of this problem has been carried out by Švestka (1972c). He has pointed out several indications that the flare radiation may be diluted, at least in some flares. If this is true, one obtains the correct physical parameters in the flare elements only when all the flare elements are optically thin in the spectral region considered. If this condition is not fulfilled (but in spite of it the flare may still seem to be optically thin), the parameters deduced are in error and the error increases with the extent of the dilution. All the errors may increase strikingly if the analysis is performed in regions which are optically thick. The parameter most strongly affected is the linear thickness of the flaring region. It is seriously underestimated. Electron and atom densities are overestimated, and the column density of hydrogen atoms is underestimated. On the other hand, even in the diluted case, the deduced source function and in consequence of it the electron temperature as well, remain unaffected if the analysis is properly carried out. This rather surprising result follows from the fact that, as described in the preceding sections, we determine the optical thickness first, after eliminating S, so that the error due to flare inhomogeneities enters into the value of τ_0 only. However, Švestka's analysis clearly shows that even in a diluted flare, reasonable estimates of the flare parameters are still possible provided that one avoids spectral regions where τ_0 approaches or even exceeds the value of 1.

2.2.8. HYDROGEN DENSITY

It is somewhat curious that with all our knowledge of the physical conditions in flares, we know essentially nothing about the density of hydrogen atoms in the flare elements. This is of course due to the fact that there is no measurement of the Lyman series in flares. The hydrogen density n_H is practically equal to the density of hydrogen atoms in the ground state and this can only be obtained from the Lyman lines or the Lyman continuum. Recently, de Feiter and Švestka (1975) discussed methods for deducing the $n_1 z$ values from the Lyman spectrum. Because of the large optical thickness of flares throughout the Lyman series it is not easy to find the correct $n_1 z$ value, particularly if the flare structure is inhomogeneous (cf. preceding section). The best chance for finding correct $n_1 z$ values would be in the optically thin gaps between low Lyman lines where, however, metal lines complicate the intensity measurements. Thus one probably cannot expect good $n_1 z$ values until high-dispersion flare spectra in EUV are available. Before that we might estimate $n_1 z$ from the highest resolvable Lyman line in a flare spectrum. But we must not forget that we are analyzing then a spectral region in which a flare is optically thick.

We have mentioned in Section 2.2.6 that the best agreement between theory and observation has been found for detailed balancing between the levels 1 and 2 (De Feiter, 1966). Therefore, we may assume $b_1 = b_2$. In this case we find the n_1 values and the degree of ionization, $n_e/(n_e + n_1)$, given in Table IX, again for the flare of 1960, August 7. If $b_1 \neq b_2$, then necessarily $b_1 > b_2$, since the deviation of the ground level population from LTE at temperature T_e must be larger than that of level 2 (cf. Section 2.2.6).

Hence the density of hydrogen atoms ($n_1 = n_H$) found from Table IX is a lower limit, and the deduced ionization degree is an upper limit to the ionization in the chromospheric part of the flare.

2.2.9. LINE ASYMMETRY

One of the most puzzling problems in flare spectra is the asymmetrical shape of flare emission lines. The asymmetry does not change the position of the center of the line, but only makes one of the wings stronger (cf. Figure 5). Usually the long-wavelength wing is the more intense and therefore the effect was first explained by absorption of the blue wing in a rising hydrogen cloud above the flare (Ellison, 1949; Švestka, 1951; Kazachevskaya and Severny, 1958). This explanation, however, became untenable as soon as the same kind of asymmetry was discovered in metal lines (Jefferies et al., 1959), with the hypothetical cloud velocities being quite different and, of course, having too low an absorbing capacity (Fritzová, 1960; Banin, 1965a). Therefore, we probably encounter here some sort of internal motion in the flare region and several models have been proposed to explain this effect: in terms of falling material (Fritzová, 1960; Hyder, 1967a, b), loop-like motions (Ballario, 1963, 1968; Banin, 1965a, 1969), moving condensations (Guseynov, 1967), contracting and expanding

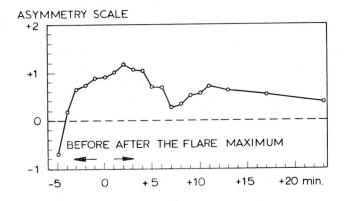

Fig. 31.   Mean time dependence of the Hα line asymmetry on the flare development (time of maximum Hα linewidth = 0), constructed from a set of 244 spectra of 92 flares. The asymmetry scale is qualitative: 0 = no, 1 = weak, 2 = moderate, 3 = strong asymmetry. Plus sign means an intensification of the long-wave wing, i.e., a red asymmetry (After Švestka et al., 1962.)

shells (Severny, 1968), or impacting electrons causing downward motion of the emitting atoms (Zirin and Tanaka, 1972). None of these models, however, can explain all aspects of the observed phenomenon.

Very often at the onset of a flare a blue asymmetry is observed (Figure 31), i.e. the short-wave wing is more intense (Švestka et al., 1962; Severny, 1968; Švestka, 1968). This blue asymmetry, however, disappears within one or just a few minutes, and subsequently a strong red asymmetry sets in which reaches its maximum approximately at the time of the flare maximum in Hα light and persists for some five minutes after this phase (Figure 25, also compare Figures 6 and 7). In one case (the large flare of 1972, August 7) Rust (1973a) observed a similar velocity scheme also in the photosphere: unusual blue shifts appeared before the flare below the flare site and these turned to red shifts after the flare onset.

If the asymmetry is explained by motion, then the highest velocity is found for hydrogen (~ 100 km s^{-1}), smaller velocities are deduced from helium and Ca II H and K lines (a few tens of km s^{-1}), whilst the velocity in metals is close to 10 km s^{-1} (Jefferies et al., 1959; Fritzová, 1960). Banin (1965a) interpreted this velocity variation as being due to the difference in the height of formation of the different lines in the solar atmosphere. In such a case the velocity decreases with the depth in the flare which indicates a downward slowing motion in the flare region. On the other hand, the deduced velocities may indicate conservation of momentum. Then, as Hyder (1967b) has pointed out, this differentiation might also be due to some kind of a mass effect (e.g. if the radiating atoms are set in motion after collisions with particles in an accelerated stream).

According to Ondřejov observations (Švestka et al., 1962; Fritzová, 1961) the asymmetry decreases from the center to the limb of the solar disc. This would indicate that if the asymmetry is associated with motion, then the motion is predominantly radial. However, Michard (1959) and Severny (1968) have also found striking asymmetries in limb flares; whilst Michard did not find any preference for either blue or red asymmetry, Severny has found that the features of the asymmetry on the limb are very similar to those discovered in disc flares. Red asymmetry is observed in 85% of cases on the limb (80% on the disc); blue asymmetry is present (mainly close to the flare onset) in 35% of cases on the limb (23% on the disc); and in 14% of cases only blue asymmetry is observed on the limb (5% on the disc). The numbers in brackets are taken from Švestka et al. (1962).

Several other quite peculiar features of the asymmetry have been discovered:

(a) Regions with different (and even opposite) asymmetries are often observed at the same time within the same flare (Švestka et al., 1962; Severny, 1968), though most frequently the same type of asymmetry is found in all flare parts. (This, for example, was the case in Figure 6.)

(b) The asymmetry appears to vanish above sunspots (Švestka et al., 1962).

(c) Sometimes on the limb the two wings do not coincide in position, and it looks as if the flare were composed of two different blue- and red-shifted components (Severny, 1968).

(d) Sometimes on the disc wings are not emitted in the same position as the line cores, and they also differ from the position of metal emissions, which coincide with the hydrogen cores. According to Michard (1959) this may be due to a difference in heights at which these emissions are formed, but according to Blaha et al. (1962) this explanation fails for many events observed.

(e) The He I emission is stronger and appears more frequently in flare regions without any asymmetry than in regions where the asymmetry is pronounced (Švestka et al., 1962); this reminds one of Leroy's (1969) observation that the $D_3/H\alpha$ ratio in stationary prominences is twice that in moving parts of a prominence.

(f) In flares with deeply reversed Balmer lines the opposite (blue) asymmetry is found more often and persists for a longer time (Švestka et al., 1962); this strongly reminds one of the behaviour of moustaches, which are characterized by strong central absorption and in which blue asymmetry strongly prevails (Koval, 1967; Severny, 1968).

Generally, however, the characteristics shared by moustaches (cf. Section 2.2.11(b)) and flares are very few. In many aspects (line asymmetry, light curves, location in a sunspot group, depth in the atmosphere, size of the event) the moustaches differ strongly from flares (Severny, 1957; Severny and Khokhlova 1959; McMath *et al.*, 1960; Bruzek, 1972). Even when a few of them seem to be associated with flares at the beginning of the flare formation (Severny, 1959, 1968), the latest very detailed study by Bruzek (1972) clearly shows that the flare elements are not identical with the moustaches, which obviously represent quite a different physical phenomenon, probably closely associated with the threads of continuous emission discussed in Section 2.2.11(b).

An explanation of the asymmetry common to both moustaches and flares was proposed by Severny (1968) who shows that such an effect would appear in lines formed by incoherent scattering in contracting or expanding shells, having a velocity gradient within. However, this hypothesis, though very attractive for moustaches, does not seem appropriate for flares which appear to be composed of arcades of loops.

It is of interest to check whether the flare lines formed at higher altitudes are also asymmetric. According to Brueckner (1974b) this is the case, but asymmetry towards

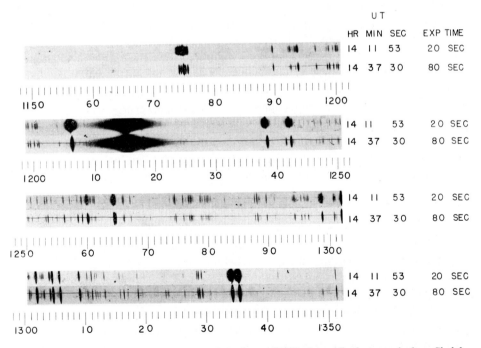

Fig. 32.   Ultraviolet lines in the spectrum of the flare of 1973, June 15, photographed on Skylab. Note the differences in the line asymmetry. (Courtesy of G. E. Brueckner, NRL, Washington.)

short or long wavelengths appears in the spectra without any particular preference. Sometimes opposite asymmetry is observed even in different lines of the same spectrum. Figure 32 shows an example of it: the λ 1206 Å line appears to be brighter towards longer wavelengths, Ly-α is rather symmetrical and all other strong lines show the opposite asymmetry. Brueckner suggests that hot plasma rises up and cold gas falls down, thus causing the differences observed. But another explanation can be proposed, too. Since the flare radiation is integrated in each line, the profile is composed of radiation in all the integrated flare areas. If there are different asymmetries in different flare positions, as we observe in the optical spectrum, the resulting asymmetry in a line is determined by the contribution of each area to the radiation in that particular line.

On the other hand, Widing and Cheng (1974) who studied flare images in iron and calcium lines still higher in the atmosphere, where temperatures exceed 5×10^6 K, found indications of a downward motion in the Ca XVII and Fe XII images, while the Fe XXIV image at an estimated altitude of 13 000 km was stationary.

We still do not know the explanation for the asymmetry. It may be connected with downward streaming in the feet of the loops composing the flare; the Skylab measurements reported by Widing and Cheng support this view, but simple loop models proposed by Ballario (1963, 1968) and Banin (1965a, 1969) leave unexplained many of the observed features, e.g. the initial phase, the short duration of the asymmetry, and the asymmetry observed on the limb, in particular. Hyder's (1967a, b) in-fall model was criticized already in Section 2.1.3.G, on the grounds that the emitting (and asymmetrical) regions do not show spatial correspondence with the falling material. Fritzová's (1960) streams do not explain the asymmetry observed on the limb. Gusejnov's (1967) moving condensations would have travelled longer paths than are available in the flare region.

The occurrence of the asymmetry during the flash phase and for only a few minutes after that rather indicates some connection with the flare energy transport from the corona to the chromosphere. However, the asymmetry is not observed solely in the impulsive kernels, hence it cannot be ascribed simply to streams of accelerated electrons as Zirin and Tanaka (1972) suggested. It is doubtful that heat conduction could produce any asymmetry in the chromospheric lines.

2.2.10. DIFFERENCES BETWEEN PROMINENCES AND LIMB FLARES

On the limb it is sometimes quite difficult to recognize whether we are encountering a flare or a flare-associated active prominence. Therefore, it seems important to mention the criteria for distinguishing these two different limb phenomena.

The first criterion, of course, is the brightness. Since however, limb events are very often overexposed, reliable photometry is difficult.

The second criterion is based on the observed motions. One can suspect that any phenomenon that shows large Doppler shifts or lateral displacements is much more likely to be a prominence than a flare.

However, we wish particularly to point out two spectral criteria that can be used to distinguish between prominences and flares. The first one was discovered by Tandberg-Hanssen (1963) when studying metal lines in a number of these phenomena. Compared with prominences, the Fe II lines in flares are always strengthened in comparison with the lines of Ti II. The intensity ratio λ 4584 (Fe II)/λ 4572 (Ti II) always exceeds 1 and may be greater than 10 in flares, whilst it is always less than one and may be as low as 0.05 in prominences.

The second criterion is based on the measurements of halfwidths of higher Balmer lines (Švestka, 1964b). In flares (Figure 22) the halfwidths decrease up to about H9 as the optical thickness decreases, and begin to increase again for lines higher than H9 due to the high electron density in flares. Since the electron density in prominences is one or two orders of magnitude smaller, this rise does not appear in prominence spectra, as one can verify in Figure 26b.

These spectral differences give evidence that at least some portions of Hα features on the limb differ in density from prominences and thus must be considered as flares projected above the limb. This is in contradiction to some evaporation theories of flares (e.g. Hirayama, 1974) which imply that Hα flare is confined only to the low chromosphere.

2.2.11. CONTINUOUS EMISSION

We can distinguish several different forms of continuous emission in the optical region of the flare spectra:

(a) A weak continuum in flares above the limb. Behind the Balmer limit it is due to free-bound transitions of hydrogen to the second quantum level (as we have already discussed in Chapters 2.2.4.B and 2.2.7), but its intensity is much less than 1% of the intensity in the disc center so that it cannot be distinguished in projection on the solar disc (Jefferies and Orrall, 1961a, b; Švestka, 1966c). At longer wavelengths the continuum (when observed) is due to Thomson scattering (Zirin, 1964; Zirin and Acton, 1967), and this can never produce any emission in projection on the disc.

(b) Narrow emission threads in the spectrum of disc flares. Often they do not coincide in position with the flare line emission (Steshenko, 1971a), they are not characteristic of large flares, and sometimes they may even be found in plages without any flare (Severny, 1958, 1964a; Khokhlova, 1958; Dolginova and Korchak, 1968). Therefore, they do not seem to be related to the flare itself, but they probably are peculiar to plages that are active in producing flares. Gurtovenko (1967) suggests that they have the same origin as bright granules, namely, they are small areas of rising overheated gas. On the other hand, the grains producing the thread-like continuous emission might be similar to the moustaches (or simply be the brightest moustaches), because these two phenomena coincide on some occasions (Severny, 1957; Severny and Khokhlova, 1959; Bruzek, 1972). According to Vorpahl and Pope (1972) all moustaches (sometimes also called Ellerman bombs) observed by them at Hα – 0.9 Å had a cospatial bright feature in the 3840 Å photospheric network.

If this association is correct, it still does not explain the physical origin of the thread-like continua, because we do not know what the moustaches are. One can describe them as bright points (Figure 33) of very small diameter close to 1000 km, that occur in high numbers in some active regions, mostly around the penumbrae of sunspots (Bruzek, 1972) and in newly emerging flux regions (Vorpahl and Pope, 1972). They are very often associated with ejections of dark material (surges, cf. Section 4.2.1; Bruzek, 1972; Roy and Leparskas, 1973). Their lifetime is close to 10 min (13 min on the average, according to Roy and Leparskas).

The moustaches differ from subflares mainly in three aspects: (1) Their light-curve does not show the flash phase and subsequent slower decay typical of flares; they show fast rise and fast decay with rather constant brightness in between (Bruzek, 1972). (2) Emission lines in their spectra seem to be broadened by the Doppler effect corresponding to velocities up to 200 km s^{-1} (Koval, 1972). (3) Predominantly the blue line asymmetry (i.e. opposite to that in flares) is observed in their spectra (Section 2.2.9).

(c) Short-lived continuous emission in major disc flares. Due to its short duration, this continuum was only recorded twice in the flare spectrum: by Grossi Galegos *et al.* (1971) and Rust (1973a). On all other occasions it was detected only through the visibility of parts of the flare in white light. Lists of approximately 30 white-light flares observed since Carrington's (1859) discovery of a flare can be found in Becker (1958b), Švestka (1966c), McIntosh and Donnelly (1972) and Slonim and Korobova

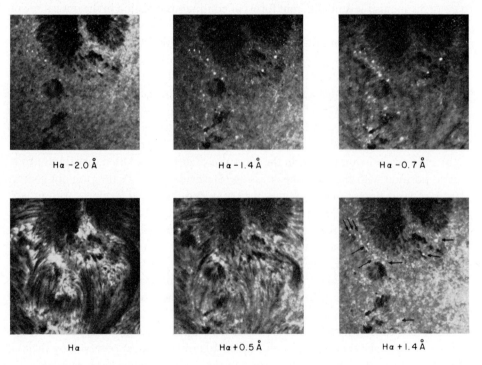

Fig. 33.   Examples of moustaches around a spot (arrows point to the brightest of them) as photographed at various distances from the Hα line center. (Courtesy of D. M. Rust, Sacramento Peak Observatory.)

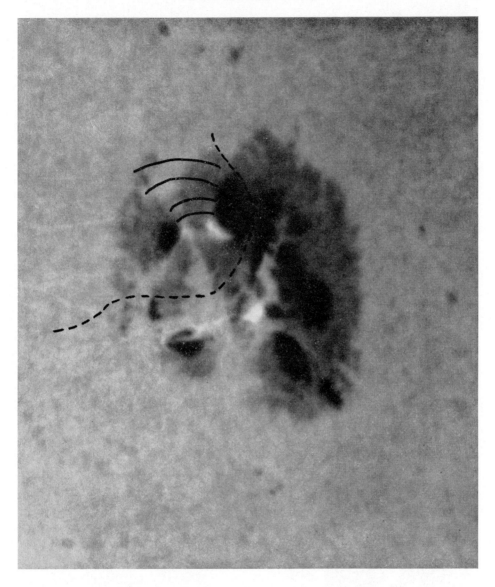

Fig. 34.   Photography of the white-light flare of 1972, August 7, with 250 Å wide passband centered at λ 4950 Å. Two pairs of bright patches were observed in sunspot penumbrae for about 4 min, obviously located at foot points of loops crossing the $H_\parallel = 0$ line (dashed in the figure). After the bright patches decayed a white-light wave was seen propagating out of the sunspot group; the curved lines mark the successive positions of this wave between $15^h 25^m$ and $15^h 36^m$ UT. (Courtesy of D. Rust, Sacramento Peak Observatory.)

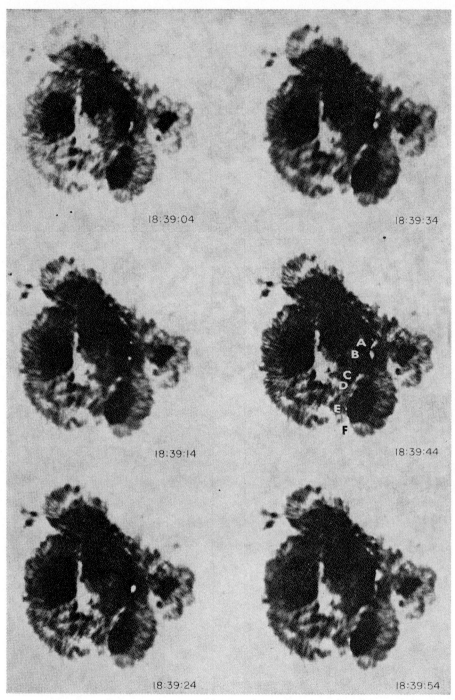

Fig. 35a. Flashes in 15 Å band centered at λ 3835 Å observed along the $H_{\|} = 0$ line during the flare of 1972, August 2. The average lifetime of the bright points was 5 to 10 s and their appearance corresponded to the peaks of the hard X-ray burst in Figure 35b. The strands D and point B rose with the first X-ray pulse at about $18^h 39^m 04^s$. A new point appeared at $39^m 24^s$ as the X-ray flux rose. It faded in 10 s and a row of points flashed at the X-ray maximum, $39^m 34^s$. The points appeared in pairs, e.g., B and D or A and E. The coupled points, even when very distant, lay on opposite sides of the $H_{\|} = 0$ line. (After Zirin and Tanaka, 1973.)

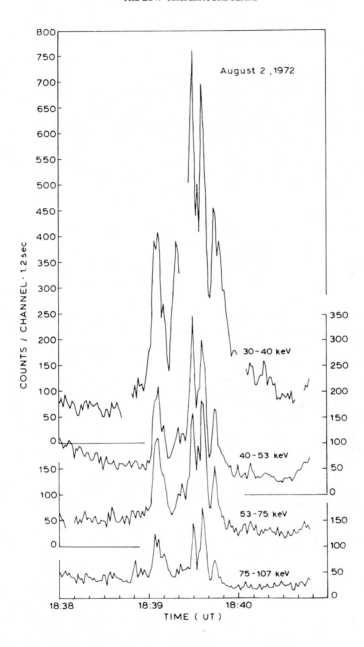

Fig. 35b.   The hard X-ray burst observed with high time-resolution during the flash phase of the flare of 1972, August 2. The individual peaks corresponded to the penumbral flashes shown in Figure 35a. The authors of this record (Van Beek *et al.*, 1973) draw attention to the similarity in shape of the individual peaks and assume that the hard X-ray enhancement is composed of 'elementary bursts' of similar properties.

(1975). Another white-light flare, not listed by these authors, was observed in 1974, July 4 (Feibelman, 1974).

   One has to emphasize that in contradistinction to the flare patrol service, which has provided nearly continuous Hα coverage for the last 15 yr, white-light observations of the Sun have been spotty. Therefore, all the existing observations of white-light flares were made only by chance. In consequence, many white-light flares might have been missed, because the duration of the white-light emission is only a few minutes. According to McIntosh and Donnelly (1972) the occurrence of strong EUV bursts, which are associated with white-light flares, indicates that the real number of white-light flares might be 5 to 6 per year in solar maximum. The majority of the observations are also very incomplete. Therefore our discussion has to be based only upon a few well-observed and described events.

   When these observations are considered, one finds that the typical appearance of the white-light emission is one, mostly two, and sometimes more bright patches (Figures 34, 36, 38) that occur very close to (or in) penumbral regions of sunspots (examples: De Mastus and Stover, 1967; Rust and Hegwer, 1975). This kind of emission exactly coincides with the impulsive phase of the flare (i.e. with hard X-ray and impulsive EUV and microwave bursts) and one can interpret it as enhancement at the feet of one or more loops which bridge the zero line (Švestka, 1970a; McIntosh and Donnelly, 1972; Rust and Bar, 1973; Slonim and Korobova, 1975). The white-light emission patches lie within the bright Hα flare ribbons (De Mastus and Stover, 1967; Švestka, 1970a; Slonim and Korobova, 1975), but in the case of the 1972 August 7 flare, well-examined by Rust and Hegwer (1975), they did not move with the ribbons, as one would expect.

   Near the limb, on the other hand, the area of white-light flares is more extensive, being in the form of a bright facula which often resembles the shape of the Hα flare (example: Becker, 1958).

   A completely different type of white-light emission has been described by Machado and Rust (1974) and Rust and Hegwer (1975) in the flare of 1972, August 7. After the bright white-light knots shown in Figure 34 disappeared, the whole edge of the moving Hα ribbon became visible in white light, moving with a velocity of ~ 40 km s^{-1}, very close to that determined by Křivský et al. (1973) for the Hα ribbons (~ 50 km s^{-1}). Occurrence of this type of continuous emission did not coincide with the impulsive phase (which coincided with the occurrence of white-light knots a few minutes earlier). Rust and Hegwer suppose that the continuum wave they observed was not unique; it was detected because a cinematic film was available for the first time.

   Finally, Zirin and Tanaka (1973) have reported short-lived flashes along the zero line over sunspot penumbra in the flare of 1972, August 2. Lifetime of these flashes (shown in Figure 35) was only 5 to 10 s, and their size was $\sim 1''$. They coincided in temporal detail with hard X-ray peaks (van Beek et al., 1974). The passband in which these flashes were seen contained the H9 line of the Balmer series of hydrogen; thus it is not clear whether Zirin and Tanaka recorded white-light flashes or tiny impulsive H9 kernels.

A. *Synchrotron Radiation*

Let us first consider white-light flares in central parts of the disc, and select the event of 1967, May 23, as an example. In this flare the white-light emission had the form of two small areas of about 10^{17} cm^2 each coinciding in position with parts of the Hα bright flare ribbons on the edges of the common penumbra of two large spots of opposite magnetic polarity (Figure 36). The emission could be seen for 7 min and its maximum coincided in time with maxima of hard X-ray and microwave bursts. The maximum flare brightness in Hα light occurred 4 min later. Similar events were observed in 1966, March 5 (M. D'Azambuja, 1947), in 1957, September 3 (Becker, 1958), in 1960, November 15 (Nagasawa *et al.*, 1961), and in 1972, August 7 (Rust and Hegwer, 1975; Figure 34), as well as possibly in some other cases photographed at Tashkent (Slonim and Korobova, 1975).

The very close time association of the white-light emission with the impulsive hard X-ray, EUV and microwave bursts indicates that its origin has to be due to the existence of accelerated particles in the flare region. These particles may be either electrons or protons. As Gordon (1954) has pointed out, relativistic electrons might produce the continuous emission by synchrotron radiation, since flares are obviously located in regions of strong magnetic field. This explanation was also applied by Severny (1958a) to thread-like continua and Stein and Ney (1963) tried to explain the whole continuous flare spectrum, from radio waves up to the visible radiation, by this process. However, as one can easily prove, the total number of electrons required for this process would give much higher X-ray intensities than are observed.

Under the most favourable approximation of a circular trajectory the spectral power of synchrotron radiation emitted by a highly relativistic electron ($E \gg mc^2$) is, in erg s^{-1} Hz^{-1} sr^{-1},

$$P(v) = 1.86 \times 10^{-23} H \, g(E) \, F(E), \tag{35}$$

where

$$F(E) = \int_{g(E)}^{\infty} K_{5/3}(x) \, dx,$$

$K_{5/3}$ is the Bessel function, and

$$g(E) = 1.6 \times 10^{-19} \, v/HE^2,$$

with v in Hz, H in G, and E in erg. According to analyses of X-ray and microwave bursts the electron distribution above 10 keV can be well represented by a power-law in energy,

$$N(E) \, dE = CE^{-\gamma} \, dE, \tag{36}$$

with $\gamma \simeq 3$ (Takakura, 1969a; Holt and Ramaty, 1969). Thus the maximum synchrotron radiation expected from such a set of electrons is

$$I(v) = 1.86 \times 10^{-23} \, HC \int_{E} E^{-3} \, g(E) \, F(E) \, dE. \tag{37}$$

In the flare discussed, the maximum observed enhancement near 5700 Å was about 4×10^{12} erg s^{-1} Hz^{-1} sr^{-1}. Then Equation (37) leads to $C = 1.6 \times 10^{37}$ for $H = 1000$ G, which may be considered as the maximum magnetic field strength to be expected in the flare region, and Equation (36) gives $N(>100 \text{ keV}) = 8 \times 10^{38}$ as the number of electrons with energies above 100 keV. Still higher values of

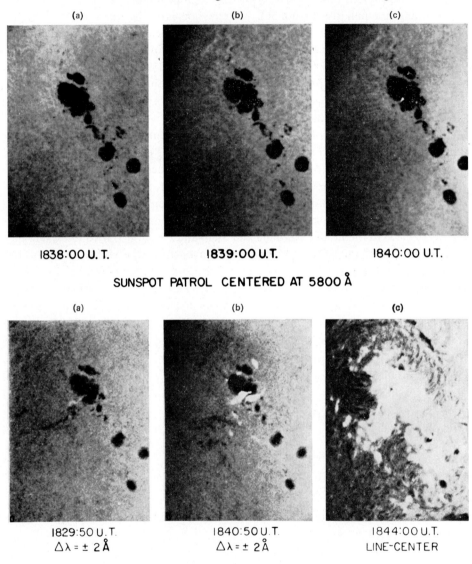

(a)   (b)   (c)

1838:00 U. T.   1839:00 U.T.   1840:00 U.T.

SUNSPOT PATROL CENTERED AT 5800 Å

(a)   (b)   (c)

1829:50 U.T.   1840:50 U.T.   1844:00 U.T.
$\Delta \lambda = \pm 2$ Å $\Delta \lambda = \pm 2$ Å   LINE-CENTER

Hα FLARE PATROL

Fig. 36. Above: Photographs of the white-light flare of 1967, May 23, in maximum brightness on picture (c) (after De Mastus and Stover, 1967). Below: The active region in the Hα light, shortly before the flare onset (a), close to the time of the white-light flare maximum (b), and at the time of the flare maximum in the Hα light (c). (Courtesy of Howard L. DeMastus, Sacramento Peak Observatory.)

$N(>100 \text{ keV})$ are obtained when more reasonable assumptions are made (lower H, since in fact only the component perpendicular to the electron velocity enters into Equation (37), and random trajectories). However, as we shall see in Section 3.2.1, even in the largest flares $N(>100 \text{ keV})$, deduced from X-rays, does not exceed the order of 10^{36}, hence it is more than two orders of magnitude lower. The approximately 10^{39} electrons needed should give rise to X-rays at energies well above 1 MeV in all white-light flares, both through bremsstrahlung (Švestka, 1970a) and through the inverse Compton effect (Korchak, 1965, 1967). This has not been observed. Only in the flare of 1972, August 4, which also produced γ-rays (cf. Section 5.4.3), did the X-ray energy exceed 1 MeV. In all other flares the highest energy of detectable X-rays was always below 600 keV.

B. *Particle Penetration to Low Atmospheric Layers*

The only other way energetic particles can give rise to an immediate increase in the white-light continuous emission seems to be through their collisions with the ambient gas (Švestka, 1970a; Najita and Orrall, 1970). If the accelerated particles deposit their energy in atmospheric layers dense enough, the heating can manifest itself by increased brightness in the continuum.

The easiest way to explain this effect is to consider high energy protons as the source particles (Švestka, 1970a). Protons with energies above 20 MeV can penetrate the normal solar atmosphere to the depth where the optical continuum is formed (Figure 37). Let us take for the proton energy distribution the power law (36) with $\gamma = 2$, which was the actual value observed by Lanzerotti (1969) in the earliest phase of the proton flux in space associated with the white-light flare discussed (as well as in another case by Heristchi et al. (1969)); then the observed emission requires some 10^{10} protons cm^{-3} in a volume of 5×10^{27} cm^3, of which 2% are accelerated to energies above 100 keV. In this case one gets $N (>100 \text{ keV}) = 10^{36}$, i.e. the same as the number of electrons deduced from the hard X-ray bursts. Najita and Orrall (1970) also included the contribution of high-energy electrons. However, as we shall see in Section 6.6.3, it is questionable whether enough electrons are accelerated to highly relativistic energies, since the particle injection into the acceleration process may be rigidity-dependent.

Nevertheless, if protons, and possibly electrons, are accelerated to high energies during the impulsive phase, their penetration to the upper photosphere can explain the white-light emission using quite reasonable assumptions. The only difficulty is that all the other effects observed during the impulsive phase only give evidence for the acceleration of electrons to energies below 1 MeV. Apart from the white-light flare, there has been only one indication of protons being accelerated at the same time, and that is the observation of γ-rays in the flare of 1972, August 4. As Chupp *et al.* (1973a, b, 1974) have shown, the γ-ray 2.2 MeV line became excited within 3 min of the hard X-ray onset, and its excitation requires colliding atomic nuclei with energies above 30 MeV (cf. Section 5.4.3).

This observation gives support to the interpretation of white-light knots by penetra-

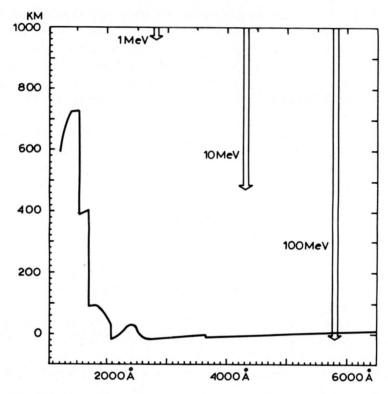

Fig. 37. The height in the solar atmosphere at which $\tau = 1$ for different wavelengths (after Gingerich and de Jager, 1968). Zero level corresponds to $\tau = 1$ at λ 5000 Å. The arrows show the depth to which protons of different energy (with zero pitch-angle) penetrate in the solar atmosphere. (After Schatzman, 1965, and Švestka, 1970a.)

tion of high energy particles into deep atmospheric layers. A theoretical problem then arises in trying to find an acceleration process efficient enough to increase the particle energy within the very short time of the impulsive phase from its thermal value up to the high energies needed (cf. Section 6.6). Many scientists now believe that the particles with energy above 1 MeV are accelerated only in a second acceleration step – a step that sets in after the impulsive phase. The occurrence of white-light and γ-ray emission in time-coincidence with the impulsive phase indicates that this second step acceleration process is accomplished almost simultaneously with the first phase. We shall see in Sections 3.2.5 and 6.6 that this actually seems to be the case.

Hudson (1972) suggests an alternative solution, namely that the white-light emission is produced by electrons of relatively low energy and that the energy deposition takes place in the chromosphere. He applies a thick-target theory in the interpretation of the X-ray spectrum and concludes that the white-light continuum arises from free-free and free-bound transitions in a medium with density 10^{12} to 10^{13} cm^{-3} at a height between 700 and 1500 km above the base of the chromosphere, and with a temperature between 10^4 and 10^5 K. However, as Korchak (1972) has pointed out,

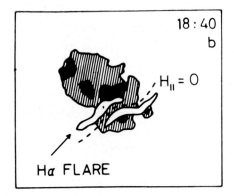

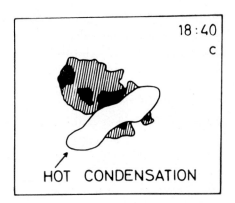

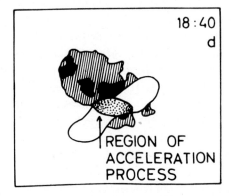

Fig. 38. A schematic drawing of the flare event of 1967, May 23 (cf. Figure 36), as it may be inter-preted at 18^{h}40^m UT when the maximum of the impulsive phase occurred. The two bright flare ribbons (b) are the visible feet of a set of loops forming a hot coronal condensation (c); in a limited part of it, in between two large spots of opposite polarity an impulsive acceleration process (d) takes place. At the feet of this (loop-like) region the two white-light flare patches (e) appear. (After Švestka, 1971c.)

Hudson has overestimated the ionization losses (and thus the energy output) by more than one order of magnitude and thus he needs more accelerated electrons than the total number of all electrons one expects to be present in the acceleration region. There is also another objection against Hudson's interpretation, namely that the proton energy spectrum of all white-light flares, as recorded in space, is unusually hard (Najita and Orrall, 1970). Thus Švestka's and Najita and Orrall's assumption of extremely hard particle energy spectra in white-light flares appears to be much nearer to the truth than Hudson's assumption of abnormally increased particle density.

As soon as we accept the interpretation that the small white-light patches in flares are produced by streams of particles accelerated during the impulsive phase (and there seems now to be general agreement on this), the position of these brightenings gives us the only information that we have on the location and size of the region where the impulsive acceleration occurs. Figures 34 through 36, as well as the few other observations available (e.g. McIntosh and Donnelly, 1972; Slonim and Korobova,

1975) prove that the impulsive acceleration process takes place in a restricted volume of the flare, in the immediate vicinity of the zero line of the longitudinal magnetic field, and usually between two large spots, where a high field gradient exists (Figure 38).

C. *Heating of Upper Photosphere*

White-light flares close to the limb, which are usually characterized by a much larger area ($> 10^{18}$ cm^2), are certainly at least partly produced by the mechanism discussed above – a bombardment of low atmospheric layers by accelerated particles. This supposition is supported by the fact that many of these close-to-the-limb white-light flares were associated with type IV bursts, polar cap absorptions, and even ground-level effects (1949, November 19; 1956, February 23; 1960, September 3). This shows that particles were accelerated to very high energies in these flares. The limb event of 1960, July 16, also coincided closely with the microwave burst in a way quite similar to the disc events (Fortini and Torelli, 1970). The only question is whether all white-light flares close to the limb can be explained entirely in this way. Many observations indicate that only a restricted part of the flare is involved in the impulsive acceleration process (see Figure 38 and also Sections 3.2.1 and 3.2.3). These observations are difficult to reconcile with the large white-light area observed in some flares near the limb. Therefore, possibly some other process is combined with the particle bombardment and may significantly increase the visible white-light area if the flare is close to the limb. As in faculae, the uppermost photospheric layers then play the dominant part in producing the outgoing radiation.

The idea that the flare continuum is due to an increase in temperature in the uppermost photospheric layers was first expressed by Mustel (1955) many years ago. He simply supposed that some flares (which are essentially chromospheric and coronal phenomena) extend down to the photosphere thus increasing its temperature. In that case, however, one would expect the maximum brightness in white light to coincide with the Hα maximum, which is not the case. Thus particle penetration during the impulsive phase must be the predominant cause of white-light emission. Mustel's mechanism, however, could explain the additional white-light emission observed within larger areas and for longer times near the limb.

It has been shown by Švestka (1966a) that the source of this additional white-light emission might be the formation of negative hydrogen ions, provided that the density of hydrogen atoms is high enough. For example, at 12° from the solar limb we get an observable continuum near 5000 Å if n_H significantly exceeds 10^{15} cm^{-3}. A comparison with Table IX clearly shows that n_H is large enough only if the electron temperature is extremely low. Therefore, the white-light emission can be explained as due to the existence of a very dense and low-temperature layer at the bottom of the flare. This is not far from Mustel's original idea of a 'penetration' of the flare into the photosphere.

Machado (1971) presented evidence that the white-light emission close to the limb is produced in the upper photosphere. In the flare of 1969, March 12 (78° W), the spectrum of which was photographed by Grossi Gallegos *et al.* (1971), Machado compared the enhancement of different Fraunhofer lines in the continuum and outside of it in the flare region; he found that different Fraunhofer lines were influenced in a different way by the continuum, in accordance with their excitation potential and oscillator strength. Hence the emission must come from the layers where the cores of these lines are formed; one could never get this result if the emission were formed higher in the atmosphere (through synchrotron emission, e.g.) and simply projected onto the disc.

We shall mention in Section 2.3.2 that in several flares Machado and Linsky (1975) have found temperature increases of the order of 100 K over the temperature distribution in the preexisting faculae at a height of 300 km above the level where $\tau_{5000} = 1$ (compare Figure 37 for estimating this altitude). The flares they studied were not visible in white light and in fact one cannot expect that they would be with such a small temperature excess. However, the temperature increase may be higher in white-light flares so that it makes the 'photospheric flare' visible when it occurs near the limb.

D. *White-Light Flare Wave*

The white-light flare spectrum obtained by Grossi-Gallegos *et al.* (1971) was not calibrated. Thus it permitted some qualitative estimates (like Machado's (1971)), but it could not be quantitatively analyzed. The only spectral analysis of flare continuum was made for the flare of 1972, August 7 (Machado and Rust, 1974), but all this analysis concerned only the white-light flare wave and not the typical white-light knots. Thus, without any doubt, Machado and Rust discussed a type of white-light emission significantly different from what is usually observed and called a 'white-light flare'.

Machado and Rust repeated Machado's (1971) analysis of enhancements in cores of Fraunhofer lines and obtained similar results: the lines with low excitation potential are more affected than those with high excitation potential. At this time, however, they had at their disposal the heights where the different lines are formed, after Lites (1973). Thus they could establish that the white-light emission does not penetrate below $h \simeq 300$ km, but extends through the chromosphere at least up to $h \simeq 1000$ km, with $h = 0$ where $\tau(5000 \text{ Å}) = 1$. By applying Equations (4) through (12) and the Griem (1962) approximation (cf. Table VII) they found $n_e = 3.0 \times 10^{13}$ cm^{-3} in the flare region that emitted continuum and $n_e = 1.1-1.7 \times 10^{13}$ cm^{-3} outside of it. They explain this difference by flare penetration of 100 to 200 km deeper in those parts where the continuum is observed.

Machado and Rust explain this continuous emission by electron penetration into the upper photosphere (or low chromosphere, according to definition); 20 to 100 keV electrons heat these atmospheric layers to ~ 8500 K and the atmosphere emits continuous radiation through free-bound transitions in hydrogen. This explanation is open to some doubt, since the wave was observed after the impulsive phase when the streaming of ~ 100 keV electrons from the corona to the photosphere must be weak. The spectrum (Figure 39) might show an increase behind the Balmer limit (as Machado and Rust's interpretation requires), but the increase appears to start at rather long

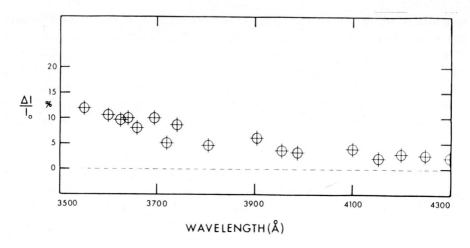

Fig. 39. Spectrum of the continuum emission of the moving white-light flare wave in the flare of 1972, August 7 (cf. Figure 34) relative to the normal photospheric spectrum. (After Machado and Rust, 1974.)

wavelengths; thus it also can be fit with continuous intensity increase towards short wavelengths (as observed, e.g., also for thread-like continua by Severny (1957)). Since the white-light emission was observed on the leading edge of the moving Hα ribbon and the ribbon moved unusually fast (40 to 50 km s^{-1}), one wonders whether perhaps the moving front rather than precipitating particles might have been responsible for the heating observed (through turbulence heating, e.g.).

2.3. Models of Chromospheric Flares

On the basis of the physical parameters deduced from the optical flare spectra (Section 2.2) several attempts have been made to model the chromospheric (low-temperature) part of a flare. In most approaches it has been assumed that the primary energy source for the flare is in the solar corona (or transition layer) and that the chromospheric flare is only a secondary product of an instability in the corona (Brown, 1971, 1973b; Shmeleva and Syrovatsky, 1973; Canfield, 1974; Kostjuk and Pikelner, 1974; Somov and Syrovatsky, 1974). For arguments supporting this view see, e.g., Kahler and Kreplin (1970) and Švestka (1973). Only Machado and Linsky (1975) created a model independent of this assumption.

Under the assumption that energy is supplied to the chromospheric flare from the corona, the energy transfer can take three different forms: by heat conduction, by streams of energetic particles, or by some mechanical mode (falling material or shock waves). We have shown already in Section 2.1.3.G that Hyder's (1967a, b) model of infalling material has not been confirmed by observations. Nakagawa et al.'s (1973) model of a downwards propagating shock does not produce emission line profiles comparable to those observed in flares (Canfield and Athay, 1974; cf. Section 2.3.1.D).

Thus only heat conduction and streams of particles remain as the two acceptable modes of energy transfer.

Many authors have assumed that streams of non-thermal particles accelerated in an active region drive all other flare phenomena and are thus the exclusive mode of energy transport in the flare (Elliot, 1969; Brown, 1971, 1973b; Kandel *et al.*, 1971; Strauss and Papagiannis, 1971; Cheng, 1972a; Syrovatsky and Shmeleva, 1972; McKenzie, 1972; Biswas and Radhakrishnan, 1973; Petrosian, 1973; Canfield, 1974; Kostjuk and Pikelner, 1974). This idea, however, as we shall see more in detail in the coming sections, contradicts several observed facts (cf. Section 3.1.3): (a) In many flares we do not see any evidence for a non-thermal acceleration process and thus for the presence of suprathermal particles in the flare region. (b) In those flares where this acceleration is observed (in the form of hard X-ray, EUV, and impulsive micro-wave bursts) it often occurs several minutes after the onset of the thermal flare (soft X-rays) and even after the first brightening in the Hα light. (c) The energy contained in the particles producing the non-thermal burst is in many flares definitely smaller than the total energy of the flare.

Therefore, as Takakura (1973), Švestka (1973), and Somov and Syrovatsky (1974) have emphasized, the basic flare phenomenon should be a thermal flare; by this we mean that all kinds of radiation we observe can be interpreted as due to a simple temperature rise and decrease. The earliest phase of flare development appears to be thermal in all events, and many flares show purely thermal (or quasi-thermal) behaviour during all their life. Only in some flares do we also observe non-thermal effects, and then it is always for a short time of only tens of seconds or a few minutes during the flash phase.

This means that the chromospheric emission must be predominantly (and at the flare onset completely) due to heat conduction. Heating by particle streams must be considered as an additional effect only.

2.3.1. MODELS OF ENERGY TRANSFER FROM THE CORONA

Brown (1973b) and Shmeleva and Syrovatsky (1973) and Kostjuk and Pikelner (1974) independently studied the distribution of temperature and density in chromospheric flares heated by energy input from the corona. Brown considered only the heating by particles and neglected the heat conductivity; Kostjuk and Pikelner also considered energetic particles as the primary source of heating, but they included heat conduction as a secondary effect. In contradistinction to it, Shmeleva and Syrovatsky found the temperature distribution to be determined only by thermal conductivity whereas the heating by particles can be neglected. Later on, Somov and Syrovatsky (1974) tried to find a middle way between these two contradictory points of view. Let us summarize briefly the main ideas of these three studies:

A. *Brown's Model*

Brown (1973b) has considered injection of a non-thermal electron beam vertically downward into a plane-parallel HSRA atmosphere (HSRA = *Harvard-Smithsonian*

Reference Atmosphere, published by Gingerich *et al.* (1971)). He adopts a power-law energy spectrum for the electrons (cf. Equation (85)) and a thick-target model (Section 3.2.1.E) for the electron energy input. Radiative losses balance this input at each point in the flare. Above a certain height the atmosphere is unable to radiate rapidly enough to balance the heating. Thus above this height (for $T > 6 \times 10^4$ K according to Brown) the atmosphere becomes unstable. This might explain the observed increase of non-thermal Doppler motions with height (Sections 2.1.3 and 2.2.3).

Brown assumes that below this height the atmosphere is in a steady state, since the time scale for electron injection (~ 100 s) is larger than the time scales for mass motion in the gas, adjustment of T_e and ionization. This, however, need not be true if the electron injection consists of a series of short $\lesssim 1$ s impulses, as high-resolution X-ray observations indicate (Van Beek *et al.*, 1974). According to Canfield (1974), Brown also underestimated the hydrogenic losses in the upper chromospheric layers, where the Ly-α dominates, and overestimated them in the lower chromosphere.

Brown's model certainly cannot explain the whole flare event in the chromosphere. His computations, however, could be relevant if the impulsive flare kernels only are considered. We shall come back to this subject in Section 2.3.1.D.

B. *Model by Shmeleva and Syrovatsky*

Shmeleva and Syrovatsky (1973) start with the energy equation

$$\frac{\partial \varepsilon}{\partial t} = \text{div}(\kappa \nabla T) - L(T, n_e, n) + P(t), \tag{38}$$

where ε is the internal energy per unit volume of plasma ($= 3 n_e kT$ in completely ionized hydrogen plasma).

The first right-hand term represents the change of heat content in the plasma due to heat conduction: κ is the coefficient of thermal conductivity and ∇T the temperature gradient. In a strong magnetic field, which is to be expected in flares, and for $T_e > 2 \times 10^4$ K (when hydrogen is completely ionized), heat conduction has only one component, $\kappa_\parallel$, along the field lines:

$$\kappa_\parallel = CT^{5/2}, \tag{39}$$

where C, of the order of 10^{-6}, depends only very slightly on the electron density and temperature. Both the contribution of ions (~ 50 times smaller) and conduction across the field lines are negligible. For temperatures below $\sim 2 \times 10^4$ K the situation becomes significantly more complicated, since one has to take into account the thermal flux of neutral atoms.

The second right-hand term in (38) is the radiative loss function that gives the amount of energy removed per unit plasma volume and per unit time by radiation. n is the number density of both neutral and ionized hydrogen atoms. This function can be written as

$$L(T, n_e, n) = L_r(T) n_e n, \tag{40}$$

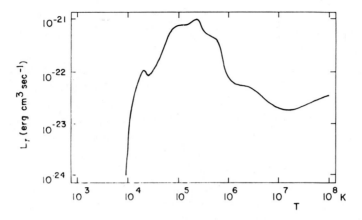

Fig. 40. Dependence of the total radiative losses in plasma on the plasma temperature T. (After Cox and Tucker, 1969.)

where $L_r(T)$ is the radiative cooling coefficient, dependent only on temperature. It was computed by Cox and Tucker (1969) for abundances which should not differ much from those in the solar corona and its variation with T is shown in Figure 40. Equation (40) is valid only when plasma is optically thin at all wavelengths, which again restricts its validity to regions where $T \gtrsim 2 \times 10^4$ K. At higher layers, with temperature in excess of this value,

$$\frac{n_e}{n} = 1 + \Delta_e, \tag{41}$$

where Δ_e is the relative electron addendum from ionisation of the other elements, representing only a small correction (for $T > 10^7$ K, $\Delta_e = 0.33$). Thus one can approximate,

$$L(T, n) \simeq L_r(T)\, n^2. \tag{42}$$

The last right-hand term in Equation (38) includes all other modes of flare heating. Effects to be considered are heating by mechanical waves coming from the convective zone and heating by non-thermal accelerated particles. As long as we suppose that the primary source of the flare is in the solar corona, the first mechanism need not be considered, since it only maintains the chromosphere and corona at their 'quiet' temperatures (but, as we shall see in Section 6.4.1, some flare theories suppose that an increased flux of Alfvén waves from below is the primary source of the flare).

When solving Equation (38), Shmeleva and Syrovatsky assume a steady-state atmosphere, similarly to Brown (1973b); i.e. they put $\partial\varepsilon/\partial t = 0$. As we mentioned in the preceding section, this may not be correct if the atmosphere is heated by streams of electrons. In their treatment of the electron energy input, Shmeleva and Syrovatsky suppose that the electron heating occurs in a thin plasma layer at an altitude h where

$$N = \int_{h}^{\infty} n(h)\, \mathrm{d}z \simeq 3 \times 10^{17} E_0^2 \ (\mathrm{cm}^{-2}), \tag{43}$$

where E_0 is the typical 'mean' energy of the accelerated electrons (after Syrovatsky and Shmeleva, 1972). The electron energy quickly transforms into a thermal flux and energy transfer to deeper layers in the atmosphere occurs by conduction. Ultimately the energy is transformed into radiation of the optical flare with a very large temperature gradient, which may explain the observed very small geometrical thickness of flares in the chromosphere. Therefore, they suggest a 'shell' structure of the flare as opposed to the usually assumed filamentary one (cf. Section 2.2.7).

Two points in this analysis have been attacked by Brown (1974). First, one cannot replace the power-law energy spectrum of accelerated electrons (Equation (85)) by a 'mean', monoenergetic value E_0 (Equation (43)). According to Brown (1973b) the electron energy flux drops off with depth as $N^{-\delta/2}$. There are about 10^{19} particles cm^{-2} at $T \simeq 10^7$ K (primary hot flare source) and about 10^{21} cm^{-2} at $T \simeq 10^4$ K (chromospheric flare). Thus, for typical $\delta \simeq 4$, the electron energy deposition falls by a factor of $(10^{21}/10^{19})^{-2} \simeq 10^{-4}$ between these layers. The conductive flux, on the other hand, varies as $T^{5/2} \nabla T$ (Equations (38) and (39)) so that, even when ignoring ∇T, it drops by a factor of $(10^4/10^7)^{5/2} \simeq 10^{-7}$ from the coronal to the chromospheric flare. It appears, therefore, that wherever streams of electrons participate in the energy transfer, they should dominate heat conduction.

Secondly, Brown's objection is that Shmeleva and Syrovatsky applied Equation (42) (which is correct in upper flare layers) down to chromospheric regions, which are optically thick in many lines. Shmeleva and Syrovatsky argue that this equation becomes invalid only for true absorption. Lines that originate through scattering (all resonance lines, e.g.), might modify the equation, but not make it invalid.

C. Models Including Both Conduction and Particle Heating

Even when quantitative theoretical approach encounters great difficulties, as we saw in the preceding two sections, at least two results seem to be well established:

(1) There must be some restricted regions in flares where Shmeleva and Syrovatsky's (1973) approach is wrong and Brown's (1974) criticism correct. This follows from the fast changes in brightness of the Hα impulsive kernels (Section 2.1.2.A) associated in time with fast changes in the transition layer and corona. Such a fast reaction of the chromosphere to coronal changes can never be accomplished through heat conduction, since the velocity of thermal (conduction) waves is in general considerably less than the velocity of sound (Bessey and Kuperus, 1970). Hence the conduction speed in the coronal flare plasma ($T \simeq 10^7$ K) would be less than 300 km s^{-1} and in the chromospheric flare ($T \simeq 10^4$ K) less than 10 km s^{-1}. On the other hand, accelerated electrons moving at about 100 000 km s^{-1} can mediate a response within a second.

(2) One can show that heat conduction is able to bring energy from the corona to the chromosphere (Švestka, 1973), hence complete neglect of heat conduction by Brown (1973a) was also incorrect.

Let us assume a vertical energy transport from the flare source in the corona to the chromosphere along magnetic field lines. Let z equal zero at the bottom of the flare source and increase downwards along the field lines. Let the whole energy gain in the $(0, z_0)$ region be due to conduction and all energy losses to radiation. Then, prior to the flare, heat conduction and radiation must have been balanced and Equation (38) can be written as

$$\frac{\partial}{\partial z}\left(\kappa \frac{dT}{dz}\right) = L(T, n_e, n), \tag{44}$$

where κ is given by (39) and L by (42).

The heat flux at depth z,

$$F(z) = -\kappa \frac{dT}{dz}, \tag{45}$$

is according to (44),

$$F(z) = F(0) - \int_0^z L(z)\, dz, \tag{46}$$

where $F(0)$ is the flux emerging from the coronal source region. The depth z_0 where $F(z)$ drops to zero is defined by

$$\int_0^{z_0} L(z)\, dz = F(0). \tag{47}$$

This is the depth where downwards conduction ceases to be effective.

Let us suppose that the temperature in the coronal flare is increased by a factor m over quiet conditions. Then, according to (39) and (45), $F(0)$ increases by a factor $m^{5/2} (dT/dz)_{z=0}$; thus the rise in T to $\sim 10^7$ K involves a rise $F(0)$ by some three orders of magnitude. Therefore, as long as n in Equation (42) can be considered constant, there is clearly.

$$\int_0^{z_{0F}} L_{rF}(z)\, dz \gg \int_0^{z_0} L_r(z)\, dz, \tag{48}$$

where the index F refers to the conditions in a flare. Figure 40 shows that L_r decreases with increasing T for $T \gtrsim 2 \times 10^5$ K. Hence, for $T \gtrsim 2 \times 10^5$ K, $L_{rF} < L_r$ in (48) and the increased flux can be compensated only by an extension of the heated region to a greater depth. This implies that the transition layer must be lowered in the flare and the increase from $\sim 10^4$ K to $\sim 10^5$ K must become significantly steeper than in a non-flaring active region. This also is the result obtained by Shmeleva and Syrovatsky (1973).

Below $T \simeq 2 \times 10^5$ K the cooling coefficient L_r increases with increasing T. However, L_r is roughly proportional to T, hence it still cannot compensate the increase in κ, which is proportional to $T^{5/2}$. Therefore, there is no obvious reason why the heat conduction would not bring the enhanced flux down to regions with $T \simeq 20\,000$ K, i.e. to the chromosphere. Also Rust and Roy (1974) in two flares of June, 1972, have given arguments supporting the view that the energy for the main phase Hα emission is supplied by conduction from a coronal source.

At still lower temperatures the situation changes drastically, since hydrogen is not completely ionized any more. The conductivity coefficient is now governed by neutral atoms, becomes less sensitive to T, and is able to conduct the heat in all directions irrespective of the magnetic field. At $T \simeq 8000$ K, which seems typical for disc flares (Section 2.2.6), $\kappa_{\parallel}$ along the field lines and $\kappa_{\perp}$ across the field lines are approximately equal. In addition, the radiative loss is governed now by optically thick lines so that Equations (40) and (42) are invalid.

The items (1) and (2) strongly support the view that both the heat conduction and streams of particles participate in the chromospheric flare heating. This point of view has been adopted also by Somov and Syrovatsky (1974), who have considered conditions under which one of these two types of heating dominates. Their conclusions can be summarized as follows:

Most of the flare particle energy is contained in the low-energy part of the particle spectrum. These particles have very short range for collision losses and rapidly transform their energy into heat. Therefore, the temperature run with depth is determined mainly by heat conduction. Only a very small fraction of the initial heat flux penetrates to the chromospheric regions below $T \simeq 2 \times 10^4$ K. When, however, the particle energy spectrum is sufficiently hard, a considerable number of particles (with high initial energies) can penetrate the transition layer at least in some restricted areas and thus contribute to the energy flux carried to the low-temperature region of the flare. Kostjuk and Pikelner (1974) have presented a similar model, but they add to it a shock which forms in the consequence of a temperature jump below the dissipated particle stream. The shock further heats and condenses the gas thus increasing the electron density in the flare region.

Somov and Syrovatsky still suppose that all flares start with acceleration of particles to suprathermal energies and that the 'quasi-thermal' flares are those events where the spectrum of accelerated particles is soft. However, they also consider it possible that the heating of the high-temperature flare region may have nothing at all to do with non-thermal particles. As an example they mention stationary dissipation of magnetic energy in a current sheet (Syrovatsky, 1972; Tucker, 1973) when there are few or no non-thermal particles produced.

D. *Models by Canfield and Athay*

Canfield and Athay (1974) have solved simultaneously the equations of radiative transfer and statistical equilibrium for a model hydrogen atom including the two lowest lines of the Lyman series, the Hα line, and the Lyman, Balmer and Paschen continua. They have considered the response of the HSRA atmosphere to a downwards propagating shock (after Nakagawa *et al.*, 1973) and to a stream of non-thermal electrons (after Brown, 1973; Canfield, 1974). Their aim was to calculate the theoretical profile of the Hα line following a shock or a particle stream through the atmosphere, and to compare it with the Hα profiles actually observed.

The intensity profile I_v is related to the source function S_v by

$$I_v = \int_0^\infty S_v \exp\left[-\int_0^{\tau} (\phi_v + r_0)\, d\tau_0\right](\phi_v + r_0)\, d\tau_0, \tag{49}$$

where ϕ_v is the profile of the line absorption coefficient ($= 1$ in the line center), τ_0 is the line-center optical thickness, $r_0 = d\tau_c/d\tau_0$ is the ratio of continuum to line-center optical thickness, and the source function S_v is composed of the line (S) and continuum (B) components,

$$S_v = \frac{\phi_v}{\phi_v + r_0} S + \frac{r_0}{\phi_v + r_0} B, \tag{50}$$

with B being the Planck function. Under the assumption of scattering with complete redistribution in frequency the line source function S can be written as

$$S = \frac{\varepsilon + \delta'}{\varepsilon' + \delta'} B + \frac{1}{\pi^{1/2} \Delta v_D(\varepsilon' + \delta')} \int_0^\infty \frac{\phi_v}{\phi_v + r_0} \frac{dH_v}{dT_0}\, dv. \tag{51}$$

Here ε and ε' are the source (i.e. excitation) and sink (i.e. deexcitation) parameters for line photons (cf. Athay, 1972), δ' is a sink parameter measuring the probability of photon absorption by continuum processes, Δv_D is the Doppler width, and H_v is the monochromatic net flux. All are functions of depth.

The transfer equation is coupled with the statistical equilibrium equations through the escape coefficient ρ_{ji} for the $j-i$ transitions,

$$\rho_{ji} = \varepsilon_{ji}(B_{ji}/S_{ji}) - \varepsilon'_{ji}. \tag{52}$$

The radiative loss per unit volume in the $j-i$ transition, $4\pi\Delta H_{ji}$, is related to ρ_{ji} by

$$4\pi\Delta H_{ji} = hv_{ji} A_{ji} \rho_{ji} N_j, \tag{53}$$

where A_{ji} is the spontaneous transition probability and N_j the population of the upper level. Natural, resonance, Doppler, and Stark broadening have been included. By using Equations (51) through (53), the line source functions S were obtained by an iterative procedure. Electron densities n_e were determined from the solution of the hydrogen ionization equilibrium, also by iteration.

This method was first applied to an atmosphere subject to a downward propagating shock as suggested by Nakagawa *et al.* (1973). The resulting Hα line profiles disagree with the profiles observed in flares. In particular they show a much stronger self-reversal and the red asymmetry is much too small. This is a particularly significant disagreement in view of the fact that Nakagawa *et al.*'s model was designed to explain the existence of a red shift.

In a second approach, Canfield (1974) applied this method to Brown's (1973b) model described in Section 2.3.1.A. The calculations again do not produce the typical Hα line profiles of a large flare, but are certainly within the range of observed characteristics of flares. With $\delta = 5$ in Equation (85) Canfield has obtained Hα line profiles similar to those observed in flares, with three differences: (a) Rather large self-reversal

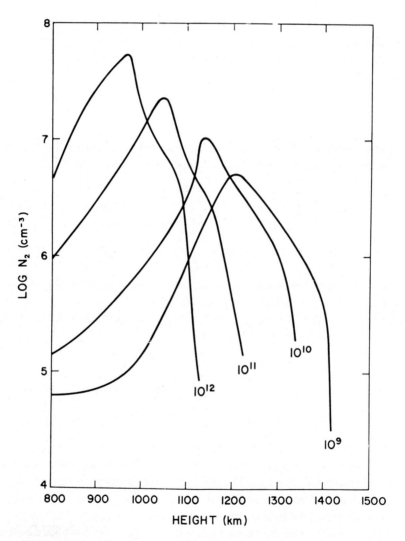

Fig. 41 a.   Calculated height distributions of the density of hydrogen atoms in the second quantum state when an electron stream coming from the corona excites the chromosphere. It is assumed that the beam electron energies follow a power-law spectrum with exponent $\delta = 5$. The numbers inside the graph give the incident flux of beam electrons of initial energy >20 keV. (After Canfield, 1974.)

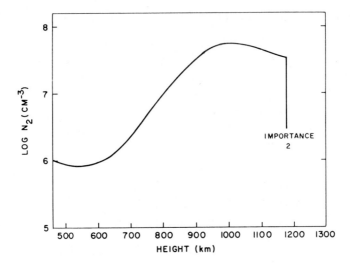

Fig. 41 b.   Calculated height distribution of the density of hydrogen atoms in the second quantum
state in a flare of importance 2, after Machado and Linsky (1975).

(but much more acceptable than for Nakagawa *et al.*'s model). (b) No asymmetry.
(c) The electron density ($< 5 \times 10^{12}$ cm^{-3}) and optical thickness ($\int n_2 \, dh < 6 \times 10^{14}$)
are less than one fourth the observed values for large flares. (For the profiles most
similar to the observed Hα line in flares the computed density and optical thickness
are 20 to 30 times too low.)

In all these computations we suffer from an ignorance of the radiative losses in the
flaring regions. Furthermore, only two discrete levels have been considered. Nevertheless, even though some error may be involved, Canfield's computation demonstrates again that streams of particles cannot produce the observed chromospheric flare.
Canfield's results seem to fit the real situation if one assumes (in agreement with
observation) that Brown's streams heat the chromosphere during the flash phase
while the quasi-thermal flare is still rising to its maximum. In that case one can expect
what Canfield has found: relatively small values of electron density and optical thickness compared to the values observed at flare maximum, which occurs a few minutes
later.

As an illustration of the energy deposition by a stream of energetic electrons,
Figure 41a shows the calculated height distribution of the population of the second
hydrogen level. The curves have been obtained for radiative balance in the Lyman
lines, $\delta = 5$ in Equation (85), and 10^9 to 10^{12} erg cm^{-2} s^{-1} as the incident energy
flux of beam electrons with initial energy > 20 keV streaming from the corona to the
chromosphere. One can see that the range of height where energy is deposited is quite
small, some 300 to 400 km; still it exceeds the values of the flare linear thickness z
deduced in Section 2.2.7.

2.3.2. MODEL BY MACHADO AND LINSKY

Machado and Linsky (1975) did not make any assumption about the mode of energy transfer in flares. They divide the low-temperature flare (which is assumed homogeneous, plane-parallel, and in vertical hydrostatic equilibrium) in two layers: photospheric flare, below the temperature minimum, and chromospheric flare, from the temperature minimum to temperatures exceeding a few times 10^4 K.

A. *The Photospheric Flare*

As a first step they derive photospheric flare models from the Ca II K line profiles in flares (photographed at Sacramento Peak Observatory) on the assumption that in the damping wings of the K line complete redistribution is valid and LTE conditions are met. The approach is essentially the same as that used for computing model photospheric faculae (Shine and Linsky, 1974).

When the K-line profiles in faculae and flares are compared, one finds temperature in the flare in excess of the pre-flare facular temperature everywhere from $h \simeq 170$ km upwards; ($h = 0$ where $\tau_{5000} = 1$; see Table X and, for comparison, Figure 37.) The temperature minimum also occurs deeper in the atmosphere than in faculae. This gives evidence that the effect of the flare can be observed at much greater depths in the atmosphere than is predicted in the model calculations made by Brown (1973 b) and Canfield (1974).

This result must be taken with some caution, since the analysis of the K-line wings under disturbed conditions is very difficult. The result certainly does not contradict other observations, because the temperature increase of some 100 deg found in the uppermost photospheric region would not produce detectable continuum in the optical flare spectrum. However, it is quite difficult to explain it as long as we assume that flares originate above the chromosphere; as we saw in Section 2.3.1, both conduction and particle streams deposit essentially all energy at much higher altitudes.

Machado and Linsky suggest that the photospheric heating might be caused by protons that penetrate to these layers (Figure 37). Nevertheless, as they admit, we don't have any evidence of protons being accelerated in the studied flares. One could also suppose that conditions in the uppermost photospheric layers possibly could be changed during a flare through radiative transfer of energy downward from the corona. The flare EUV and X-ray radiation illuminates the lower atmospheric layers. Thus some energy, which certainly is not negligible, must be deposited there through absorption of this radiation and this energy may be re-radiated at completely different wavelengths. Physical conditions in the photosphere under the flare may thus change slightly. This process was mentioned many years ago (Jefferies, 1957; Švestka, 1957), and recently Somov (1975) made some quantitative estimates of the radiative heating produced by thermal X-ray emission of flares in lower atmospheric layers. According to his result essentially all X-ray energy is deposited in the chromospheric region; only for $T > 3 \times 10^7$ K a slight part of the energy is deposited in atmospheric layers with densities $n_H > 10^{16}$ cm^{-3}. This may not be enough for explaining the heating observed; if this explanation fails, the found temperature excess might be considered a strong argument against all theories that place the flare origin above the chromosphere; it would then support Unsöld's (1958) and Piddington's (1974) suggestion that flares are due to enhanced dissipation of waves propagating from the convection layer of the Sun through the photosphere and chromosphere into the corona.

B. *The Chromospheric Flare*

As the second step Machado and Linsky have calculated models of chromospheric flares taking the photospheric model, obtained before, as the lower boundary condition. From there they assumed a rise in temperature up to $T = 8500$ K; above that T jumps to the coronal values. The rise is linear with mass column density. Thus the upper boundary condition must determine at which height (i.e. mass column density) the value $T = 8500$ K is reached.

In the HSRA atmosphere (Gingerich *et al.*, 1971) mass column density at $T = 8500$ K is $m_0 = 5 \times 10^{-6}$ g cm^{-2}. Machado and Linsky suppose that in a flaring region density is enhanced by a factor of more than 100 above the quiet Sun in the transition region and upper chromosphere. They base this supposition on measurements in the EUV spectral range (Noyes, 1973; see Section 3.2.2), but we have seen that electron densities determined from the optical region also indicate similar increase (de Feiter, 1974; see Section 2.2.4.B). Therefore, Machado and Linsky adopt as the upper boundary condition (rather arbitrarily, but there is obviously no other way) m_0 (flare) = 60, 180, 300, and 600 times m_0 (HSRA) at $T = 8500$ K for flares of importance S, 1, 2, and 3, respectively.

When computing the model flare chromospheres, Machado and Linsky represented

TABLE X

The model of low-temperature flare by Machado and Linsky (1975) for a flare of Hα importance 2

h(km) (above $\tau_{5000} = 1$ level)	T(K)	n_H(cm^{-3})	n_2(cm^{-3})	n_e(cm^{-3})	P(dyn cm^{-3})
	rise to corona				
1177	15 000	9.5×10^{12}	3.0×10^6	9.5×10^{12}	41.4
1176	8 397	2.3×10^{13}	3.5×10^7	1.1×10^{13}	41.5
1168	8 376	2.3×10^{13}	3.5×10^7	1.2×10^{13}	42.8
1159	8 353	2.4×10^{13}	3.6×10^7	1.2×10^{13}	44.1
1098	8 190	3.3×10^{13}	4.5×10^7	1.3×10^{13}	55.1
1020	7 935	5.2×10^{13}	5.4×10^7	1.2×10^{13}	75.9
941	7 530	8.8×10^{13}	4.8×10^7	9.8×10^{12}	110
763	6 139	3.4×10^{14}	6.0×10^6	1.7×10^{12}	315
560	5 171	1.8×10^{15}	8.6×10^5	4.3×10^{11}	1.41×10^3
397	5 010	7.2×10^{15}	1.7×10^6	9.3×10^{11}	5.51×10^3
316	5 013	1.4×10^{16}	3.4×10^6	1.7×10^{12}	1.09×10^4
240	5 043	2.9×10^{16}		2.8×10^{12}	2.00×10^4
185	5 173	4.4×10^{16}		4.4×10^{12}	3.17×10^4
	facular conditions				

the hydrogen atom by three discrete levels and continuum, with detailed balance in the Lyman transitions. They followed the approach used by Shine and Linsky (1974b) in computing plage models. Table X shows one of their models, for a flare of Hα importance 2. Table XI gives basic characteristics of all models, for importances S, 1, 2, and 3. They are in fairly good agreement with flare parameters actually observed and reported in Sections 2.2.4, 2.2.5, 2.2.6, and 3.2.2. The transition layer shifts downwards with increasing flare importance, as the computations of heat conduction from the corona predict (Sections 2.3.1.B and C).

Figure 41 b shows the n_2 distribution with height, according to Table X. It is quite similar to that obtained by Canfield (1974) for Brown's (1973b) model (Figure 41a), with an extension towards the photosphere which Canfield's results do not show. The region that gives the main contribution to the flare radiation is again narrow, some 400 km.

TABLE XI

Characteristic parameters of Machado and Linsky's flare models

Hα importance	$n_{e\ max}$ (cm^{-3})	Column density $N_{2\ max}$ (cm^{-2})	$\dfrac{P_{flare}}{P_{quiet\ Sun}}$	h (8500 K)
S	4.0×10^{12}	2.8×10^{14}	54	1430 km
1	8.2×10^{12}	5.6×10^{14}	179	1205 km
2	1.2×10^{13}	1.5×10^{15}	276	1177 km
3	2.3×10^{13}	4.1×10^{15}	546	1096 km
$\geqslant 2$ observed (Sections 2.2 and 3.2.2)	$1-4 \times 10^{13}$	3×10^{15}	150?	no data

Machado and Linsky then computed the Ca II K line profiles that would be produced by their models. Equations of statistical equilibrium for a five-level plus continuum calcium ion were considered and complete redistribution was assumed. However, the resulting K profiles differ considerably from the observed ones in several respects, which are surprisingly similar to those found by Canfield and Athay (1974) and Canfield (1974) in their models for the Hα line. Two of them are most apparent:

(1) The profiles show a very strong central reversal that is not observed in flares. Machado and Linsky, like Canfield and Athay, try to remove the reversal by assuming large (macro-turbulent) non-thermal motions in chromospheric flares. But we have emphasized in Section 2.2.1 that the chromospheric flare spectrum with broad hydrogen and narrow metal lines contradicts this assumption. Thus the central reversal must be removed in some other way. If flares have filamentary structure (Section 2.2.7), we observe the interflare matter in the line center while all the models considered a homogeneous flare with conditions proper to the flare elements only; the interflare matter can fill in the central part of the profile. Another possibility is that the reversal is due to the neglection of all levels in the hydrogen atom higher than 3. As de Feiter (unpublished) showed many years ago, by neglecting higher levels one gets wrong b

values (cf. Section 2.2.6) for the levels taken into account. An incorrect ratio b_3/b_2 might produce a fictitious central reversal in the Hα line.

(2) The line cores are more intense than those observed in flares. This may be an indication that flares have filamentary structure and that the radiation is diluted.

Machado and Linsky conclude that the flare must be basically a thermal response to heating occurring at the same time over a wide range of heights in the solar atmosphere. The fact that the flare is first observed in transition region lines does not necessarily mean that the flare originates close to the transition layer; this region is simply most sensitive to temperature changes and after the flare onset it quickly moves to more dense layers lower in the atmosphere due to an inability to radiate the increasing energy input. In some flares the heating seems to trigger the acceleration process which produces the non-thermal bursts. This happens above the chromosphere and only in this case does the energy transport through particle streams become significant, causing impulsive kernels in the chromosphere. In a few cases, when the particle energy spectra are unusually hard, the particles produce an overheating of the photosphere as well and cause visibility of some flare parts in white light.

CHAPTER III

THE HIGH-TEMPERATURE FLARE

Under the term 'high-temperature flare' we shall include not only the thermal flare phenomena that are two or three orders of magnitude hotter than the chromospheric part of the flare discussed in Chapter II, but also the impulsive non-thermal phenomena observed in the X-ray, EUV and radio spectral regions. The reason for this is that thermal and non-thermal effects are coupled in the high temperature flare regions in the solar corona; non-thermal effects arise from pre-heated plasma and eventually become thermalized.

3.1. Thermal Phenomena

3.1.1. CORONAL CONDENSATIONS

As soon as an active region forms on the solar surface, the brightness of the red (Fe x, 6374 Å) and green (Fe xiv, 5303 Å) coronal lines above the active region increases, and we speak about *permanent coronal condensations* or *coronal enhancements* above the sunspot groups. In soft X-rays these enhancements bridge the zero line of the longitudinal magnetic field ($H_\parallel = 0$), being brightest along the $H_\parallel = 0$ line (Vaiana *et al.*, 1973). If the group is very active, a hot core is occasionally observed embedded in this coronal enhancement and is characterized by emission in the yellow coronal lines of Ca xv (5445 and 5694 Å). Also emission in the 4086 Å line of Ca xiii and in the continuum are typical for these cores which are called *sporadic* or *high-temperature condensations*.

These sporadic condensations are closely associated with the occurrence of flares. Since however, optical observations of the condensations are restricted to the solar limb, and can be associated with events in front of, as well as behind the limb, a one-to-one correlation between flares and the occurrence of yellow coronal line emission can not be established. Most probably, high-temperature condensations are formed in flares, but they survive much longer than the low-temperature flare in the chromosphere, as do the loop prominence systems observed for about ten hours after some major flares (cf. Section 4.4). As a matter of fact, the structures observed in yellow coronal lines often also take the shape of loops.

In soft X-rays the regions with high flare activity show one or more bright and very hot cores which de Feiter and de Jager (1973) call 'superthermal plasma nodules'. They may precede some flares, but as it seems, they are mostly remnants of flares, similar to the yellow coronal line emission.

In Section 2.2.4 we mentioned Zirin's (1964) observations of the Ca xv lines in a flare more than 20 000 km above the limb. Measurements of the Doppler width of these lines give temperatures around 4×10^6 K, and according to the coronal ionization theory, a temperature of about the same order is required for equal concentrations of Ca XIII and Ca XV, which are the ions that produce the lines most characteristic of a sporadic condensation. This temperature is about twice as high as that observed in the permanent coronal condensations, where the observed emission in the green line leads to $T_e = 2.4 \times 10^6$ K, and in the red line to $T_e = 1.7 \times 10^6$ K (Nishi and Nakagomi, 1963). The electron density is in excess of 5×10^9 cm^{-3} in the dense parts of sporadic condensations (Waldmeier, 1963, Nishi and Nakagomi, 1963; Saito and Billings, 1964; Zirin, 1964), and it has been found to be as high as 10^{11} cm^{-3} from the continuum (Zirin, 1964; cf. Section 2.2.4). Since however, the low-temperature parts are still mixed with the high-temperature condensation at these heights, the n_e value deduced from the continuum may refer to these, possibly more dense, flare regions, as well. Therefore, to an order of magnitude, 10^{10} cm^{-3} seems to be a good estimate of the electron density in the flare-associated high-temperature condensations. This value also is in good agreement with density estimates made from observations of thermal X-rays (Section 3.1.4).

Some observations indicate that an increased density in the flare-associated condensation may survive for many days, particularly after large proton flares which eject material into the corona. Newkirk et al. (1969) have shown that in 1966, July 10 (after proton flare on July 7) and on September 5 (after proton flare on September 2) unique, low elevation coronal condensations were observed on the western limb, with anomalously increased density up to 1.5 solar radii. Similarly the large flare of 1959, June 9, showed an expanding condensation 2 to 3 days after the flare occurrence. If the flare was actually its source, one gets expansion velocities of about 2 km s^{-1} (Newkirk et al., 1969), very close to the velocities in the loop prominence systems (Bruzek, 1964b). This again points to a close association between loop prominences and high-temperature condensations, which may represent different aspects of the same phenomenon.

Observations in X-rays (Vaiana et al., 1973; Chase et al., 1975) show that many coronal condensations (i.e. active regions) are interconnected with loops, in particular when they belong to the same complex of activity (as defined by Bumba and Howard, 1965). Existence of these loops is also confirmed by white-light corona observations on the solar limb (Hansen et al., 1972). Some of them cross the equator, connecting condensations on opposite hemispheres (trans-equatorial arches in white-light corona), others connect active regions on the same hemisphere (intermediate helmets in white-light corona). In X-rays the longest interconnecting loops, observed during the Skylab mission, extended over 37 heliographic degrees. It is not yet clear whether this is the actual limit of the length of these interconnections, or whether we just cannot see the longer loops.

These interconnections can be strongly influenced by activity that occurs in one of the interconnected regions (Chase et al., 1975; AS & E, 1975). One some occasions (as it seems, in relation to flares) the loops brighten, and even new loops may become

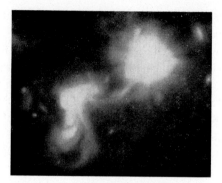

**7 AUGUST 1973
12:08 UT**

**7 AUGUST 1973
15:41 UT**

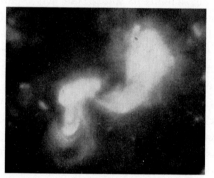

**7 AUGUST 1973
18:09 UT**

5 arc. min.

Fig. 42. Transequatorial loops interconnecting active regions McM 12472 (10°S) and 12474 (07°N), visible in soft X-rays. A subflare occurred in McM 12474 at 13^h07^m UT; after that (maybe in association with the subflare) the interconnecting loops brightened for about five hours, but the other interconnected region looks unaffected (central frame). Note that another, fairly twisted loop, becomes visible at 18^h09^m UT. (Skylab photograph, courtesy of AS & E, Cambridge, Mass., U.S.A.)

visible (Figure 42). Nevertheless, so far no evidence has been found that flare-associated enhancement of a loop produced a sympathetic flare (cf. Section 4.3) in the
other interconnected region.

3.1.2. SOFT X-RAY EMISSION

As we mentioned in the preceding section, the increased temperature and density in a
permanent coronal condensation also manifest themselves by an increased flux in
soft X-rays (i.e. X-rays with energy $\lesssim 10$ keV) above the active regions. The flux and
the hardness of the X-ray spectrum varies within wide limits above individual active
regions indicating large differences in the temperatures and/or densities. The highest
flux and the hardest spectra are observed in those active regions which are most
productive in flares (de Feiter and de Jager, 1973; AS & E, 1975) and in emission of
particles into the corona and into interplanetary space (cf. Section 5.1.1.B).

A. Soft X-Ray Bursts

The X-ray flux from an active region varies greatly with time. These variations may
be fast, lasting from a few minutes up to several hours, as well as gradual, lasting for
hours up to several days (Krieger et al., 1972; Parkinson, 1973a). The fast variations
are mostly associated with flares. According to Hudson et al. (1969) and Drake (1971)
about 70% of these X-ray enhancements can be identified with Hα flares, but this
percentage decreases if more sensitive X-ray receptors are used. On the other hand,
if the X-ray receptors are sensitive enough, essentially every solar flare is associated
with a soft X-ray burst. Out of 283 flares studied by Thomas and Teske (1971) only
one (of importance 1f) appeared to be unaccompanied by soft X-rays between 8 and
12 Å. In addition, about 75% of subflares are associated with soft X-ray bursts (Drake,
1971). One is very much inclined to believe that even many of the X-ray enhancements
for which no optical counterpart can be found in the published lists of flares are
associated with flare-like brightenings too small to be reported by the observing stations
as subflares. This, e.g., was the case with the brightening shown in Figure 11. Other
non-flare X-ray increases have been identified with surge-like spikes on the solar limb
(Teske, 1971c), eruptive prominences (Hyder, 1973), loop prominences (Teske, 1971a;
Brinkman and Shaw, 1972; Machado et al., 1973; Waldmeier, 1973; AS & E, 1975),
type III radio bursts (Kane et al., 1974) or with gradual non-flare temperature variations in the active region (Krieger et al., 1972; Parkinson, 1973a).

X-ray photographs from rockets and particularly from Skylab also show a number
of X-ray brightenings which are not related to active regions. The most conspicuous
of them are the flaring bright points (Golub et al., 1974, 1975) mentioned in Section
2.1.2 and demonstrated in Figure 10. But also some loop-structures outside of active
regions show short-lived brightenings; some of them are associated with an activation
of quiescent filaments; others brighten for reasons that we do not yet understand
(AS & E, 1975).

A typical flare-associated soft X-ray burst has a time profile roughly similar to the
time development of the Hα intensity in the brightest point of the flare and similarly,

is characterized by a fast rise and much slower decay (Figures 2, 53 and 54). According to Thomas and Teske (1971) the soft X-ray burst begins before the optical Hα event (about 2 min on the average), reaches its peak flux later (the mean difference is about three minutes), and ends several minutes after the optical flare is no longer visible. On the other hand, Datlowe et al. (1974) find the onsets of Hα flares and soft X-rays approximately simultaneous. Statistically, the X-ray bursts' strength increases with the increasing Hα importance of the flare, but there are great deviations from this rule in individual cases. Teske and Thomas (1969) and Spangler and Shawhan (1974) also find a correlation between intensities of microwave and soft X-ray bursts. Generally, very active regions usually produce relatively strong X-ray bursts.

The height of the soft X-ray emission can be estimated by two methods: from study of flare events that occur close behind the solar limb and from simultaneous observations by two spacecraft in markedly different heliocentric longitudes.

The first method was first applied by Beigman et al. (1969) and Teske (1971 b) who found $h \lesssim 20\,000$ km and $\simeq 10\,000$ km, respectively, above the base of the chromosphere. More recently Roy and Datlowe (1975) analyzed 37 X-ray bursts beyond the limb in energy range of 5.1 to 6.6 keV. For 27 of the bursts the minimum height was below 8000 km above the chromospheric base. In two events, however, the minimum height was about 25 000 km, and in other two cases 100 000 km. Thus, unless the identification was wrong, these cases indicate that measurable soft X-ray emission in flares can occasionally extend very high into the corona.

Observations from two spacecraft were used by Catalano and Van Allen (1973) who compared the soft X-ray data from Explorers 33 and 35 near the Earth and from Mariner 5 deep in space. During the 117 day period 607 bursts were recorded. For those fully visible from one spacecraft but partly occulted by the solar disc for the other, the dependence of soft (~ 1 to 3 keV) emission I on altitude h above the base of the chromosphere was found well represented by

$$\frac{dI}{dh} \sim \exp[-h/h_0] \tag{54}$$

with

$$h_0 = (11 \pm 3) \times 10^3 \text{ km}.$$

Roy and Datlowe (1975) find their measurements (mentioned above) consistent with the scale height of 11 000 km in (54). On the other hand, a few anomalous cases in Catalano and Van Allen's analysis seem to confirm Roy and Datlowe's conclusion than some X-ray flares can extend to unusually large altitudes.

B. X-Ray Flares

In 1968 the first direct photograph of an X-ray flare was obtained between 3 and 14 Å (Vaiana et al., 1968; Vaiana and Giacconi, 1969) and in 1973 many X-ray subflares and several flares were photographed during the Skylab mission (Kahler et al., 1975; Petrasso et al., 1975; Vorpahl et al., 1975; Pallavicini et al., 1975;

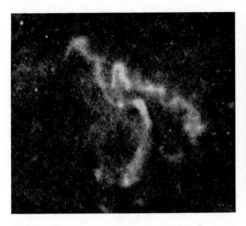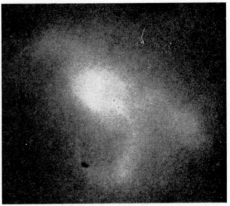

Fig. 43a. The flare of 1968, June 8 in the Hα light (left) and in 3–14 Å X-ray light (right), at 17^h43^m UT. (After Vaiana and Giacconi, 1969.)

Vorpahl, 1975). Images with lower resolving power ($\sim 20''$ against $<5''$ on Skylab) were also obtained on OSO 7 (Neupert *et al.*, 1974). Examples of X-ray flares are shown in Figures 11, 20b, and 43a, b.

The X-ray flare of Hα importance 1 n photographed during a rocket flight by Vaiana *et al.* (1968) showed a filamentary structure with characteristic dimensions of the order of 10^4 km. The general correspondence between bright regions in X-rays and bright regions in Hα is quite apparent (Figure 43a). However, apart from the bright regions visible both in the X-rays and Hα, part of the bright X-ray emission corresponds to a bridge between the two Hα emission areas located on opposite sides of the zero line of the longitudinal magnetic field (Vaiana *et al.*, 1973). This again indicates (cf. Sections 2.1.3.E and 3.2.2) that the basic shape of flares is a loop, or a system of loops, of which only the feet are visible in the Hα line, the top being too hot to produce Hα emission.

The loop-like structure of flares has been confirmed also by Skylab and OSO-7 measurements. According to Kahler *et al.* (1975) and Vorpahl *et al.* (1975), who studied many subflares recorded on Skylab, spatial configurations vary widely from event to event, exhibiting features as small as 3600 km as well as large structures up to 100 000 km length. Generally, the subflares (and presumably flares as well) appear to be composed of two basic kinds of structures: X-ray knots (which also might be unresolved tiny loops) and extensive, often diffuse, loop-like structures. The small features are usually brightest during the rise phase, but some of them keep their entity and stay clearly visible also during later phases of a flare. On the other hand, the big loop structures prevail during the maximum and decay phases, but in some flares they are the only structure one can detect even in the early flare phase. Both groups of authors tried to find a correlation between the occurrence of X-ray knots and the impulsive flare bursts, but without any obvious success. It might be that some of the knots are X-ray impulsive kernels, but the vast majority of them occurs independently of the impulsive phase.

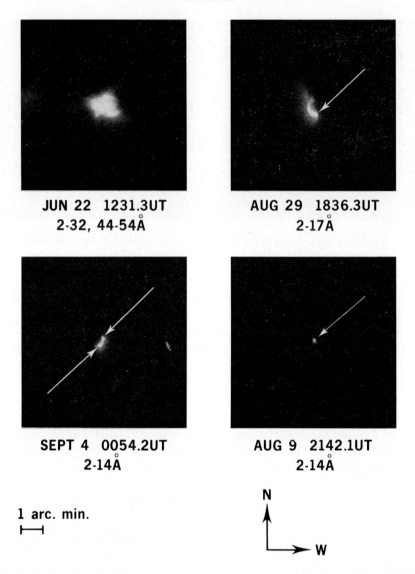

JUN 22   1231.3UT
2-32, 44-54Å

AUG 29   1836.3UT
2-17Å

SEPT 4   0054.2UT
2-14Å

AUG 9   2142.1UT
2-14Å

1 arc. min.
⊢━━┥

N

W

Fig. 43b. Subflares photographed in soft X-rays by the AS & E experiment on Skylab. All were associated with microwave bursts, and those in 1973, August 29 and September 4 also with a radio type III burst. Arrows point to bright X-ray knots that might indicate the sites where the particle acceleration occurred. (Courtesy of S. W. Kahler, AS & E, Cambridge, Massachusets, U.S.A.)

Many times the flare core could be seen in longer X-ray exposures taken before the flare actually started, indicating that some gradual heating or energy release occurs before the more catastrophic event takes place (Vorpahl *et al.*, 1975). In one subflare, studied in detail (Figure 11), Petrasso *et al.* (1975) found that a gradual point brightening occurred in a pre-existing loop crossing the $H_{\parallel} = 0$ line about 10 min prior to the striking increase in brightness. But similar gradual brightening of the loop was also observed a few hours earlier without any associated flare event. Thus a preheating need not necessarily give rise to a flare.

In the decay phase of most flares the flare features merely become fainter and more diffuse. However, in some cases also significant changes in structure occur (Vorpahl et al., 1975). In one case, studied by Silk et al. (1975) both cooling and mass loss contributed to the flare decay.

Bigger flares were studied by Vorpahl (1975) and Pallavicini et al. (1975) on Skylab, and by Neupert et al. (1974) on OSO-7. The flare studied by Vorpahl (on 5 September 1973) had the shape of an arcade of loops. Sequential brightening of the loops indicated that some excitation moved perpendicular to the magnetic field of the arcade at velocities of 180–280 km s^{-1}, and Vorpahl suggests a magnetosonic wave as the propagating mechanism. The flare of June 15, 1973, discussed by Pallavicini et al., was highly structured and exhibited complex systems of loops which changed during the event: as it seems, loop systems at successively increasing heights formed during the decay phase, finally leading to the large loops observed in the postflare phase. A brightening over an extended portion of the active region preceded the flare onset.

With lower spatial resolution of OSO-7, Neupert et al. (1974) found that images of the flare in several XUV lines (corresponding to lower temperatures) all had similar spatial distributions and matched the most intense region of Hα emission. On the other hand, the X-ray emission near 1.9 Å was displaced relative to the other emitting regions. This displacement suggests greater height for the X-ray emission, tentatively at the top of an arch, the feet of which emit radiation in the EUV lines.

Quite a similar arch-like structure was observed by Widing and Cheng (1974) when comparing pictures of the 1973, June 15 two-ribbon flare in high-temperature and low-temperature XUV lines (cf. Section 3.2.2). The Fe xxiv emission formed a bridge-like structure across the $H_{\parallel} = 0$ line, whereas less ionized ions appeared cospatial with the Hα bright ribbons (Figure 59).

C. *Indirect Observations of Solar X-Rays*

Direct observations of the X-ray radiation from flares date from 1959 (Peterson and Winckler, 1959; Chubb et al., 1960, 1961). Before that time all our information on the X-ray emission from flares was based only on observations of the flare effects on the ionosphere, the 'sudden ionospheric disturbances' (SID). Since most SID's are caused by an increased ionization in the ionospheric D layer, and this layer is mainly affected by X-rays between 1 and 10 Å, the SID effects give fairly good information on the flux of soft X-rays from flares. More sophisticated SID measurements can even give information on the incident X-ray and EUV flux in different energy ranges. Since in earlier decades these were the only X-ray measurements available, they are certainly worth a brief mention.

The most widely used SID types can be summarized as follows:

SWF–*Short Wave Fadeout:* a sharp decrease in the field strength of a distant short-wave transmitter due to attenuation of the reflected signal by enhanced *D*-region ionization. Frequencies used: 1 to 30 MHz.

SCNA – *Sudden Cosmic Noise Absorption:* a decrease in the intensity of the galactic radio noise due to increased attenuation of the signal passing through the enhanced *D* region. Frequencies used: 15 to 60 MHz.

SEA – *Sudden Enhancement of Atmospherics:* an increase in the signal strength of atmospherics (i.e. noise from distant thunderstorms) due to increased reflection of very low-frequency signals by the enhanced *D* layer. Frequencies used: mostly 22 to 30 kHz.

SPA – *Sudden Phase Anomaly:* A change in the phase between the ground wave from a frequency standard and the wave reflected from the *D* layer due to the changing height of the reflection with time. Frequencies used: 16 to 150 kHz.

SFD – *Sudden Frequency Deviation:* an increase in the frequency of the reflected signal from a short wave transmitter due to enchancement of the electron number density in the *F* layers. In contradistinction of the other effects the SFD is sensitive mostly to radiation above 10 Å and up to about 1030 Å which penetrates into the *F* layers (Donnelly, 1971). As soon as the flux of X-rays below 10 Å increases, the effect is lost in the heavy absorption (SWF) in the *D* layer. Finally, one more effect is observed in the records of the Earth's magnetic field variations:

SFE – *Solar Flare Effect* (magnetic crochet): a sudden short-lived variation mainly in the *H* component ($\leq 30\,\gamma$) due to enhanced conductivity of the ionosphere. Usually only large flares (with very strong other types of the SID) produce this geomagnetic effect.

3.1.3. FLARE SPECTRUM IN THE X-RAY REGION

The radiation in the soft X-ray region is composed of continuum and emission in lines of highly ionized ions. The continuous emission consists of two parts: free-free (bremsstrahlung) and free-bound (radiative recombination) transitions. Below about 5×10^6 K also the two-photon decay of the $2s$ states of hydrogen-like and helium-like ions makes a contribution (Tucker and Koren, 1971), but above 2×10^7 K the bremsstrahlung fully dominates in the formation of the continuum spectrum. Toblin (1972) has drawn attention to the fact that Compton backscattering can also contribute significantly to the observed X-ray flux below 2 Å.

The line emission results from downward radiative transitions in highly ionized, almost completely stripped ions, following the occupation of an excited level either by recombination or by inelastic collisions. Collisional excitation is prevalent, but dielectronic recombination can make a very significant contribution in some spectral lines (Gabriel and Jordan, 1969a; Doschek and Meekins, 1970; Tucker and Koren, 1971; Doschek *et al.*, 1971; Grineva *et al.*, 1973a, b).

With very few exceptions (Landini *et al.*, 1972, 1973), the radiation in soft X-ray region ($\lesssim 10$ keV) is generally considered to be of thermal, or quasi-thermal, origin. Thermal electron distribution can result from either direct heating of the flare plasma, e.g. by Joule dissipation of an electric current, or from collisional energy transfer from non-thermal electrons to the electrons of the ambient plasma.

As we mentioned in Section 2.3, many authors have adopted the latter interpreta-

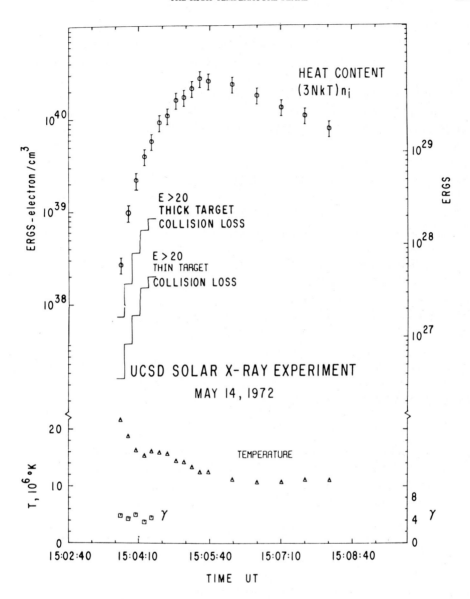

Fig. 44. Energetics of the X-ray burst of 1972, May 14. The top trace gives the heat content of the thermal plasma, the product of the total thermal energy ($3NkT$) and the ambient density (n_i). The peak value is 3×10^{40} erg electron cm^{-3}, so that if the ambient density is 3×10^{10} cm^{-3} the thermal energy is 10^{30} erg. The next two traces give the cumulative thick and thin target collision losses in ergs, obviously insufficient to produce all the flare energy. When the collisional energy input stopped the heat content had reached only 25 % of its final value. The triangles show the temperature and the boxes the spectral index γ variations. (After Datlowe et al., 1974.)

tion: they suppose that electrons are accelerated first to suprathermal energies and the thermal soft X-ray emission occurs after thermalization of these particles (see references in Section 2.3). However, this interpretation contradicts several observed facts:

(a) In many flares (Figure 54c) we do not see any impulsive burst that would indicate a non-thermal acceleration process (Kane, 1969; Kane and Anderson, 1970; Vorpahl, 1972; McKenzie et al., 1972; Švestka, 1973; Takakura, 1973; Datlowe et al., 1974).

(b) When the impulsive burst is observed, it often occurs several minutes after the onset of the thermal flare (Figure 54a; see also examples in Kane and Anderson, 1970; Vorpahl and Zirin, 1970; Kahler and Kreplin, 1970; Kelly and Rense, 1972; Kane et al., 1972; Vorpahl, 1972; de Feiter, 1973; Anderson and Mahoney, 1974). Spangler and Shawhan (1974) who correlated impulsive microwave bursts at 15.4 GHz with soft X-rays, found that in 23% of the cases the soft X-rays started 8 min or more before the impulsive microwave burst.

(c) According to Peterson et al. (1973) and Datlowe et al. (1974) there are definitely flares in which there is not enough energy in the non-thermal electrons to provide the heat input to the thermal plasma (Figure 44). Also the time history of the soft X-ray burst contradicts this supposition (cf. Section 3.1.4).

Therefore, though the non-thermal electrons can contribute in many flares to the plasma heating through collision losses, the basic source of the soft X-ray emission in flares must be direct heating, e.g. through stationary dissipation of magnetic energy (Syrovatsky, 1972; Tucker, 1973).

A. *The Continuous Emission*

Following Culhane (1969), let us first consider bremsstrahlung with a Maxwellian distribution of velocities in the non-relativistic case. Then the energy emitted per unit time, volume, and wavelength interval due to encounters of electrons at a temperature T_e with ions of nuclear charge Z is given by

$$E_{ff}(\lambda, T_e) = 2.04 \times 10^{-22} \, \lambda^{-2} \, T_e^{-1/2} \, n_e n_H \exp(-144/\lambda T_e) \times$$

$$\times \sum_Z (n_Z/n_H) \, Z^2 \, g(\lambda, Z, T_e) \quad \text{erg cm}^{-3} \, \text{s}^{-1} \, \text{Å}^{-1}, \qquad (55)$$

where n_e, n_H, and n_Z represent the number density of electrons, hydrogen ions, and ions of nuclear charge Z, respectively, and g is the free-free Gaunt factor. T_e is expressed in 10^6 K and λ in angströms. No serious error is involved if one takes

$$\sum_Z (n_Z/n_H) \, Z^2 \, g(\lambda, Z, T_e) = g(\lambda, 1, T_e) + (n_{He}/n_H) \, 4g(\lambda, 2, T_e) + K, \qquad (56)$$

where K is a constant independent of wavelength and temperature, amounting to about 6% of the total sum, and $n_{He}/n_H \simeq 0.2$ (Tucker and Koren, 1971). The values of the hydrogen and helium Gaunt factors can be taken from the graphs published by Karzas and Latter (1961).

Under the same assumptions, the emission due to radiative recombinations, i.e. captures into the state n of an hydrogen-like ion Z, is

$$E_{fb}(\lambda, T_e) = 6.52 \times 10^{-23} \, \lambda^{-2} T_e^{-3/2} n_e n_H \, \exp(-144/\lambda T_e) \times$$

$$\times \sum_Z (n_Z/n_H) \sum_i (n_{i+1}/n_Z) \sum_n G_n \quad \text{erg cm}^{-3} \text{s}^{-1} \text{Å}^{-1}, \quad (57)$$

where, with the Gaunt factor set equal to unity,

$$G_n = r_n X_{i,n}^2 n^{-1} \exp(0.012 \, X_{i,n}/T_e). \tag{58}$$

The first sum extends over the elements, the second one over the ionization stages, and the last one over all levels n to which recombination is allowed for a given photon energy, i.e. for which the ionization potential in electron volts exceeds $12\,398/\lambda$, for λ in angströms. r_n represents the number of vacancies in the nth shell which can accept a captured electron, $X_{i,n}$ is the potential in electron volts of an electron in the nth shell of an ion of ionization stage i; n_{i+1}/n_Z is the fraction of an element in the $(i+1)$th stage of ionization, and T_e is given again in 10^6 K.

Assuming that the X-ray emission region is isothermal, the total X-ray flux at the Earth is then

$$F_c(\lambda, T_e) = \frac{E_{ff}(\lambda, T_e) + E_{fb}(\lambda, T_e)}{4\pi R^2 n_e n_H} \int n_e^2 \, dV \quad \text{erg cm}^{-2} \text{s}^{-1} \text{Å}^{-1}, \quad (59)$$

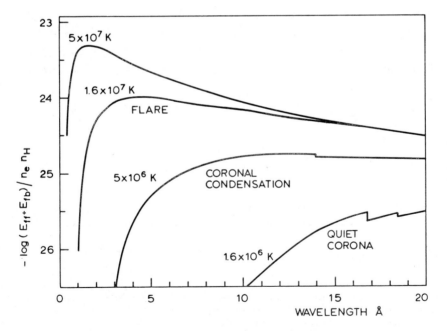

Fig. 45. Intensity of the X-ray continuum below 20 Å as computed by Tucker and Koren (1971) for different temperatures corresponding to quiet corona, permanent coronal condensation, an average flare, and the maximum phase of a large X-ray flare. The intensity is given in erg cm^{-3} s^{-1} Å^{-1}.

where R is the Earth-Sun distance, and the integration is extended over the emitting volume on the Sun. The quantity $\int n_e^2 \, dV$ is called the *emission measure*, and it is a basic and widely used parameter of X-ray bursts. A simplified empirical formula for F_e has been given by Culhane and Acton (1970).

Using formulae (55) and (57) the thermal continuous emission in the soft X-ray region has been computed in the last few years by several authors (Culhane, 1969; Landini and Monsignori Fossi, 1970a, b; Tucker and Koren, 1971; Mewe, 1972b), and Figure 45 shows the resulting X-ray flux between 1 and 20 Å for different T_e values according to Tucker and Koren. From it, one can see that the quiet corona (with $T_e \simeq 1.6 \times 10^6$ K) does not emit any observable X-ray flux below 10 Å. In high temperature condensations ($T_e \simeq 5 \times 10^6$ K) the flux is greatly increased and the observable limit shifts to some 3 Å. In order to give rise to an observable flux near 1 Å, as recorded in flare-associated X-ray bursts, T_e must exceed 10^7 K.

For X-rays below 1 Å, which occur in many flares as well, T_e should be in excess of 10^8 K. However, as we shall see in Section 3.2.1, X-rays below 1 Å can hardly be produced by a thermal mechanism. Instead, anisotropic streams of impulsively accelerated electrons are the main source of hard X-ray emission, the electron velocity distribution not being Maxwellian and Equations (55) and (57) no longer being valid. Kahler and Kreplin (1971) have demonstrated that in some flares the non-thermal component can be distinguished clearly even in the 3 to 10 keV (i.e. 1.2 to 4 Å) energy channel.

B. *The Line Emission*

When the first spectra of flares in the soft X-ray region were obtained in 1967 aboard the satellites OV 1–10 (Rugge and Walker, 1968), OSO 3 (Neupert *et al.*, 1967), and OSO 4 (Meekins *et al.*, 1968), the identification of the many new lines observed

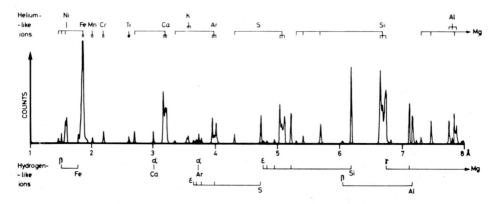

Fig. 46.   Schematic drawing of spectral lines observed in the region between 1 and 8 Å in a large flare. The positions of the helium-like lines of different ions are indicated in the upper part of the figure, starting with the three-component line $1s^2\,{}^1S - 1s2p\,{}^1P$, $1s2p\,{}^3P$, and $1s2s\,{}^3S$. Higher transitions in the same series are indicated when observed. In the lower part the positions of the hydrogen-like lines are indicated, starting with Ly-α. The highest line observed is marked with α, β, γ, or ε, respectively. (Based on same observations as Table XII.)

appeared to be quite a difficult task and the individual authors differed greatly in their interpretation of the X-ray spectra. After a few years, however, even though we still cannot be sure of our identification of all lines in the X-ray region, we began to understand a lot of what we observed, in particular due to the theoretical work of Gabriel and Jordan (1969 a, b) and the improved spectral resolution aboard OSO 5 (Neupert and Swartz, 1970; Neupert, 1971), OSO 6 (Doschek *et al.*, 1971) and OV 1–17 (Walker and Rugge, 1970).

As one can see in Figure 46, the most prominent lines in the X-ray region are formed in helium-like ions (Figure 47) by transitions to the ground $1s^2\ {}^1S$ term from $1s2p\ {}^1P$ (resonance line), $1s2s\ {}^3S$ (forbidden line), and $1s2p\ {}^3P$ (intercombination line). These three lines form strong complexes in the X-ray spectrum for Fe xxv, Ca xix, Ar xvii, S xv, Si xiii, Al xii, Mg xi, and O vii, which can be resolved into the three components if the resolving power is good enough. Weak lines of the same type also seem to be present for Mn xxiv, Cr xxiii, Ti xxi, and K xviii, as well as for Ni xxvii, Na x, and Ne ix, where these lines are blended with other possibly important lines of iron. In addition, for most abundant elements higher lines to the ground term are observed, possibly up to $1s^2 - 1s5p$ for Fe, Si, and Mg.

Next most important lines in the X-ray region are the lines in the Lyman series in hydrogen-like ions. The Ly-α line is clearly observed for Fe xxvi, Ca xx, S xvi, Si xiv, Al xiii, Mg xii (particularly strong), and O viii, most probably for Na xi and Ne x, where it is blended, and possibly for Ar xviii. Lyman lines up to Ly-γ are clearly seen for sulphur, silicon and probably magnesium (blended).

Lines of a third type, which are fairly weak, but very important from the theoretical point of view, are produced through innershell transitions in ions with three or more electrons. It is fairly difficult to observe them; nevertheless, there is now good evidence

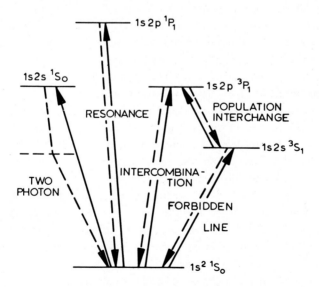

Fig. 47.   Schematic energy level diagram for helium-like ions.

that these innershell (satellite) lines also are present in the flare spectra. This evidence is based upon the following facts:

(a) At the long-wave side of the complex helium-like line of Fe xxv at $1.850 + 1.855 + 1.865$ Å one observes a low-intensity tail which corresponds to the wavelengths of innershell transitions in Li-like $(1.860 + 1.875$ Å$)$, Be-like $(1.868$ Å$)$, and less stripped $(1.88–1.94$ Å$)$ Fe ions (Meekins et al., 1968, 1970; Neupert and Swartz, 1970; Doschek et al., 1971, Neupert, 1971). According to Neupert and Swartz, 53% of the intensity in the 1.9 Å line complex comes from helium-like Fe xxv, 20% from lithium-like Fe xxiv, 11% from Fe xxiii, and 7% from lower stages of ionization. (According to Toblin (1972) these intensities should be corrected for Compton backscattering which produces a red line shift, thus fictitiously enhancing the long wavelength wing of the line complex).

More recently, a very detailed analysis of this group of Fe lines has been presented by Grineva et al. (1973a), with a resolution of about 4×10^{-4} Å. Between 1.850 and 1.870 Å the authors identify seven lines of Fe xxiv and one line of Fe xxiii which can be resolved from the main complex of the helium-like Fe xxv lines. The most prominent Fe xxiv line at 1.866 Å attains half of the Fe xxv resonance line intensity. In the range 1.89 to 1.92 Å (Grineva et al., 1973b) one cannot easily separate individual lines, but increased intensity there points to existence of lines emitted from iron ions of still lower ionization (down to Fe xix).

(b) Similar satellite lines have been detected in the helium-like lines of other elements as well (Meekins et al., 1968, 1970; Neupert and Swartz, 1969; Neupert, 1971; Doschek et al., 1971, 1972; Walker and Rugge, 1969, 1971). Doschek et al. have performed a detailed analysis of the helium-like lines of Ca xix and have shown in a convincing way that a good fit between the observed and theoretical profiles can be obtained only after taking the innershell transition $1s^2 2s^2 S - 1s2s2p\,^2P$ in lithium-like Ca xviii into account. For the ratio [(Li + Be) ion satellite lines/helium-like lines of the same element] the different authors give values 0.07 for Ne, 0.11–0.16 for Si, 0.20 for S, 0.14 for Ar, 0.23–0.33 for Ca, and 0.7 for Fe.

Apart from these basic lines which fully dominate the spectrum below 10 Å, many other lines are observed at longer wavelengths. Most originate in various states of ionization of the iron atom. Here the identification is still uncertain and if, for example, one compares the theoretical line intensities predicted by Tucker and Koren (1971) and by Mewe (1972b), respectively, sometimes it is not clear which of several alternative transitions is responsible for the line observed.

The line at 1.51 Å and the somewhat uncertain line at 1.46 Å, both ascribed to transitions from higher levels to the ground state of the helium-like Fe ion, are the shortest wavelength lines ever observed in the solar spectrum. Although highly stripped heavy ions with $Z > 27$ can give rise to emission lines at still lower wavelengths, the low abundance of such elements makes any observable occurrence of these lines highly improbable (Neupert, 1969).

Using the assumptions that the concentration of ions in various ionization stages is determined by the equilibrium between collisional ionization and combined dielectronic and radiative recombination, that the line emission is produced only as a consequence of collisional excitation, and that the velocity distribution again is Maxwellian, one can compute the energy emitted in the various lines of highly ionized atoms for different values of T_e. Generally the calculated line fluxes involve errors which are due mainly to inaccuracies in the excitation cross-sections, the relative ion concentrations, and the element abundances. Also the neglect of recombination processes in line formation may lead to significant errors in some of the lines. In addition, in flares, i.e. for high values of T_e, the plasma parameters can change so rapidly that the starting assumption about ionization equilibrium may be questionable, and in the flash phase of a flare, non-thermal electron streams may also contribute to the process, thus making the assumption about Maxwellian distribution invalid. Nevertheless, such computations give us at least a general idea of which lines can be expected to appear in the soft X-ray spectrum of a solar flare, and what range of temperatures is indicated by the occurrence and intensity of the different lines.

Computations of this kind have been made by Landini and Monsignori Fossi (1970; for ions of C, N, O, Ne, Mg, Si, S, Ca, and Fe, region 1 to 100 Å, $T_e = 10^6$ to 10^8 K), Tucker and Koren (1971; the same elements plus Ni, region 1 to 70 Å, $T_e = 6 \times 10^5$ to 10^8 K), and Mewe (1972b; the same elements plus Na, Al, Ar, K, Ti, Cr, and Mn, region 1 to 60 Å, $T_e = 10^5$ to 10^9 K). In Table XII we have tried to summarize all the spectral lines observed in the soft X-ray region of flare spectra and to compare them with these theoretical results. The list is based on spectra published by Neupert et al. (1967), Meekins et al. (1968, 1970), Rugge and Walker (1968), Walker and Rugge (1969, 1970), Neupert (1969, 1971), Doschek and Meekins (1970), Meekins and Doschek (1970), Neupert and Swartz (1970), and Doschek et al. (1971, 1972). The lines have been identified according to these authors, or according to theoretical line intensities published by Mewe (1972b) if the original identification appeared to be improbable from the theoretical point of view. The readers' attention is drawn to the fact that after this table had been completed, lists of X-ray lines were also prepared by Neupert et al. (1973, 6 to 25 Å), Doschek et al. (1973, 8.5 to 16 Å) and Beigman et al. (1974, 7.8 to 17.4 Å). These lists do not alter the basic information given in Table XII. However, they contain more lines than Table XII, many with tentative, or without any, identification. Some identifications at lines above 10 Å are different.

In order to give the reader better orientation in the table, we have marked with a cross the lines where the identification is most probably correct, and with two crosses those lines where the identification is believed to be sure. Circles denote lines that are seen clearly in the solar spectrum even in the absence of a flare, i.e. which originate in permanent condensations above the active regions. We have also tried to give a rough estimate of intensity to each line (in the column I_{obs}), but the reader must be aware of the fact that due to the rapidly changing efficiency of the spectrometers with wavelengths these estimates are mutually comparable only within relatively small wavelength ranges. The same note applies to the intensities in Figure 46. The reality of several lines (OW in column I_{obs}) is open to some doubt and a few lines reported in only one source have been omitted.

TABLE XII

X-ray lines observed in spectra of solar flares

λ(Å)	Ion	Transition[a]	T_M (10^6K)	I_M[b]	$I_{1.5}$[c]	I_{obs}[d]	References[e]
1.46	Fe XXV	He 5p	60	3.0	0.02	OW	1, 8, 10
1.51	Fe XXV	He 4p	60	6.5	0.03 ⎫	W	ibid.
1.51	Fe XXVI	H Lβ	110	11	0.001 ⎭		
+ 1.57	Fe XXV	He 3p	60	20	0.10 ⎤	M	ibid., 5, 6
1.59	Ni XXVII	He 2p, s	60	14	0.05 ⎦		
+ 1.78	Fe XXVI	H Lα	110	87	0.001	W	1, 5, 8, 10
+ + 1.85	Fe XXV	He 2p1	60	140	1.4 ⎤		
+ 1.86	Fe XXIV	Li Kα	30	6	1.1 ⎢		
+ + 1.86	Fe XXV	He 2p3, s	50	80	1.8 ⎬	S	ibid., 2, 6, 9, 12[f]
1.87	Fe XXIII	Be Kα	20	1.7	1.4 ⎢		
1.89	Fe XXII	B Kα	13	1.5	1.2 ⎦		
⎮	possibly						
1.94	lower ion.	Kα					
2.01	Mn XXIV	He 2p, s	55	0.5	0.02	OW	1, 8, 10
2.19	Cr XXIII	He 2p, s	50	1.0	0.04	W	ibid., 6
2.62	Ti XXI	He 2p, s	40	0.4	0.07	W	1, 8, 10
+ 2.70	Ca XIX	He 3p	35	1.3	0.25	W	ibid.
+ + 3.02	Ca XX	H Lα	47	4.5	0.03	W	ibid.
+ + 3.17	Ca XIX	He 2p1	35	10	2.5 ⎤		
+ + 3.19	Ca XIX	He 2p3 ⎫					
+ + 3.21	Ca XIX	He 2s ⎬	33	6.3	2.4 ⎬	S	2, 6–11
3.21	Ca XVIII	Li Kα ⎭			⎦		
3.35	Ar XVII	He 3p	24	1.5	0.74	OW	10
3.57	K XVIII	He 2p, s	25	0.5	0.23	W	10
3.65	S XVI	H Lε	24	0.2	0.07	OW	12
3.70	S XVI	H Lδ	24	0.4	0.14	OW	2, 9, 12
+ 3.73	Ar XVIII	H Lα	29	7.1	0.36	blend	6, 8, 9, 10, 12
+ 3.78	S XVI	H Lγ	24	1.0	0.34	W	9, 12
+ + 3.95	Ar XVII	He 2p1	23	11	6.4 ⎤		
+ + 3.99	Ar XVII	He 2p3, s	23	7.6	5.6 ⎬	M	2, 6, 8, 10
+ 3.99	S XVI	H Lβ	24	3.5	1.3 ⎦		
+ 4.31	S XV	He 3p	17	6.3	6.1	W	8, 10
+ + 4.73	S XVI	H Lα	24	27	12	M	ibid., 2, 6
4.77	Si XIV	H Lε	13	0.5	0.49	OW	10
4.83	Si XIV	H Lδ	13	1.0	1.0	OW	8, 10
+ 4.95	Si XIV	H Lγ	13	2.4	2.4	W	ibid., 12
+ + 5.04	S XV	He 2p1	17	47	47 ⎤		
+ + 5.07	S XV	He 2p3 ⎫			⎬		
+ + 5.11	S XV	He 2s ⎭	17	35	35 ⎬	S	ibid., 2, 6, 7
5.09	S XIV	Li Kα	13	0.7	0.65 ⎦		
+ 5.22	Si XIV	H Lβ	13	9.1	9.1	M	2, 6, 8, 10, 12
5.30	Si XIII	He 5p	9			OW	12
5.41	Si XIII	He 4p	9	3.2	1.4	W	6, 10, 12
+ 5.68	Si XIII	He 3p	9	8.4	4.2	MW	ibid., 2
6.05	Al XIII	H Lβ	10	0.3	0.28	OW	6
+ + 6.18	Si XIV	H Lα	13	78	75	MS	2, 6, 8, 10
○ + + 6.65	Si XIII	He 2p1	9	70	30 ⎤		
○ + + 6.68	Si XIII	He 2p3 ⎫			⎬		
○ + + 6.74	Si XIII	He 2s ⎭	9	56	19 ⎬	S	ibid., 7
6.74	Mg XII	H Lγ	9	1.3	0.82 ⎢		
6.75	Si XII	Li Kα	6	0.6	⎦		

Table XII (continued)

	λ(Å)	Ion	Transition[a]	T_M[b] $(10^6 K)$	I_M[b]	$I_{1.5}$[c]	I_{obs}[d]	References[e]
	6.81	Si XI	Be Kα	3.5	0.04		W	2, 6
	+ 7.11	Mg XII	H Lβ	9	5.0	3.0	MS	*ibid.*, 8, 10
	+ 7.17	Al XIII	H Lα	10	2.8	2.3	M	*ibid.*
	7.23	Fe XXIV	Li 5p	17	3.5	2.8	OW	2, 6
	7.31	Mg XI	He 5p	6	0.6	0.09	W	*ibid.*
	7.47	Mg XI	He 4p	6	1.4	0.19	M	*ibid.*, 10
O	+ 7.76	Al XII	He 2p1	7.5	2.4	0.48 ⎫		
O	+ 7.81	Al XII	He 2p3	⎱ 7.5	2.1	0.28 ⎪	MS	*ibid.*
O	+ 7.88	Al XII	He 2s	⎰		⎪		
	+ 7.85	Mg XI	He 3p	6	4.2	0.55 ⎪		
	7.88	Al XI	Li Kα			⎭		
	7.99	Fe XXIV	Li 4p	17	9.1	7.8	OW	2, 6
	8.08	?					W	*ibid.*
	8.16	?					W	*ibid.*
	8.21	?					W	*ibid.*
	8.31	Fe XXIV	Li 4d	17	29	2.7	W	*ibid.*
O	+ + 8.42	Mg XII	H Lα	8.5	43	23	S	*ibid.*, 1, 3, 4
	9.00	Fe XXII	B 4d	12	20	15	OW	1
O	+ + 9.17	Mg XI	He 2p1	6	35	3.6 ⎫		
O	+ + 9.23	Mg XI	He 2p3	⎱ 6	33	1.9 ⎬	MS	1, 3, 4
O	+ + 9.32	Mg XI	He 2s	⎰		⎪		
	9.27	Mg X	Li Kα	5	0.3	⎭		
	9.48	Ne X	H Lδ	5	0.6	0.14	W	1,3
	9.71	Ne X	H Lγ	5	1.4	0.33	OW	3, 4
	10.0	Fe XXV	He s3p	60	1.5	0.02 ⎫	W	*ibid.*, 1
	10.03	Na XI	H Lα	6	4.8	1.7 ⎭		
	10.24	Ne X	H Lβ	5	5.4	1.2 ⎱	OW	*ibid.*
	10.3	Fe XXV	He p3d	60	1.5	0.02 ⎰		
	10.7	Fe XXIV	Li 3p	17	63	51	WM	1, 4
	11.00	Na X	He 2p1	4.5	4.2			
	11.08	Na X	He 2p3	⎱ 4.5	3.5		WM	1, 3, 4
	11.18	Na X	He 2s	⎰		⎫		
	11.1	Fe XXIV	Li 3d	17	110	100 ⎭		
	11.2	Fe XXIII	Be 3p	13	170	150	OW	3
	11.3	Fe XXIV	Li 3s	17	46	45	W	1, 3, 4
	11.5	Fe XXIII	Be 3d3	13	85	77 ⎱	W	*ibid.*
	11.56	Ne IX	He 3p	3.5	3.3	0.03 ⎰		
	11.8	Fe XXIII	Be 3d1	13	85	76 ⎱	M	*ibid.*
	11.8	Fe XXII	B 3d	12	220	160 ⎰		
O ⎰	12.12	Fe XVII	Ne 4d1P	5.5	83	0.01 ⎱	M	*ibid.*
⎱	+ 12.13	Ne X	H Lα	4.5	54	9.1 ⎰		
O	12.26	Fe XVII	Ne 4d3D	5.5	75	0.01 ⎱	M	*ibid.*
	12.3	Fe XXI	C 3d	10	350	95 ⎰		
	12.4	Fe XXII	B 3s	12	5.6	4.9 ⎱	W	3, 4
	12.42	Ni XIX	Ne 3d1P	6.5	110	1.4 ⎰		
	12.64	Ni XIX	Ne 3d3D	6.5	45	0.54	OW	*ibid.*
	12.9	Fe XX	N 3d	9	500	23	M	1, 3, 4
⎰	13.44	Ne IX	He 2p1	3.5	35	0.18 ⎫		
⎪	13.5	Fe XIX	O 3d	8.5	420	11 ⎪		
O ⎨	13.55	Ne IX	He 2p3	⎱ 3.5	35	0.08 ⎬	M	*ibid.*
⎪	13.70	Ne IX	He 2s	⎰		⎪		
⎩	13.7	Fe XX	N 3s	9	28	0.65 ⎭		

Table XII (continued)

	λ(Å)	Ion	Transition [a]	T_M [b] $(10^6$K)	I_M [b]	$I_{1.5}$ [c]	I_{obs} [d]	References [e]
○ ⎧	13.77	Ni XIX	Ne $3s1P$	6.5	1.7	0.02 ⎫		*ibid.*
⎨	13.82	Fe XVII	Ne $3p1$	5.5	160	0.03 ⎬ W		*ibid.*
⎩	13.89	Fe XVII	Ne $3p3$	5.5	190	0.03 ⎭		*ibid.*
	14.09	Ni XIX	Ne $3s3P$	6.5	2.3	0.03	OW	*ibid.*
○	14.25	Fe XVIII	F $3d2D$	6.5	960	1.5	M	*ibid.*
○	14.40	Fe XVIII	F $3d2D$	6.5	650	0.91	W	*ibid.*
○	14.54	Fe XVIII	F $3d2P$	6.5	600	0.82	W	*ibid.*
	14.67	Fe XVIII	F $3d4P$	6.5	330	0.49	OW	3, 4
○ +	15.01	Fe XVII	Ne $3d1P$	5.5	960	0.13	S	*ibid.*, 1
○ +	15.26	Fe XVII	Ne $3d3D$	5	280	0.04	WM	*ibid.*
○	15.45	Fe XVII	Ne $3d3P$	5	4.5		OW	3, 4
○	15.62	Fe XVIII	F $3s2D$	6.5	530	0.78	W	*ibid.*, 1
○	15.88	Fe XVIII	F $3s2P$	6.5	7.8	0.01	W	*ibid.*
○ ⎧	16.01	O VIII	H $L\beta$	3.5	4.0	2.7 ⎫	M	*ibid.*
⎩	16.01	Fe XVIII	F $3s4P$			⎭		
○ +	16.77	Fe XVII	Ne $3s1P$	5	28		M	*ibid.*
○ +	17.05	Fe XVII	Ne $3s3P$	5	31		S	*ibid.*
○	17.4	Fe XVI	Na $3s$	4	4.2		OW	3
○	17.6	?					W	1, 3
○	18.63	O VII	He $3p$	2.2	24		W	3, 4
○ ++	18.97	O VIII	H $L\alpha$	3	350	21	M	*ibid.*, 1
○ ++	21.60	O VII	He $2p1$	2.2	200		WM	3, 4
○ ++	21.80	O VII	He $2p3$ ⎫				W	*ibid.*
○ ++	22.09	O VII	He $2s$ ⎭	2	210		WM	*ibid.*

○ The line is also observed in the spectrum of active regions without flares.
+ Pretty safe identification.
++ Safe identification.
[a] The transitions are abbreviated as follows:
H Ln = nth line of the Lyman series in the hydrogen-like ion
He np = $1s^2 - 1snp$
He $2pn$ = $1s^2 - 1s2p$, term $^nP^0$
He $2s$ = $1s^2 - 1s2s$
He $2p, s$ = $2p1 + 2p3 + 2ps$, unresolved
He $a3b$ = $1s2a - 1s3b$
Li np = $2s - np$
Li nd = $2p - nd$
Li $3s$ = $2p - 3s$
Be $3p$ = $2s^2 - 2s3p$
Be $3dn$ = $2s2p - 2s3d$, term ^nD
B na = $2p - na$
C $3d$ = $2s^2 2p^2 - 2s2p3d$
N $3a$ = $2p^3 - 2p^2 3a$
O $3d$ = $2p^4 - 2p^3 3d$
F $3anb$ = $2p^5 - 2p^4 3a$, term ^nb
Ne $manb$ = $2p^6 - 2p^5 ma$, term ^nb
Ne $3pn$ = $2s^2 2p^6 - 2s2p^6 3p$, term $^nP^0$
Na $3s$ = $2p^6 3s - 2p^5 3s^2$
Kα denotes innershell transitions
(for details see Mewe (1972b))

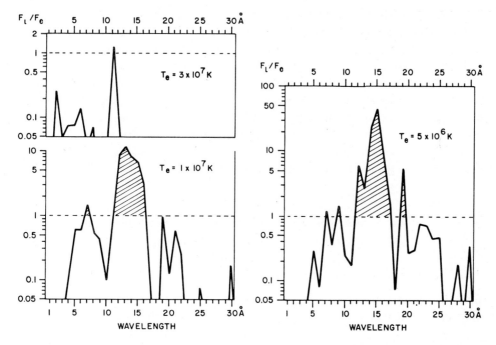

Fig. 48. The ratio of the line (F_l) and continuum (F_c) contribution to the total X-ray emission at different wavelengths (for 1 Å intervals). Importance of the line emission decreases with increasing temperature, and the region where the lines become important shifts to shorter wavelengths. (After Mewe, 1972b.)

In columns 4 and 5 we give the temperature T_M at which the line in question should reach its maximum intensity I_M, after Mewe (1972b). These values have been interpolated from Mewe's tables and therefore, they should be considered estimates only. In column 6, again after Mewe, we give the intensity at $T_e = 1.5 \times 10^7$ K, a temperature that seems to be typical for most of the life of large flares (Meekins *et al.*, 1970).

The continuous emission prevails in most X-ray spectral regions. However, the line emission becomes a significant (and even prevalent) contributor to the flux

---

Notes to Table XII (continued)

[b] Temperature (T_M) at which the line intensity should be maximum (I_M). The intensity is given in 10^{-4} erg cm^{-2}s^{-1}. The values of T_M and I_M have been interpolated from Mewe's (1972b) tables.
[c] The line intensity at $T = 1.5 \times 10^7$ K, in 10^{-4} erg cm^{-2}s^{-1}, after Mewe (1972b).
[d] An estimate of the line intensity observed; the scale proceeds from the weak to the strong lines as OW – W – M – MS – S, but one should be aware of the fact that the intensities vary with the flare development, differently for different lines.
[e] References: 1 – Neupert *et al.* (1967); 2 – Meekins *et al.* (1968); 3 – Rugge and Walker (1968); 4 – Walker and Rugge (1969); 5 – Neupert (1969); 6 – Meekins *et al.* (1970); 7 – Doschek and Meekins (1970); 8 – Neupert and Swartz (1970); 9 – Meekins and Doschek (1970); 10 – Neupert (1971); 11 – Doschek *et al.* (1971); 12 – Doschek *et al.* (1972).
[f] Detailed table of fourteen Fe xxv–xxiii lines within $\lambda = 1.850$ to 1.870 Å observed in a flare can be found in Grineva *et al.* (1973a).

between approximately $\lambda = 30/T_e$ and $100/T_e$, where λ is in Å and T_e in 10^6 K (Mewe, 1972b). Thus the region where line contribution is most important shifts towards shorter wavelengths as T_e increases, as one can see in Figure 48, where the theoretical ratio F_l/F_c is plotted for 1 Å intervals for three different values of T_e.

On OSO 5 Kastner *et al.* (1974) obtained several flare spectra in the range from 66 to 171 Å, all emitted from large flares of Hα importance 2B. In this spectral range one can expect transitions of the type $2s^r2p^k - 2s^{r-1}2p^{k-1}$ in highly ionized Fe ions possessing less than 10 electrons. Theoretical knowledge of these transitions, however, is not yet available and therefore, identification of the observed lines is difficult. The most outstanding lines in flare spectra in this range are

$$\lambda\ 132.83 \text{ of Fe xxiii } [2s^2(^1S_0) - 2s2p(^1P_1)],$$

$$\lambda\ 135.73 \text{ of Fe xxii } [2s^22p(^2P_{1/2}) - 2s2p^2(^2D_{3/2})], \text{ and}$$

$$\lambda\ 117.18 \text{ of Fe xxi } [2s^22p^2(^3P_1) - 2s2p^3(^3P_1)].$$

3.1.4. TEMPERATURE IN THE X-RAY FLARE

Both the continuum and the line spectrum allow us to estimate the electron temperature in the flare region where the soft X-ray emission is produced.

A. *Temperature and Emission Measure*

Meekins *et al.* (1968, 1970) tried to determine T_e from the energy distribution in the observed continuum. Using the assumption that the continuum is due entirely to bremsstrahlung, they found, from the slope of the continuum above 2.4 Å in two large flares, T_e between 1.4 and 2.0×10^7 K and in excess of 2.0×10^7 K if still shorter wavelengths were considered. These estimates, however, may be seriously influenced by flare intensity variations during the time when the spectral scan was made.

This source of error disappears if one compares simultaneous fluxes in two different energy channels, but then, of course, one integrates the contribution of both the continuous and line emission within the energy range considered. Figure 48 shows that the total sum of the line contribution decreases with increasing temperature when shifting to the shorter wavelengths. Therefore, one can make the simplifying assumption that the line contribution can be completely neglected between 1 and 8 Å for any temperature. According to observations (Friedman, 1969), lines never contribute more than 10% in this spectral region, but of course, any narrowing of the spectral bands may make this relative contribution higher. In particular, the Fe line near 1.9 Å may give rise to errors, since it becomes temporarily very strong in the initial phase of a flare (cf. Figure 49).

With the line contribution neglected one can compute the temperature and density in the volume emitting the soft X-ray burst utilizing Equations (55) and (57). As an example (Horan, 1971), let us suppose that we have measured the X-ray flux with ionization chambers in two different energy channels 1 and 2, both with upper wavelength limits so low that the line contribution can be neglected. According to Kreplin

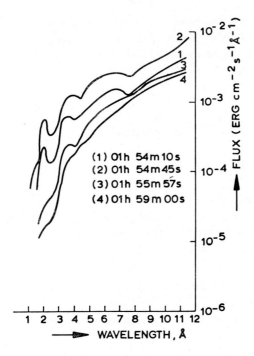

Fig. 49.   Time variation of the X-ray spectrum of the flare of 1967, December 21 (after Culhane *et al.*, 1969). Note the fast decay of the peak near 2 Å, which corresponds to the strong but short-lived line of Fe xxv.

(1961) the current generated in an ionization chamber is given by

$$J = e\omega A \int \varepsilon(\lambda)\, F_c(\lambda, T_e)\, d\lambda, \tag{60}$$

where e is the electronic charge, ω is the number of ion pairs produced in the gas per unit of absorbed energy, A is the effective window area of the detector, $\varepsilon(\lambda)$ is the efficiency of the detector and the incident continuous flux F_c is given by Equation (59) which can be rewritten as

$$F_c(\lambda, T_e) = (4\pi R^2)^{-1}\, C(\lambda, T_e) \int n_e^2\, dV \tag{61}$$

with

$$C(\lambda, T_e) = E_{ff}(\lambda, T_e) + E_{fb}(\lambda, T_e)/n_e n_{\mathrm{H}}, \tag{62}$$

where E_{fb} and E_{ff} are given by Equations (55) and (57).

Equations (59) and (61) involve the assumption that the X-ray emission region is homogeneous in temperature and density. Assuming in addition that the emission measure $\int n_e^2\, dV$ is identical for both the spectral bands studied, the current ratio J_1/J_2 becomes a function of T_e only,

$$J_1/J_2 = \omega_1 A_1 \int \varepsilon_1(\lambda)\, C(\lambda, T_e)\, d\lambda \bigg/ \omega_2 A_2 \int \varepsilon_2(\lambda)\, C(\lambda, T_e)\, d\lambda, \tag{63}$$

and if the detector characteristics ω, A, and $\varepsilon(\lambda)$ are known for both bands, one can determine T_e and its time variations from the measured ratio of the two fluxes.

Once the electron temperature is known, one can compute the detector current per unit emission measure,

$$J_0(T_e) = e\omega A \int \varepsilon(\lambda) \, C(\lambda, T_e) \, d\lambda / 4\pi R^2 \tag{64}$$

and the emission measure M is then given by

$$M = \int n_e^2 \, dV = J/J_0(T_e) \tag{65}$$

from either of the channels.

By this very simplified method several authors determined the electron temperature and emission measure for various X-ray bursts (Blocker *et al.*, 1971; Horan, 1971; Deshpande and Tandon, 1972; Peterson *et al.*, 1973; Datlowe *et al.*, 1974). The approach can be improved by taking the line-emission into account, by using theoretical computations of the expected line intensities mentioned above. Such an improvement has been tried for several flares by Culhane *et al.* (1970) and Catalano and Van Allen (1973). Kahler *et al.* (1970) have taken the 1.9 Å line complex into account.

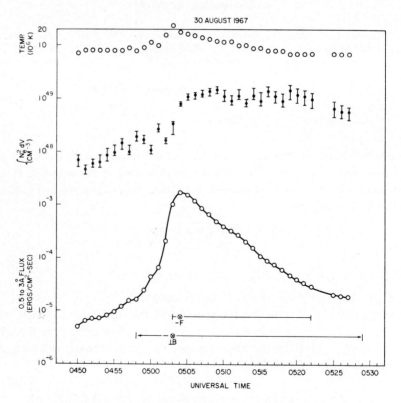

Fig. 50a.

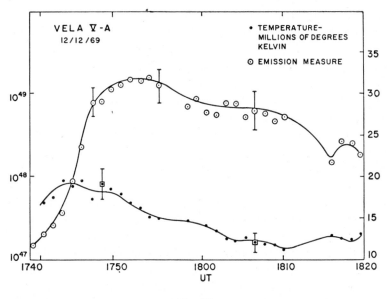

Fig. 50b.

Figs. 50a–b. Two typical examples of temperature and emission measure time variations as deduced from flare-associated soft X-ray bursts: (a) Flare of 1967, August 30. From above: temperature and emission measure deduced from 0.5 to 3 Å and 1 to 8 Å bands, total energy flux in 0.5 to 3 Å, and duration of the associated flare and subflare (after Horan, 1971). (b) Flare of 1969, December 12. Emission measures and temperature deduced from energy ranges >5, >8 and >15 keV (after Blocker *et al.*, 1971).

Of course, our knowledge of the line contribution at different temperatures is imperfect; by including theoretically computed line intensities we probably improve the result, but to what extent is unknown.

All the results show that generally there is a fast rise both in temperature and emission measure in the initial phase of the burst. However, whilst T_e peaks sharply in the few minutes after the flare onset and decays approximately exponentially after that, the increase in emission measure, though slowing down at about the time of the T_e peak, continues to increase further and reaches its maximum value much later in the burst development. Usually the emission measure stays fairly constant for ten minutes or more, irrespective of the simultaneous fast decrease in T_e and falls only in the later phase of development of the burst, sometimes slowly, sometimes fairly fast.

Examples of this behaviour are shown in Figure 50. According to Horan (1971) the peak temperature never coincides in time with the flux peak, but occurs when the flux is still rising. The flux peak appears when the decreasing temperature forces the flux to decrease even though the emission measure is still increasing. Thus the time sequence of maxima always keeps to the schedule 'temperature – flux – emission measure'.

The peak temperature exceeds 1×10^7 K and usually reaches values between 2×10^7 K and 4×10^7 K. After the temperature decrease T_e stays at a value between

0.8 and 1.5×10^7 K in most of the flares. The characteristic value of the maximum (and fairly constant) emission measure is of the order of 10^{49} cm^{-3} for flares and of 10^{48} cm^{-3} for subflares. Only Culhane *et al.* (1970) give the emission measure about one order of magnitude lower (less than 4×10^{48} cm^{-3} for flares). According to Toblin (1972), emission measure and temperature may be overestimated if one uses wavelengths below 2 Å where the Compton backscattering begins to contribute significantly to the observed X-ray flux.

However, there are two more factors that make the determination of temperature and emission measure uncertain: possible contribution of non-thermal electrons to the coronal heating, and inhomogeneities in the temperature distribution in X-ray flares.

As we mentioned several times in Chapter II and will discuss more in detail in Section 3.2, there are flares in which non-thermal electron streams heat the solar atmosphere during the flash phase. As Landini *et al.* (1973) demonstrated, we have no means for distinguishing between thermal and non-thermal heating and line excitation in the X-ray spectral range. Thus one cannot exclude the possibility that the sharp rise in temperature and the peak values of temperature are fictitious; at least in some flares they may be produced by high-energy electrons streaming through the corona which itself stays at significantly lower temperature (Deshpande and Tandon, 1972).

In any case the X-ray pictures from Skylab show that even a perfectly thermal corona is greatly inhomogeneous. Further evidence for inhomogeneity is the simultaneous occurrence of spectral lines corresponding to different temperatures in the flare spectrum (Section 3.1.4.B). Also we usually find higher values for T_e and smaller values for the emission measure when going to smaller wavelengths. This indicates that the X-ray emission comes from a hot core, or from a number of separated hot cores, surrounded by a larger amount of material at a lower temperature (de Feiter, 1973).

Herring and Craig (1973), Craig (1973) and Herring (1974) have proposed a two-component thermal model of X-ray sources: with high temperature and small emission measure in the hot component (typically $(1.5$ to $2.5) \times 10^7$ K and $(5$ to $8) \times 10^{48}$ cm^{-3}), and with low temperature and large emission measure in the cold component $(2$ to $3) \times 10^6$ K, $> 10^{51}$ cm^{-3}). This is in fact an analog of the 'flare filaments' and 'interflare matter' in the analysis of optical flare spectra (cf. Sections 2.2.1 and 2.2.7). In both cases it is a crude approximation. The flare certainly is not isothermal, but neither does it consist of two components. Thus we substitute one unrealistic assumption for another, and although the second may be nearer to the truth, it is still far from reality. At the optical flare we had no other choice. In the X-ray region, however, Batstone *et al.* (1970), Chambe (1971), and Dere *et al.* (1974) have suggested an approach that seems more promising: a multi-thermal analysis. Let us follow the approach proposed by Dere *et al.*

We will assume that the flare X-ray emission was measured in several wavelength bands: 1, 2, ..., n. (In practice, e.g., 0.5 to 3 Å, 1 to 5 Å, 1 to 8 Å, 8 to 16 Å, 1 to 20 Å and 44 to 60 Å aboard the NRL Solrad 10 satellite.) Then in the j-th band Equation

(60) can be written as

$$J_j = e\omega_j A_j \int_\lambda \varepsilon_j(\lambda) \int_{T_e} \frac{dF}{dM}(\lambda, T_e) \frac{dM}{d \ln T_e} d \ln T_e \, d\lambda, \tag{66}$$

where $F(\lambda, T_e)$ is the energy flux of the coronal plasma with electron temperature T_e per unit wavelength interval, taking both the continuous and line emission into account. The quantity $(dM/d \ln T_e)$ is related to the differential emission measure dM/dT_e:

$$\frac{dM}{d \ln T_e} = \frac{T_e \, dM}{dT_e} = T_e \frac{d}{dT_e} \int n_e^2 \, dV. \tag{67}$$

The procedure for finding $dM(T_e)/d \ln T_e$ starts with the assumption that it can be adequately represented by a simple function of several free parameters. This is a weakness of this method, because the choice of this function is rather arbitrary. Dere et al. tried several functions and found the best fit with the form

$$\frac{dM(T_e)}{d \ln T_e} = C_1 \exp[-C_2 X^{C_3}], \tag{68}$$

where

$$X = \ln[T_e/(2 \times 10^6 \text{ K})] \tag{69}$$

and C_i's are free parameters. In order to determine them one has to find those values which yield the minimum for

$$\varphi = \sum_{n \text{ bands}} [J_j(\text{measured}) - J_j(\text{expected})]/\Delta_j^2, \tag{70}$$

where Δ_j is the estimated error in band j.

The value of $T_e = 2 \times 10^6$ K in (69) has been chosen because Chambe (1971) and Walker (1972) in previous analyses of active regions have found that dM/dT_e has a maximum at about 2×10^6 K. Dere et al. suppose that the same happens in flares. Thus also the function (68) always has a maximum at $T_e = 2 \times 10^6$ K (Figure 51) and this may not be true for all flares. Analyses of X-ray line emissions by Neupert et al. (1973, 1974) and the two-component analysis by Herring and Craig (1973) both indicated a maximum of dM/dT_e in flares at higher temperatures. Dere et al. believe that this was not the case in the events they studied and they may be right because they studied only subflares. In larger flares one cannot exclude the possibility of a shift of the maximum to higher values of T_e.

Figure 51 shows typical differential emission curves at four instants during the evolution of a typical 'thermal' subflare. The differential emission measure corresponding to lower and intermediate temperatures increases up to the time of the peak flux while it begins to decrease at high temperatures after the maximum temperature is reached.

The fact that the emission measure still rises while temperature is already decreasing (cf. Figures 50 and 51) is not easy to interpret. Increasing emission measure implies

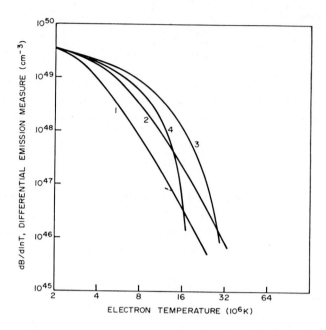

Fig. 51.   Typical differential emission curves at four instants during the evolution of a thermal flare. Curve 1: early during the rising phase, before the peak temperature is reached. Curve 2: at the time of the peak temperature. Curve 3: later on, near time of peak flux in 0.5 to 8 Å detectors. Curve 4: during the decaying phase. The curves are normalized to give the same value of the differential emission measure at 2×10^6 K. (After Dere *et al.*, 1974.)

that either n_e, or V, or both, increase. There are flares for which Skylab photographs in X-rays do not show any expansion in the corona (AS & E, 1975). Therefore, it seems to be the electron density that should increase. n_e can increase either through contraction or through bringing additional plasma into the radiating volume. Contraction would lead to heating, but the contrary is observed. Hence, new matter must be added to the radiating volume.

Datlowe *et al.* (1974) suggest that different loops are heated to the flare temperature successively. This might have been observed for some flares on Skylab X-ray photographs, but all the other flares without any apparent expansion cannot be interpreted in this way. Therefore, it appears more realistic to suppose that the flare slowly penetrates downwards (i.e. pushes the transition layer between chromosphere and corona deeper into the chromosphere, cf. Section 2.3.1.C). The lower layers are dense and thus they offer the most efficient way to increase M with only a slight increase of V. Similar model has been proposed by Zaumen and Acton (1974).

B. *Line Excitation*

Temperature also can be estimated from the line spectrum. To begin with, columns 3 through 6 in Table XII show clearly that any assumption of an isothermal region producing the X-ray spectrum is very far from being realistic. At short wavelengths we observe many lines which indicate T_e at least as high as 2×10^7 K, and the Ly-α line of Fe xxvi resolved by Neupert *et al.* (1967; Neupert, 1971) at the short-wavelength wing of the 1.9 Å line, needs T_e close to 3×10^7 K or higher. On the other hand, the strong lines of Fe xvii near 17 Å, as well as the O vii lines (still increased by some 50% in intensity in the flare spectra (Walker and Rugge, 1969)), indicate T_e below 7×10^6 K.

Meekins *et al.* (1970) tried to deduce T_e from the intensity ratios of the Ly-α lines in the hydrogen-like ions of Mg, Si, and S, and from the ratios of the resonance lines of the hydrogen- and helium-like ions in sulphur and silicon. They found T_e between 1.0 and 1.9×10^7 K, but in fact the lower limit is too high, since the ratio 2.7 found by them for the Mg xiii/Si xiv Ly-α lines corresponds to only 8×10^6 K according to the new computations by Tucker and Koren (1971) and Mewe (1972b). Mandelshtam (1974) compared intensities of various lines in a subflare and derived temperatures as high as (4 to 5) $\times 10^7$ K from the intensity ratio [Fe xxvi (Ly-α)/Fe xxv] and 2×10^7 K from ratios of Fe xxv lines. In the same subflare at the same time the ratios [Mg xi/Mg xii], [Ne ix/Ne x] and [Fe xvii/Fe xviii] gave $T_e \simeq 5 \times 10^6$ K (also see Beigman *et al.*, 1974). Landini *et al.* (1973) found from Si, S, Ar, Ca, and Fe lines T_e varying between 2×10^7 K and 4×10^7 K in a flare, for which they offered a non-thermal inter-pretation. Grineva *et al.* estimated $T_e = 1.6 \times 10^7$ K and 1.8×10^7 K, respectively, in two flares, from the energy distribution among the various components of the 1.9 Å line complex. Some line ratios at longer wavelengths clearly correspond to much lower temperatures; for example, if one compares the very weak 14.67 Å line of Fe xviii with the very strong 15.01 Å line of Fe xvii (Walker and Rugge, 1969), their ratio, according to Mewe's tables, points to T_e less than 5×10^6 K.

One can also determine temperature and emission measure from the observed line intensities; in that case the emission measure is representative of the amount of material that is at the temperature at which the line is produced. Thus, e.g., Widing and Cheng (1974) found $M = 3 \times 10^{48}$ cm^{-3} and $T_e = 1.6 \times 10^7$ K for the Fe xxiv emission region in the flare of 15 June, 1973. On the other hand, from satellite lines of Fe xvii–xx at λ 1.93 Å Phillips and Neupert (1974) estimated in an importance 2b flare $M \simeq 3 \times 10^{50}$ cm^{-3} and $T_e \simeq 1.1 \times 10^7$ K. From the continuum, a two-com-ponent model gave for the same flare $T_{hot} = 3.3 \times 10^7$ K, $M_{hot} = 3.5 \times 10^{49}$ cm^{-3} $T_{cold} = 7 \times 10^6$ K, $M_{cold} = 6 \times 10^{51}$ cm^{-3}. A more detailed analysis has been made by Neupert *et al.* (1974) for the 1b flare of 1972, August 2. They find $M = 3.2 \times 10^{47}$ cm^{-3} for $T_e = 2.3 \times 10^6$ K (Fe xiv), $M = 3.9 \times 10^{48}$ cm^{-3} for $T_e = 9.4 \times 10^6$ K (Mg xi) and $M = 7.2 \times 10^{46}$ cm^{-3} for $T_e = 3.0 \times 10^7$ K (Fe xxv). (Hence, maximum M was found near 10^7 K and not for $T_e \simeq 2 \times 10^6$ K as Equation (69) assumed.)

Obviously the X-ray flare must be considered as a highly heterogeneous medium as far as temperature (and probably density as well) is concerned, with local temper-ature variations at least within the limits of 5×10^6 to 3×10^7 K and possibly higher.

The time variation of T_e, deduced earlier from the continuum measurements, is also evident in the line spectrum. The highest stages of ionization are found to increase most rapidly at the onset of the flare, while lower stages are observed later in the event. One example has been shown in Figure 49, demonstrating the short duration of the

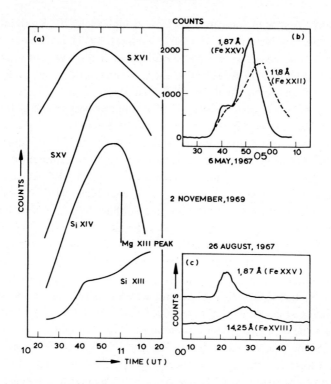

Fig. 52.   Examples of the flare light curves in different X-ray lines. Generally, the higher ionization states peak earlier in the flare development: (a) Light curves of the flare of 1969, November 2, from S XVI to Si XIII (after Doschek *et al.*, 1972); (b) Light curves of the flare of 1967, May 6, in Fe XXV and Fe XXII lines (after Neupert *et al.*, 1969); (c) Light curves of the flare of 1967, August 6, in Fe XXV and Fe XVIII lines. (After Neupert, 1969.)

1.9 Å line emission in the flash phase of the flare (which of course also may be due to a short-lived occurrence of a non-thermal burst component at that time). In the flare of 1967 August 26, Neupert (1968, 1969) has found that the 14.3 Å line of Fe XVIII peaked about 7 min after the peak of the 1.9 Å line of Fe XXV, when the intensity of this high-temperature line had already almost decreased to zero. Similarly, in the limb flare of 1969, November 2, Doschek *et al.* (1972) have found that the Ca XIX helium-like line peaked a few minutes after the Fe XXV line, and the Mg Ly-α line only 20 min later. Examples of this behaviour are shown in Figure 52. In another flare, the Al XII/Mg XI ratio changed from about 2:1 to 1:1 during less than 10 min of the flare development (Meekins *et al.*, 1968). In some other observations emission from Fe XVI (near 15 and 17 Å) has remained enhanced over its preflare level for up to four hours, whilst the lines of higher ionization stages decayed much earlier (Neupert *et al.*, 1968). The plausible explanation is that the plasma is cooling, therefore progressively favouring emission from ions that exist in equilibrium at lower temperatures.

    Another effect, which is probably thermal in origin, is the increase in intensity of the innershell satellite lines during the flare decay. Such an increase has been found by

Doschek *et al.* (1971) for calcium in the flare of 1969, November 2. Since, according to Gabriel and Jordan (1969 a), these satellite lines are not formed by collisional inner-shell excitation by electron impact of the lithium like ion, but predominantly by dielectronic recombinations to the helium-like ion, the enhancement indicates an increased recombination rate. Decrease in temperature is a plausible explanation for it.

C. *Flare Cooling*

We shall discuss later on what might be the source of the fast heating of the flare plasma. Let us now, however, summarize those processes which can take part in the subsequent cooling of the flare material.

Presumably the ions do not acquire energy as efficiently as electrons in the original input of energy to the plasma. Therefore, at the time of the temperature peak one can expect the electron temperature to be much larger than the ion temperature (Friedman, 1969). When this is the case, the electrons in the plasma subsequently lose energy to the ions by collisions, as has been suggested by Hudson *et al.* (1969). According to Culhane *et al.* (1970) the best fit to the temperature decay is of an exponential form,

$$T = T_{\text{Max}} \exp(-t/\tau), \tag{71}$$

and the cooling time constants τ observed lead to n_e value about 2×10^9 cm^{-3}. Thus, with the emission measure values observed, one obtains $V \gtrsim 10^{30}$ cm^3. These resulting n_e values appear too small and the V values too large when compared with results of other studies. Therefore, Culhane *et al.* conclude that a different cooling process is to be looked for.

Let us next suppose that the region cools by radiation. For high temperatures existing in flares this loss will be predominantly in the continuum. Then, with the τ values observed, Culhane *et al.* find n_e values about 2×10^{11} cm^{-3}. Such densities seem to be too high.

Therefore, Culhane *et al.* suggest thermal conduction as a third possible cooling mechanism. Oster and Sofia (1966) have discussed conduction cooling in connection with flare phenomena and arrived at the conclusion that radiative processes were more likely to dissipate energy in flares. However, since the coefficient of thermal conductivity is proportional to $T^{5/2}$, the situation looks different in regions with T above 10^7 K. The decay does not follow the exponential law (71) in this case, but a good fit with observations can still be found, with n_e equal to $1-2 \times 10^{10}$ cm^{-3}.

Thus, summarizing the analysis presented by Culhane *et al.*, with the temperatures observed in X-ray flares the cooling should be predominantly due to collisions with ions at lower temperature if n_e is of the order of 10^9 cm^{-3}, due to conduction if n_e is close to 10^{10} cm^{-3}, while radiative cooling should be the main process if n_e approaches or exceeds 10^{11} cm^{-3}.

More recent discussions (Craig, 1973; Craig *et al.*, 1973; Neupert *et al.*, 1974; Zaumen and Acton, 1974; Rust and Roy, 1974; Roy and Datlowe, 1975; Silk *et al.*,

1975; Pallavicini *et al.*, 1975) have generally confirmed Culhane *et al.*'s conclusions and emphasized heat conduction as the predominant mode of cooling, with two slight modifications:

In the 1972, August 2 and 1973, June 15 flares conduction should have caused a faster drop in T_e than was actually observed. Neupert *et al.* consider this as an indication that the hot plasma may not be coupled directly to the chromosphere but may rather be contained in a partially leaky magnetic bottle which permits a conduction cooling rate about 4 times less than otherwise would be the case. Another possibility is, of course, that there was an additional energy supply to the flare during the cooling period (Pallavicini *et al.*).

In some flares it appears to be difficult to explain the whole process of cooling by conduction only. Craig *et al.* (1973) describe two events in which the first (~ 7 min) period after $T_{e\,\mathrm{Max}}$ shows a faster decay than the following period. They offer two alternative explanations: (a) The first phase is conductive cooling, which is stopped after a few minutes, either because of decrease in dT/dh (Craig, 1973), or for some other reason. The following phase may be radiative cooling. (b) The first phase may be cooling due to energy exchange between electrons and protons before $T_e = T_i$; the second phase is then conduction. From Skylab observations, Silk *et al.* (1975) have found also material loss as another reason for the flare decay.

3.1.5. ELECTRON DENSITY IN THE X-RAY FLARE

There are in principle three ways to deduce electron density from the X-ray spectra: (a) From the emission measure, after the flare volume is known. (b) From the cooling time (cf. preceding section). (c) From intensity ratios of density-sensitive spectral lines.

The method (a) is fairly unreliable. Since we do not know the exact shape of the X-ray flare, all estimates of its volume V in the corona are only rough approximations. Besides, even if we were able to estimate V from X-ray photographs, there still may exist unresolved condensations within this volume with n_e much higher than $\bar{n}_e = [(1/V) \int n_e^2 \, dV]^{1/2}$. Nevertheless, some authors used this method and found reasonable results: $n_e = 5 \times 10^{10}$ cm^{-3} for the region emitting Fe XXIV (Widing and Cheng, 1974), 2.2×10^{10} cm^{-3} for Fe XIV, 1.1×10^{11} cm^{-3} for Mg XI and 1.5×10^{10} cm^{-3} for Fe XXV (Neupert *et al.*, 1974). Rust and Roy (1974) deduced $n_e = 3 \times 10^{10}$ cm^{-3} from broadband X-ray measurements.

The method (b) leads to n_e close to 10^{10} as soon as we accept heat conduction as the dominant cooling mechanism (Culhane *et al.*, 1970). From the time profile of the Fe-lines at 1.9 Å Phillips *et al.* (1974) found 10^{10} cm^{-3} as the lower limit of the electron density. The value of 3×10^{10} cm^{-3} obtained by Rust and Roy (1974) by method (a) is also in good agreement with conductive cooling. Craig (1973), who considered a two-component model, found from conductive cooling of the hot component values one order of magnitude higher: $n_e = (2\text{--}3) \times 10^{11}$ cm^{-3}. Volume corresponding to these high densities would be quite small, about 10^{26} cm^3. Nevertheless this represents still reasonable dimensions, e.g. a filament $1''$ wide and $\sim 5 \times 10^9$ km long.

The method (c) is rather controversial. The whole story started when Gabriel and Jordan (1969b) pointed to the interesting fact that collisional interchange can take place between the levels ^3S and ^3P in the helium-like ion (cf. Figure 47) so that the relative intensities of the intercombination $1s2p\,^3P - 1s^2\,^1S_0$ and forbidden $1s2s\,^3S_1 - 1s^2\,^1S_0$ lines are density dependent. They defined

$$R = \frac{\text{forbidden line strength}}{\text{intercombination line strength}}$$

and gave an expression for R in the form

$$R = R_0 \, A_f/[(1+F)\,Cn_e + A_f], \tag{72}$$

where A_f is the $^3S_1 - {}^1S_0$ transition probability and C the $^3S - {}^3P$ collisional excitation rate. F is the ratio between the $^1S - {}^3S$ and $^1S - {}^3P$ collisional excitation rates, and

$$R_0 = [(1+F)/B] - 1, \tag{73}$$

where B is the effective branching ratio (i.e. a function of the transition probabilities from the ^3P states to the ground state). The value of B varies from 0.11 for C v to 0.46 for Si XIII, whilst the value of F is fairly uncertain. Estimates range from 0.35 (Gabriel and Jordan, 1969b) to 0.55 (Walker and Rugge, 1970).

As long as n_e is so small that $(1+F)\,Cn_e \ll A_f$ in Equation (72), R is constant and equal to R_0. If n_e increases above a certain limit, R becomes less than R_0. Thus the value of n_e can be estimated from the R value provided it is high enough to cause a recognizable difference $R_0 - R$.

The results of such an analysis were rather surprising. Both in the non-flaring active regions (Gabriel and Jordan, 1970; Freeman et al., 1971; Neupert, 1971) as well as in flares (Neupert, 1971) n_e was found to increase with the degree of ionization from $<10^{10}$ cm^{-3} for O VII to $>10^{13}$ cm^{-3} for Si XIII, and even to $\sim 10^{14}$ cm^{-3} for S XV, due simply to the fact that R was always found to be somewhat lower than R_0 for essentially all the ions. Gabriel and Jordan (1970) and Neupert (1971) have concluded from this that these helium-like lines are formed in small condensations with high densities and that the higher the density is, the higher also is the temperature so that higher ionization degrees are observed in emission.

However, one hesitates to accept these results as correct, since they are not confirmed by any other method. This does not involve only the methods (a) and (b) mentioned above, but also results obtained from the optical flare spectra (Section 2.2.4.B; Figure 29) and from the microwave and hard X-ray measurements (Sections 3.1.6 and 3.2.1). In order to maintain pressure equilibrium, the densities in hotter elements should be smaller than in the cold ones, as in fact was indicated in the optical spectra (Section 2.2.4.B). The opposite result, found from the X-ray lines of the helium-like ions, implies striking deviations from the pressure equilibrium, and one should postulate the existence of very strong magnetic fields high in the corona in order to confine the high-temperature and high-density condensations.

Therefore, one strongly suspects that the application of Equations (72) and (73) leads to erroneous results. The reason for it is probably to be sought both in the measurements as well as in the theory. The effects of blends and low spectral resolution, as well as problems in interpretation of the detector characteristics, make accurate measurements of R difficult. For example, the R values published by Doschek and Meekins (1970) are in error (Doschek and Meekins, 1973), due to an incorrect evaluation of the absorption edges in the detector characteristics. Similarly, the intensities of the intercombination line as given by Rugge and Walker (1971) are admittedly overestimated by some 15% which makes the R value lower. Unresolved satellite lines may further confuse the situation.

However, the theory is also open to question mainly on the F value in Equation (73). It has been taken as a constant in the n_e computations. However, it depends on the atomic number Z (Walker and Rugge, 1970; Rugge and Walker, 1971; Mewe, 1972a) and what is more important, also on temperature (Blumental *et al.*, 1972). Acton *et al.* (1972), who studied the helium-like lines in Ne IX and O VII ions in active regions, confirmed a dependence of R upon temperature in agreement with Blumental *et al.* and concluded that none of the regions producing these lines had $n_e > 3 \times 10^9$ cm^{-3}. Bonnelle *et al.* (1973) studied the helium-like Mg XI lines which also did not reveal the presence of high electron densities in active regions (giving $n_e < 7 \times 10^{11}$ cm^{-3}). The application of the theory to the non-steady state in flares has already been doubted by Gabriel and Jordan (1969b) themselves. Thus, as Blumental *et al.* said, 'it is premature at this time to conclude that density effects in R are in fact being observed'.

Summarizing, one can say that most probably the electron density in the soft X-ray flare region is $\gtrsim 10^{10}$ cm^{-3}, with a typical value of 3×10^{10} cm^{-3}; however, one cannot exclude the existence of plasma condensations of small dimensions ($\sim 10^{26}$ cm^3) with densities up to one order of magnitude higher.

3.1.6. GRADUAL MICROWAVE BURSTS

A great majority of flare-associated radio bursts are of impulsive non-thermal origin and we shall discuss them in the next Chapter. Only the gradual microwave bursts, called 'gradual rise and fall' and 'post-burst increases' are supposed to be thermal and basically of the same origin as the soft X-ray bursts, i.e. due to thermal bremsstrahlung (Kawabata, 1960, 1966; Kundu, 1963, 1964; Elwert, 1964). These thermal radio bursts are generally weak, mostly below 4×10^{-21} W m^{-2} Hz^{-1} in the range between 3 mm and 20 cm.

Hudson and Ohki (1972) and Shimabukuro (1972) have carried out a detailed comparison of thermal X-ray and gradual microwave bursts, and have confirmed a good coincidence between them. Figure 53 shows the time development of the enhancements at 2 to 12 Å and 3 mm, respectively, as observed in association with a 1b flare in 1969, March 29. According to Hudson and Ohki the time development of a thermal microwave burst roughly follows that of the emission measure deduced from the soft X-ray variation as one should expect from the thermal model.

Shimabukuro has tried to obtain the temperature and emission measure from three thermal microwave bursts recorded in March and April 1969. The free-free-emission flux in the radio range, expressed in W m^{-2} Hz^{-1}, can be written as

$$F(\lambda, T_e) = 2kT_e\lambda^{-2}\,\omega_b[1 - \exp(-\alpha z)], \tag{74}$$

where ω_b is the solid angle subtended by the burst region, z is the linear thickness of the emitting region, and α is the absorption coefficient,

$$\alpha = 2.0 \times 10^{-23}\,g\lambda^2\,n_e n_H\,T_e^{-3/2}, \tag{75}$$

with λ in centimetres. g is the Gaunt factor which may greatly exceed unity in the radio range.

At longer wavelengths ($\lambda \geqslant 10$ cm) the emission region is optically thick and thus the flux observed does not depend on the emission measure:

$$F(\lambda, T_e) = 2kT_e\,\omega_b\,\lambda^{-2}. \tag{76}$$

Since thermal radio bursts last for several tens of minutes, one can make radio maps of the Sun during the time of their existence, thus being able to estimate the actual size of the burst source. From the 3.3 mm maps, for example, Shimabukuro has found the half-power widths of the burst regions equal to 1.1' to 2.8'. Thus the quantity ω_b can be estimated, and Equation (76) yields the value of T_e.

On the other hand, for $\lambda \leq 3$ cm, $1 - \exp(-\alpha z) \simeq \alpha z$, and thus one obtains a relation analogous to Equation (55), with the exponential term equal to unity, simplified for a hydrogen atmosphere, and in units of W m^{-2} Hz^{-1},

$$F(T_e) = 4 \times 10^{-23}\,kgn_e n_H\,T_e^{-1/2}\,\omega_b z = \text{const. } T_e^{-1/2}\int n_e^2\,dV, \tag{77}$$

hence the flux should be independent of λ. Therefore, with T_e known, one can determine the emission measure from the flux value below 3 cm.

As an example, Figure 53 shows the result of such a calculation for the flare of 1969, March 29. The curve corresponds to $T_e = 6 \times 10^6$ K and $\int n_e^2\,dV = 5.4 \times 10^{49}$ cm^{-3}. For two other flares Shimabukuro has found $T_e = 2 \times 10^6$ K and $\int n_e^2\,dV \simeq 3 \times 10^{49}$ cm^{-3}. Generally, the T_e values obtained by him are a factor 2 to 8 lower than those deduced from the soft X-ray bursts (cf. Section 3.1.4.A) whilst the emission measure is of about the same magnitude.

The size of another thermal microwave burst, in 1967, December 16 was determined by Enomé et al. (1969) from interferometer observations. During the thermal phase of the burst they found its half-power width equal to 1.3' at 3 cm wavelength, close to the values found by Shimabukuro for three other bursts. This diameter corresponded to 5.5×10^4 km on the Sun. The emission measure was determined for the same event by Horan (1971) from the associated X-ray burst and it was found to be equal to 7×10^{49} cm^{-3} when the radio size was 1.3'. Thus, if the same emission measure is also adopted for the radio burst, one finds (Hudson and Ohki, 1972)

$$n_e \simeq 2 \times 10^{15} z^{-1/2},$$

(78)

using the assumption of a homogeneous emitting region. If one sets z equal to the lateral dimension of the region, which is roughly in agreement with the altitude extension of X-ray regions observed below the limb (Krieger *et al.*, 1972), one obtains $n_e \simeq 3 \times 10^{10}$ cm^{-3}. Shimabukuro has assumed z equal to 1 to 2×10^4 km and arrived at similar densities between 1 and 6×10^{10} cm^{-3}. These are the same values as deduced from the soft X-rays in the preceding section.

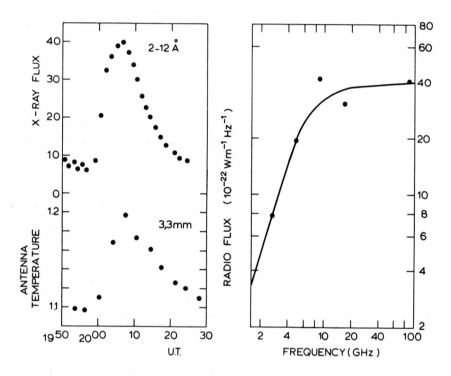

Fig. 53. The soft X-ray and gradual microwave bursts associated with the 1 b flare of 1969, March 29. The 3.3 mm peak value corresponds to 41×10^{-22} W m^{-2} Hz^{-1}. On the right: The burst spectrum of this radio event at the time of its peak flux. The solid curve corresponds to free-free emission at $T_e = 6 \times 10^6$ K and emission measure equal to 5.4×10^{49} cm^{-3}. (After Shimabukuro, 1972.)

If there are inhomogeneities the actual density in the flaring elements may be higher than these average values. However, one cannot expect them to be much in excess of 2×10^{11} cm^{-3}, since for higher n_e values microwaves at 10 cm do not easily propagate (Hudson and Ohki, 1972). This is again in agreement with the previous section. In order to obtain n_e in excess of 10^{13} cm^{-3}, as has been deduced from Equation (72), z in Equation (78) should be smaller than 100 m,

3.2. Non-Thermal Phenomena

3.2.1. HARD X-RAY BURSTS

De Jager (1965) was the first to suggest that one should distinguish two components in solar X-ray bursts: one thermal, or quasi-thermal, which we have discussed in the preceding chapter, and another one of impulsive, non-thermal character, probably caused by directed streams of accelerated electrons, with non-Maxwellian velocity distributions.

This has been fully confirmed in later years, as X-ray measurements have been extended to higher energies. However, as Kane (1969) showed and many others confirmed, the non-thermal component is not present in all X-ray bursts (cf. 3.1.3(a)). According to Vorpahl (1972) only 10 to 15% of flares give any indication of an impulsive component; on the other hand, Datlowe *et al.* (1974) have found detectable hard X-ray emission in 2/3 of soft X-ray bursts with peak fluxes above 10^3 photons $(cm^2 \, s \, keV)^{-1}$. In any case the non-thermal component is missing in many flares, so that it either does not develop there at all, or it is so weak (or short-lived) that one cannot recognize it in the record.

A. *Characteristic Properties of the Hard X-Ray Bursts*

Figures 54a, b and c show examples of two X-ray bursts that do, and another one that does not, contain the non-thermal component. When this component is present, it occurs in the initial phase (flash phase) of the flare, it has a short duration of a few tens of seconds to a few minutes, and it extends to much higher energies than the thermal X-ray burst. The thermal burst usually ceases to be discernible above a few tens of keV (and in fact it disappears at significantly lower energies, because the enhancements at > 10 keV are mostly or entirely due to a pile-up, as it is explained in the caption to Figure 54c). In contrast to it, the short-lived impulsive burst of non-thermal nature can be seen up to energies of hundreds of keV in strong events.

It is usually difficult to recognize the impulsive component in X-ray records below 10 keV. Nevertheless, there are cases when it manifests itself by a steepening in the rising part of the time-development curve of the burst even at these low energies, and Kahler and Kreplin (1971) have shown positively that the impulsive component extends in energy downwards as far as the range between 3 and 10 keV. On the other hand, the highest energy ever recorded in a hard X-ray burst was about 2 MeV in the flare of 1972, August 4 (Figure 55); we are speaking here about continuous emission produced through bremsstrahlung, not about the γ-lines that have different origin (cf. Section 5.4)). However, such a high energy is quite exceptional. Apart from this flare, the highest energy ever recorded in a hard X-ray burst was about 600 keV in the event of 1966, July 7 (Cline *et al.*, 1968).

There are many thermal X-ray bursts which do not have a non-thermal component (Figure 54c), but the opposite, an impulsive non-thermal burst without any thermal tail in the soft X-ray region, has never been observed. The hard X-ray burst always starts either simultaneously with the soft X-ray enhancement (Figure 54b) or after the

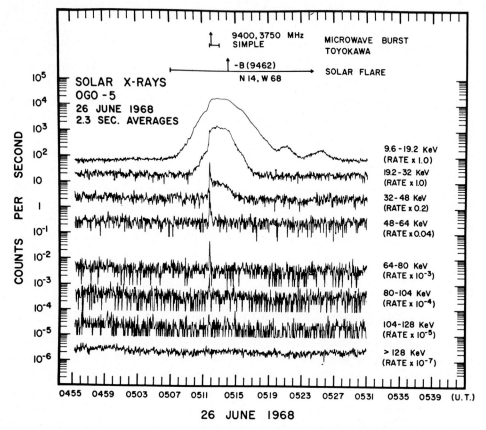

Fig. 54a. Example of an X-ray burst with the impulsive hard component occurring 4 min after the onset of the soft X-ray burst. (After Kane, 1969; *Astrophysical Journal*, University of Chicago Press, Copyright by the American Astronomical Society.)

soft X-ray onset (Figure 54a, cf. Section (3.1.3(b)). The time difference between the soft X-ray and hard X-ray onsets may be more than 8 min (Anderson and Mahoney, 1974; Spangler and Shawhan, 1974). Therefore, the impulsive non-thermal burst evidently is not the source of the whole flare phenomenon as many authors are suggesting. It develops as a secondary effect somewhere in or near the region of the thermal burst; it sets in at or after the thermal burst onset, but only when some favourable conditions are fulfilled. Or, alternately, it develops in all cases, but it is so weak in many of them that we cannot see any trace of it in the X-ray records.

There are several lines of evidence that the site of the impulsive burst is relatively small, much smaller than the coronal condensation which gives rise to the soft thermal bursts:

(a) The location and size of an impulsive hard X-ray burst associated with the flare of 1969, September 27, have been determined in one direction to a considerable precision with a balloon-borne X-ray collimator by Takakura *et al.* (1971). The center of the X-ray source was found on the line passing through the center of the flare region,

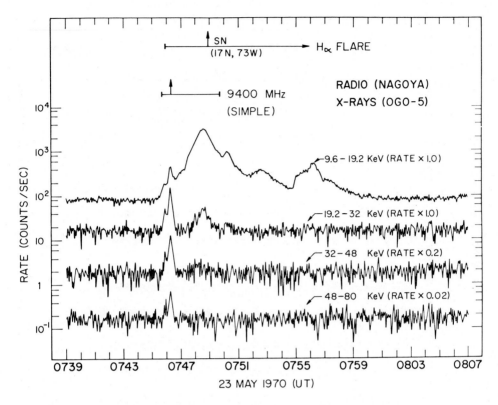

Fig. 54b.   Example of an X-ray burst with the impulsive hard component occurring at the onset of the soft X-ray burst. (Courtesy of S. R. Kane, Berkeley.)

and the size of the source was one arc minute or less, much smaller than that of the Hα flare region. Since Vaiana *et al.* (1968) have shown that the structure of a soft X-ray burst resembles that of an Hα flare (cf. Section 3.1.2), it also implies that the sources of hard and soft X-rays differed in their sizes.

(b) As we shall show later on in more detail, hard X-ray bursts are closely correlated to impulsive microwave bursts. Interferometer measurements of several radio bursts of this type have been carried out by Enomé *et al.* (1969) who find the size i.e. the half-power width of these impulsive bursts ≲ 0.5′. Kundu *et al.* (1974) have observed still smaller diameters. The gradual rise-and-fall bursts, which are believed to be of thermal origin (cf. Section 3.1.6), show much larger half-power widths, always exceeding one arc minute.

(c) De Jager (1967), Vorpahl and Zirin (1970) and Vorpahl (1972) have identified the hard X-ray bursts with small flare kernels which brightened strikingly in Hα light for the period of the burst observed. The diameter of the bright kernels is 3000 to 6000 km only, representing less than 1/8 to 1/2 of the main flare for subflares (Section 2.1.2.A).

(d) As we mentioned in Section 2.2.11.B, white-light flares are believed to be due

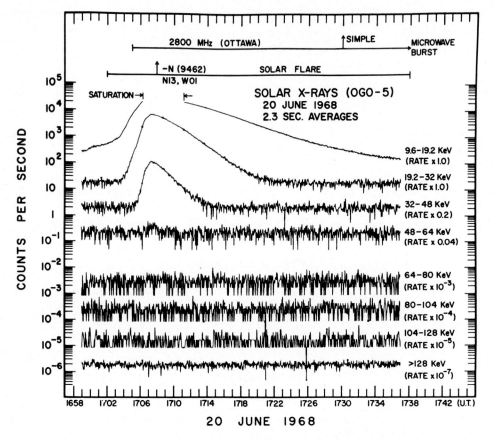

Fig. 54c.   Example of an X-ray burst without any impulsive hard component. (After Kane, 1969; *Astrophysical Journal*, University of Chicago Press, Copyright by the American Astronomical Society.) In fact, the burst does not reach the high energies (>19.2 and >32 keV) shown in the record. Very large thermal spectrum counting rates cause high probability that two pulses will arrive within the detector resolving time of 1.6 μs and be counted as one large pulse (Kane and Hudson, 1970). Thus, a 'pile-up' results, leading to fictitious fluxes at high energies.

to streams of particles accelerated in the same region as the electrons producing a hard X-ray burst, because both of these phenomena coincide closely in time. Most recently, this has been confirmed by Rust and Hegwer (1975) in the flare of 1972, August 7: The 2s time resolution was the highest ever recorded and the light curve for the white-light knots correlated with the 60 to 100 keV burst profile. But the areas of white-light flares are very small, less than 10^8 km^2 (DeMastus and Stover, 1967; McIntosh and Donnelly, 1972). According to Švestka (1970a) the size of the region where the non-thermal acceleration in the flare of 1967, May 23, took place was only about 1/20 of the whole volume of the high-temperature flare which gave rise to the thermal burst (Figure 38).

   Detailed analyses of hard X-ray bursts have been performed by Kane and Anderson (1970), Kane (1971, 1973, 1974), Peterson *et al.* (1973) and Datlowe *et al.* (1974).

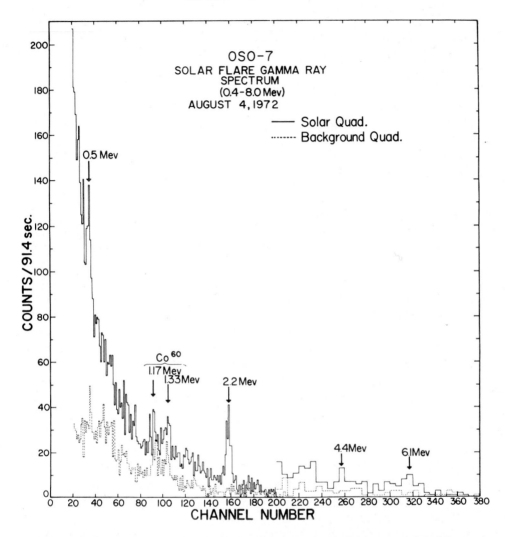

Fig. 55.   Spectrum of the Sun above 400 keV during the impulsive phase of the flare of 1972, August 4. The solid histogram shows the spectrum coming from the Sun, while the dotted histogram shows the background spectrum. γ-ray lines at 0.5 and 2.2 MeV are superposed on a hard X-ray continuum that extends beyond 2 MeV. This is the hardest X-ray spectrum ever recorded. (Courtesy of E. L. Chupp.)

The results, generally referring to small flares or subflares, can be summarized as follows:

(1) The hard X-ray burst occurs during the flash phase and peaks usually 0.5 to 3 min before the flare maximum observed in the Hα light.

(2) The X-ray flux increases and decreases are roughly exponential, with e-folding rise times of 2 to 5 s and e-folding decay times of 3 to 10 s et 40 keV. The rise time is always shorter or equal to the decay time. However, with higher time resolution,

many bursts can be decomposed into several short-lived spikes ('elementary flare bursts' after Van Beek *et al.* (1974)), with the rise and decay times as low as 1 to 2 s. In some cases (Frost, 1969; Beigman *et al.*, 1971) indications of fine structure even less than one second have been reported. Figure 57 shows an example of a highly structured hard X-ray burst, after Anderson and Mahoney (1974), with spikes of 2 to 20 s duration. Another example was shown in Figure 35.

(3) The peak flux of X-ray energy at the Earth, above 10 keV, is usually 10^{-7} to 10^{-5} erg cm^{-2} s^{-1}.

(4) The X-ray spectrum, which fits the observations at the burst maximum is of the power-law form

$$\frac{dJ(E)}{dE} = CE^{-\gamma} \text{ photons cm}^{-2} \text{ s}^{-1} \text{ keV}^{-1} \tag{79}$$

with $2.5 \lesssim \gamma \lesssim 5.0$ (median $\gamma \simeq 4.0$) in the range $10 \text{ keV} < E < E_0$, and much steeper for $E > E_0 \simeq 60$ to 100 keV. dJ/dE is the differential photon flux and E is the photon energy. γ may be as high as 8 (Peterson *et al.*, 1973), but almost never below 2.5 (Kane, 1971, 1973b), which appears to be an actual limit to the hardness of the spectrum.

(4) According to Anderson and Kane (1970) and Kane (1973) the spectrum hardens as the flux increases and softens during the decay time so that the hardest spectrum is obtained at the time of the peak flux. The hardening during rise time has not been confirmed by McKenzie *et al.* (1973); Datlowe *et al.* (1974) found it in some events, but in numerous cases the spectrum softened continuously throughout the burst. These differences may be due to the above mentioned fine time structure of the bursts and the associated spectral variations (Kane, 1974).

On the other hand, the softening of the X-ray flux during the burst decay has been confirmed by essentially all authors. Exceptions in some large flares are most probably due to the complex structure of the X-ray emission composed from tens of 'elementary flare bursts'. As we shall see in Section 3.2.1.D and E, the time variation of the X-ray spectrum is of particular interest because it gives information about the acceleration and energy loss processes in the flare region.

As was first pointed out by Peterson and Winckler (1959) and by Kundu (1961), there is a very good time correlation between the hard X-ray bursts and impulsive microwave bursts, which are often quite similar even in the fine details of the time development curves (Anderson and Winckler, 1962; Kundu, 1965; Arnoldy *et al.*, 1967; 1968a, b; Cline *et al.*, 1968). As an example, Figure 56 shows the X-ray and 8800 MHz microwave records for the flare of 1969, October 24 (Zirin *et al.*, 1971). The correlation is usually best at very high frequencies (Anderson and Winckler, 1962; Parks and Winckler, 1971), and very hard X-ray bursts (above 50 keV) are most similar to radio bursts at frequencies of 1 GHz or more (Arnoldy *et al.*, 1968b).

Parks and Winckler (1969) and Frost (1969) have detected simultaneous intensity variations in the hard X-ray and microwave flux which appear to have a quasi-periodic structure. In the case studied by Parks and Winckler (an importance 2b flare in 1968, August 8) five successive maxima were observed in the flux of >20 keV and

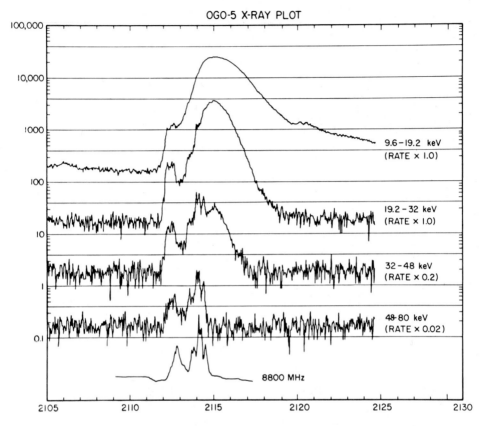

Fig. 56. This comparison of X-ray records in different energy channels with 8.8 GHz microwave burst demonstrates the great similarity between the hard X-ray (>50 keV) and microwave flux variations. The great rounded peak is mostly due to pile-up from lower energies (cf. Kane and Hudson, 1970, and caption to Figure 54c) so that it represents in fact the soft X-ray burst. (After Zirin *et al.*, 1971.)

>50 keV X-rays, with a periodicity of 16 s. In the microwave records these periodic variations were best seen at high frequencies, above 1 GHz. The energy spectrum is harder at X-ray peaks than in valleys. In the case studied by Frost (an importance 3 eruptive limb event of 1969, March 1) three successive maxima were observed in the flux of 28 keV X-rays and >2800 MHz radio records, with a period of about 35 s.

However, more recent observations have cast serious doubt on the supposed periodicity of these variations. Anderson and Mahoney (1974) observed two hard X-ray bursts composed of many 'flashes' (one of the bursts is demonstrated in Figure 57) and could not find any periodicity in their occurrence. In the flare of 1972, August 2, Zirin and Tanaka (1973) observed fast flashes along the zero line in sunspot penumbra (cf. Section 2.2.11 and Figure 35). A comparison of the time variations of the hard X-ray flux (Van Beek *et al.*, 1974) with Zirin and Tanaka's photographs suggests a clear association of the flashes in both wavelength regions (Figure 35; de

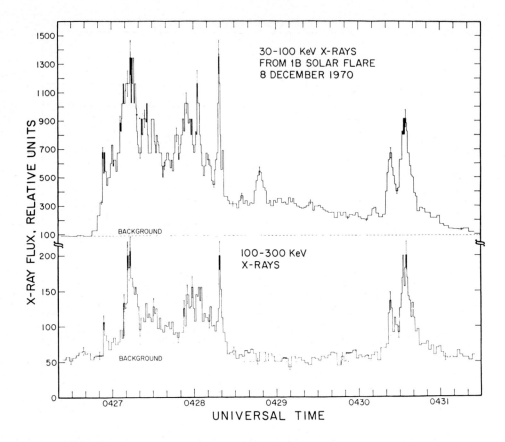

Fig. 57.   Record of hard X-rays in 1970, December 8 with high time-resolution. Similarly to Figure 35b the enhancement is composed of many 'elementary bursts'. One could suggest from the histogram a 'periodicity' of ~7.5 s; as we have seen in Figure 35, however, the individual flashes need not have identical sources. (After Anderson and Mahoney, 1974.)

Feiter, 1974). Thus the 'quasi-periodic pulsations' appear to be just successive energy releases in different areas of the flare.

There are now several more observations of similar kind. In the flare of 1972, August 7, Rust and Hegwer (1975) have found that flare continuum emissivity correlated closely in time with > 60 keV X-ray burst intensity. Three hard X-ray (and microwave) peaks in this flare coincided with brightenings in different positions (Zirin and Tanaka, 1973). Simultaneous rapid intensity fluctuations in the 1.9 Å emission and Hα have been suspected also by Neupert et al. (1969). The Hα brightenings occurred in a number of different chromospheric patches.

Thus most probably in all these cases we do not encounter any pulsations in the acceleration region as Parks and Winckler (1969) suggested, but subsequent short-lived acceleration processes in different parts of the flare. Irrespective of the interpretation, these fast variations, simultaneous in the hard X-ray and impulsive microwave flux, prove two important facts:

(1) That the energy spectrum of solar X-rays can change drastically during a few seconds and

(2) that the hard X-rays and the microwave impulsive bursts have a common process of acceleration.

B. *Evidence on the Non-Thermal Nature of the Hard X-Ray Bursts*

The physical origin of hard X-ray bursts has been argued for over a decade. The first interpretation was made by Peterson and Winckler (1959), who explained the hard X-ray burst as bremsstrahlung produced from non-thermal electrons. However, Chubb *et al.* (1966) interpreted their hard X-ray observations as emission from thermal plasmas with temperatures exceeding 10^8 K, and showed that the earlier observations of Peterson and Winckler could have been interpreted in the same way. Chubb (1970) and Milkey (1971) have pointed out that all X-ray emission from the Sun is fully interpretable as thermal plasma emission provided that one assumes sets of hot plasmas at different temperatures in the flare region.

However, at the present time the non-thermal nature of hard X-ray bursts appears to have been generally accepted, due to the increasing number of facts which are much more favourable to the non-thermal explanation than to the thermal one:

(1) The energy spectra of hard X-ray bursts generally follow a power law (Cline *et al.*, 1968; Hudson *et al.*, 1969; Kane, 1959, 1972, 1974; Frost, 1969; Kane and Anderson, 1970; Frost and Dennis, 1971). This is easy to explain if the non-thermal electrons in the solar source also have a power-law energy distribution (Kane and Anderson, 1970; Lin and Hudson, 1971). It might also be explicable by a thermal plasma with a greatly inhomogeneous temperature distribution (Chubb, 1970; Milkey, 1971), but then one has to accept temperatures greatly in excess of 10^8 K. In the event studied by Cline *et al.* (1968), temperatures up to 2×10^9 K are needed in order to explain the whole energy spectrum.

It is fairly difficult to believe in temperatures of this order of magnitude. For example, the energy density of solar gas at 10^9 K is equal to the magnetic energy density for 1000 G if $n_e = 10^{11}$ cm^{-3}. What then confines these hot inhomogeneities in the solar atmosphere? Chubb and Milkey argue that no non-thermal processes are required if one takes account of heterogeneous temperature structures with T_e in excess of 10^8 K. It seems more reasonable to argue, as Acton (1968) did, that no temperatures much in excess of 10^7 K are required if one considers non-thermal processes in the flash phase of a flare.

(2) The fact that high-time resolution receptors show a fine structure of the bursts, with life time of a few seconds (Van Beek *et al.*, 1974; Anderson and Mahoney, 1974), favours the non-thermal origin of the hard X-rays.

(3) The hard X-rays occur in very close time coincidence with impulsive microwave bursts, which are of non-thermal origin (cf. Section 3.2.3). The time coincidence in detailed structure (e.g. Frost, 1969; Parks and Winckler, 1969; also see Figure 56)

strongly indicates that both the microwaves and hard X-rays are produced by particles accelerated by the same process at the same time.

(4) Strong hard X-ray bursts also coincide in time with the short-lived white-light flares (cf. Section 2.2.11.B), which are believed to be produced by streams of highly accelerated particles, certainly of non-thermal nature, impinging onto the lowest atmospheric layers (Švestka, 1970a; Najita and Orrall, 1970; Hudson, 1972).

(5) It is believed that the flash phase, when the hard X-ray bursts occur, is the time when the particle acceleration at least to energies of ~ 1 MeV, but possibly higher, is accomplished (cf. Section 6.6). The strong particle fluxes in space are certainly due to non-thermal acceleration processes in the solar atmosphere; and type III radio bursts, associated with electron events in space, are a manifestation of non-thermal streams of electrons accelerated to velocities in excess of 10^5 km s^{-1} exactly at the time when the hard X-rays are observed (Sections 3.3.6A and 5.3.1.).

Besides these five points there are two other criteria that the non-thermal bursts should fulfill: the radiation produced by them should be polarized (Korchak, 1967, Haug, 1972) and it should show a centre-to-limb variation in the burst intensity (Elwert and Haug, 1971, Shaw, 1972).

The polarization of hard X-rays was indeed detected by Tindo et al. (1970, 1972a, b, 1973) in the wavelength range of 0.6 to 1.2 Å and confirmed by Nakada et al. (1974). However, this polarization is not confined only to the short period of the impulsive non-thermal burst, but it lasts for a significantly longer time, when the flare radiation already should be thermal and thus unpolarized. Tindo et al. (1972b) explain it by a continuous injection of accelerated electrons into the emitting region. Korchak (1974), however, has shown that continuous injection cannot explain the long existence of the polarized component in the observed energy range (10 to 20 keV) due to the fast increase of thermal radiation that would soon prevail. A non-thermal character for the X-ray emission could be probably better maintained through short-term but repeated injections.

On the other hand, there is also a possibility that the polarization (or at least a part of it) is produced through scattering of isotropic (and thermal) X-ray emission by the solar photosphere. The scattering cross-section depends on the angle between the electrical vectors of the initial and scattered photons and therefore, the reflected photons can be partially polarized. According to Beigman (1974) the polarization degree depends on the height of the X-ray flare above the photosphere; it should be zero in the center of the solar disc and maximum at $\cos \theta = 0.2$ to 0.5, depending on the height.

The polarization caused by non-thermal X-rays also depends very significantly on the geometry of the electron streams. For example, vertical streams should give zero polarization in the disc center and yield a maximum polarization near the limb (Brown, 1972b). So far only a few events have been observed in polarized light (and most of them with defective detectors, thus giving incomplete data) so that one hardly can decide which explanation is correct.

Another characteristic of bremsstrahlung from a directed beam of electrons should be an anisotropy in the emission produced. According to Elwert and Haug (1971) the most intense emission above 10 keV should be expected from flares situated in the intermediate region between the center and the limb of the Sun, provided that the electrons move predominantly parallel to the surface and to the equator. With increasing energy the maximum should shift towards the limb. Shaw (1972) who has assumed a random distribution for the orientation of the electron stream, but still parallel to the surface, finds maximum emission in the center of the disc for 10 keV X-rays, a broadening of the peak with increasing energy and splitting into two peaks at about 40° for energies above 40 keV. However, as soon as the electron paths are not parallel to the surface, the maximum is brought towards the limb. Thus in fact the situation is very complex and without a better knowledge about the electron trajectories in the flare region we can get essentially any result we like (Korchak, 1971); and, if the trajectories are different in different flares, the intensity distribution may be essentially independent of the position in a statistical analysis.

The observational results are unfortunately in good agreement with Korchak's pessimistic view. First studies of the longitudinal distribution of the X-ray bursts on the solar disc were made by Ohki (1969) and Pintér (1969). Astonishingly enough they obtained quite different results from exactly the same set of events. But apart from it, the number of observations used by them (46 events) was apparently too small to lend statistical significance to the results.

A few years later a more extensive study was performed by Pizzichini *et al.* (1974) who found some indication that the maximum of visibility shifts towards the disc center as one proceeds from > 10 keV to > 20 keV bursts. However, Datlowe *et al.* (1974), who studied 123 hard X-ray bursts, did not find any statistically significant longitude variation in the burst intensities.

C. *Theoretical Interpretation of the Hard X-Ray Bursts*

From now onwards let us assume that hard X-rays are of non-thermal origin. Then another problem arises: namely, whether there is any mechanism other than bremsstrahlung which might contribute to their production.

Only two other mechanisms might be considered: synchrotron radiation and Compton scattering. We have already shown in Section 2.2.11.A that an extention of a recognizable synchrotron radiation from the radio region to the optical one needs a very high number of relativistic electrons, which is incompatible with other kinds of observations. Our problems multiply if we try to extend the synchrotron emission up to the X-ray range. As far as the Compton effect is concerned, Korchak (1965, 1967, 1971) has shown that it might be effective only in regions with low densities $(n_i \lesssim 10^8 \text{ cm}^{-3})$ so that its contribution to the X-ray emission above 10 keV is highly improbable. Even when Compton backscattering from the photosphere might deform to some extent the hard X-ray spectra observed (depending on the iron abundance and on the degree of anisotropy in the X-ray production (Tomblin, 1972; Santangello *et al.*, 1973)), one can consider it to be fairly well established that hard X-ray bursts are produced by bremsstrahlung from directed beams of electrons accelerated to energies at least equal to, or exceeding, the highest X-ray energy recorded.

The bremsstrahlung originates in some volume in the solar atmosphere where accelerated electrons collide with ambient atoms and ions. We shall neglect all atoms other than hydrogen and assume that at some point in this volume there are n_i protons (free and in atoms, per cm^3) and n_e electrons accelerated to energy E (per cm^3 and per unit energy range), with differential energy spectrum $dn_e(E)/dE$. Then the X-ray flux is

$$\frac{dJ}{d\varepsilon} = \int_\varepsilon^\infty \frac{d\sigma(\varepsilon, E)}{d\varepsilon} v(E) \left\{ \int_V n_i \frac{dn_e(E)}{dE} dV \right\} dE \text{ photons s}^{-1} \text{ keV}^{-1}, \quad (80)$$

where v is the electron velocity corresponding to E, and $\sigma(\varepsilon, E)$ is the bremsstrahlung cross-section for electrons of energy E emitting X-rays of energy ε.

As Brown (1971) has pointed out, n_i and $n_e(E)$ are spatially non-uniform in the emitting region. Usually, however, one simplifies the problem by assuming that the particle distribution is uniform through the volume V (Takakura and Kai, 1966; Holt and Cline, 1968; Takakura, 1969a; Holt and Ramaty, 1969; Kane and Anderson, 1970; Korchak, 1971; Cheng, 1972b). Then, at the Earth's distance R, after neglecting the directivity of the bremsstrahlung process, one obtains the photon count rate

$$\frac{dJ(\varepsilon)}{d\varepsilon} = \frac{n_i}{4\pi R^2} \int_\varepsilon^\infty \frac{d\sigma(\varepsilon, E)}{d\varepsilon} v(E) \frac{dn_e(E)}{dE} V \, dE \text{ photons cm}^{-2} \text{ s}^{-1} \text{ keV}^{-1}. \quad (81)$$

Let us denote

$$V \int_\varepsilon^\infty \frac{dn_e(E)}{dE} dE = N_e(\varepsilon). \quad (82)$$

Then the quantity $n_i N_e(\varepsilon)$ is analogous to the quasi-thermal emission measure introduced in Section 3.1.3.A and some authors refer to it as the *non-thermal emission measure*. However, whilst the thermal emission measure takes into account all the electrons in the volume (since their number is approximately equal to n_i, one can also present it in the form $\int n_e n_i dV \simeq n_e^2 V$), the non-thermal emission measure depends upon the cutoff energy and $n_e(\varepsilon)$ is generally much smaller than n_i. Due to the dependence on ε, $n_i N_e(\varepsilon)$ decreases as the non-thermal electron spectrum decays. For constant V, for example, $n_i N_e(\varepsilon)$ decreases whilst $n_e^2 V$ remains the same during the decay phase of the X-ray burst (Holt and Ramaty, 1969).

As we have already mentioned, the differential X-ray flux $dJ(\varepsilon)/d\varepsilon$ below ~ 70 keV can generally be well represented by a power-law spectrum given by Equation (79), with γ varying between 2.5 and 5.0 in individual events (Kane, 1971). Thus, as soon as the cross-section $\sigma(\varepsilon, E)$ is known, one can estimate the energy spectrum $dn_e(E)/dE$ of the electrons producing the burst, as well as the non-thermal emission measure, by integrating the flux from the instrumental cutoff ε to infinity.

However, there is no 'exact' expression for the bremsstrahlung cross-section in the nuclear Coulomb field because it is not possible to solve the Dirac wave equation in closed form. One has to use an approximation, and the most widely used one is the Born approximation which takes a fairly simple form for the extreme cases of the non-relativistic and highly relativistic energies (Heitler, 1954; Koch and Motz, 1959; Cheng, 1972b). For the mildly relativistic range a simplified formula for $d\sigma$ has been proposed by Holt and Cline (1968) and used by Holt and Ramaty (1969). Though this 'intermediate' formula can be used throughout the whole energy spectrum without inducing serious errors, it seems to overestimate the $d\sigma$ value for all energies above a few keV's (Cheng, 1972b).

As long as our considerations are restricted only to X-rays in the energy range of a few tens of keV one can use the non-relativistic Born approximation, i.e. the Bethe-Heitler formula (Takakura, 1969a; Kane and Anderson, 1970; Lin and Hudson, 1971; Brown, 1971)

$$d\sigma(\varepsilon, E)/d\varepsilon = 1.58 \times 10^{-24} (\varepsilon E)^{-1} \ln\{(E/\varepsilon)^{1/2} + [(E/\varepsilon) - 1]^{1/2}\} \qquad (83)$$

in cm^2 per ion and keV, with ε and E in keV. This approximation is favoured by the fact that the electron energy spectrum is fairly steep and thus the contribution of electrons of kinetic energy E to X-ray radiation of energy ε decreases rapidly with increasing difference $E-\varepsilon$ (Brown, 1971). At these energies also the electron-electron bremsstrahlung can be neglected (Koch and Motz, 1959) and the screening effects are small so that Equation (83) applies both to ionized and neutral hydrogen atoms. Of course, as soon as one extends the discussion to energies close to, or above, 100 keV (as, for example was the case for the very strong hard X-ray burst of 7 July, 1966, discussed by Holt and Cline (1968) and Holt and Ramaty (1969)), Equation (83) cannot be used without taking the risk of very serious errors in the results. In such a case either the relativistic approximation (Cheng, 1972b), or Holt and Cline's intermediate formula should be used.

Under the assumption that the approximation (83) can be used for electron energies lower than 100 keV, and that electrons with higher energies do not contribute in any significant way to the production of <70 keV X-rays, Equation (81) takes the non-relativistic form (Kane and Anderson, 1970; Lin and Hudson, 1971)

$$\frac{dJ(\varepsilon)}{d\varepsilon} = 1.05 \times 10^{-42} n_i V \frac{1}{\varepsilon} \int_\varepsilon \frac{1}{E^{1/2}} \frac{dn_e}{dE} \times$$

$$\times \ln\{(E/\varepsilon)^{1/2} + [(E/\varepsilon) - 1]^{1/2}\} dE \text{ photons cm}^{-2} \text{ s}^{-1} \text{ keV}^{-1}.$$

$$(84)$$

This equation can be solved numerically for different electron energy spectra dn_e/dE to obtain the best fit with to the observed X-ray spectrum $dJ(\varepsilon)/d\varepsilon$. This has been done by Kane and Anderson (1970) and Lin and Hudson (1971) who have assumed that the accelerated electrons have a power-law energy spectrum of the form

$$\frac{dn_e}{dE} = KE^{-\delta}. \qquad (85)$$

They have found that this assumption fits Equation (84) well within the range $20 < \varepsilon < 70$ keV with $\delta = \gamma - \alpha$, where α decreases from ~ 1.3 at $\gamma = 3.0$ to ~ 0.9 at $\gamma = 5.0$. A more general solution of this problem has been given by Brown (1971; cf. next two sections D and E).

By inserting Equation (85) into (81) one can find the value of $n_i V K$ which best fits the observations, and from Equations (85) and (82)

$$N_e(\varepsilon) = V \frac{K}{\delta - 1} \varepsilon^{-(\delta - 1)}, \tag{86}$$

hence $N_e(\varepsilon)$ can be substituted for VK and we find the non-thermal emission measure $n_i N_e(\varepsilon)$. By this or similar methods different authors have found the following non-thermal emission measures in hard X-ray bursts: 5×10^{47} for $\varepsilon > 6$ keV (Acton, 1968); 9×10^{44} to 5×10^{46} for $\varepsilon > 10$ keV (Arnoldy et al., 1968; Kane and Anderson, 1970); $\sim 10^{46}$ for $\varepsilon > 15$ keV (Acton, 1968); 3×10^{44} to 6×10^{45} for $\varepsilon > 20$ keV (Takakura, 1969a; Lin and Hudson, 1971); and in a very strong burst, 6×10^{45} for $\varepsilon > 80$ keV (Holt and Cline, 1968).

D. Models of Impulsive and Continuous Acceleration

Many authors (Takakura and Kai, 1966; Holt and Cline, 1968; Takakura, 1969a; Holt and Ramaty, 1969) have made the assumption that the acceleration process is accomplished at the time of the burst maximum and that the subsequent decay in the X-ray flux is due to energy losses of the assembly of accelerated electrons.

Generally the energy loss could be caused by four different processes: collisions of the electrons with the ambient plasma, bremsstrahlung, synchrotron radiation, and escape of the accelerated electrons from the emitting region. In the energy range of a few tens of keV, which we are discussing, the synchrotron loss is not important. For 10 to 100 keV electrons in a magnetic field $H < 1000$ G the time τ in which the energy is reduced by a factor of e due to synchrotron radiation is > 200 s (Snijders, 1968; Kane and Anderson, 1970), i.e. one or two orders of magnitude longer than the value observed. The bremsstrahlung energy loss is about four orders of magnitude lower than the loss due to collisions. Hence only electron escape and collisions can be responsible for the decay observed.

Let us first assume that escape is also negligible and that collisions are the only source of the energy loss. Then the rate of energy loss for 10 to 100 keV electrons is given approximately by (Takakura, 1969a, Kane and Anderson, 1970)

$$\frac{dE}{dt} = -4.9 \times 10^{-9} \, n_0 E^{-1/2} \quad \text{keV s}^{-1}, \tag{87}$$

where n_0 is the density of the ambient plasma. Then the time $\tau(E)$ in which the initial energy E of an electron decreases by a factor e, is

$$\tau(E) \simeq 10^8 \, n_0^{-1} E^{3/2} \text{ s} \ (E \text{ in keV}). \tag{88}$$

As we mentioned before, τ for ~ 40 keV electrons has been found to be equal to

3 to 10 s. Thus $n_0 = 3 \times 10^9$ to 10^{10} cm^{-3}, and since $n_0 \simeq n_i$, we can estimate the total number N_e of electrons with energies in excess of ε from the emission measure $N_e n_i$. The results give $N_e(>10 \text{ keV}) \simeq 10^{35}$ (Kane and Anderson 1970), $N_e(>20 \text{ keV}) \simeq 3 \times 10^{34}$ to 6×10^{35} (Takakura, 1969a; Lin and Hudson, 1971) and, in the very strong event of 7 July 1966, $N_e(>80 \text{ keV}) \simeq 2 \times 10^{36}$ (Holt and Cline, 1968).

New observations (Van Beek et al., 1974; Anderson and Mahoney, 1974; cf. Section 3.2.1.A and Figures 35 and 57) reveal that large hard X-ray bursts also possess a fine structure with τ as small as one or two seconds; it is possible that even smaller τ values will be found with future high-time resolution observations. Thus, according to (88), the ambient plasma density is as high as $n_0 \simeq 3 \times 10^{10}$ cm^{-3} at least in some cases, and it may be even higher. N_e is then smaller, in proportion to the increase in $n_0 (\simeq n_i)$ but it does not represent any more the total number of accelerated electrons, which is (and may be significantly) larger.

If the escape of electrons to regions of lower density also played a significant role during the decay time, n_0 would be smaller and N_e larger than the values given above. A comparison of N_e with the number of electrons observed in interplanetary space (cf. Section 5.3.1) indicates that the fraction of electrons escaping from the Sun is small, generally less than 1% (Arnoldy et al., 1968a; Holt and Ramaty, 1969; Lin and Hudson, 1971; Kane and Lin, 1972). Of course, many more electrons escaping from the dense region can be trapped in the corona close to the Sun.

Because of the $E^{3/2}$ dependence in Equation (88) we should expect the hard X-ray bursts to last longer the higher the photon energy is. Contrary to this expectation the observations show a decrease of the burst duration with photon energy (e.g. Kane and Anderson, 1970; McKenzie et al., 1972) or a constant decay time (van Beek et al., 1974). This discrepancy can be reconciled by assuming that the higher energy electrons penetrate deeper into the atmosphere (Brown, 1972a, 1973c). Then n_0 in Equation (88) becomes energy-dependent, and if it increased with electron energy E more rapidly than $E^{3/2}$, a softening spectrum would result.

However, an alternate explanation is also possible: that the X-ray production takes place deep in the atmosphere, at such high densities that τ is smaller than the observed burst decay time at all energies. Then, however, the model of impulsive acceleration, as described above, cannot be held any more, and the time profile of the hard X-ray burst reflects in this case the time variation of the acceleration process (or injection process) itself.

Acton (1968) and Arnoldy et al. (1968a) were the first who proposed this other alternative, namely a continuous acceleration mechanism which lasts throughout the whole life-time of the burst, and whose modulations determine the time variations of the hard X-ray emission. This model of continuous acceleration has also been discussed by Kane and Anderson (1970), Lin and Hudson (1971), Brown (1971, 1972b, 1973a), Kane and Lin (1972), McKenzie et al. (1972), Kane (1973), Petrosian (1973), Datlowe et al. (1974), Vorpahl and Takakura (1974), Van Beek et al. (1974), Roy and Datlowe (1975), and McKenzie (1975).

If the X-ray burst directly mirrors the acceleration process, the lifetime of the accelerated electrons must be much shorter than the time constants of the observed intensity variations i.e. τ in Equation (88) must be significantly smaller than 1 s. Consequently, the density of the emitting region must be much higher in this case, with $n_0 > 10^{11}$ cm^{-3} according to Equation (88). Thus the problem of whether the acceleration process is impulsive or continuous, is also very close to the problem of whether the thin-target or the thick-target theory should be applied to the discussion of the flare-associated hard X-rays (Brown, 1971; Hudson, 1972).

E. *Thin-Target and Thick-Target Emission*

In order to make perfectly clear what we are talking about, let us make the following definitions:

τ = characteristic time for energy exchange, given by Equation (88);

τ_a = characteristic time of the acceleration process (during which energy increases by factor e);

τ_e = characteristic time for electron escape from the region where the emission is produced (during which the number of electrons decreases by factor e).

We speak about an *impulsive acceleration model* when

$$\tau \gg \tau_a. \tag{89}$$

In that case the number of electrons producing the X-rays at burst maximum ($= N_e$ in our previous discussion) is equal to the total number of electrons produced. The electron energy spectrum deduced at burst maximum is the same as the spectrum of the accelerated electrons at the ultimate end of the acceleration process.

We speak about a *continuous acceleration* (or continuous injection) model when

$$\tau \ll \tau_a. \tag{90}$$

In that case the total number of electrons produced is greater than the number in the source region at any instant of time (i.e. also greater than the above mentioned N_e values deduced from the X-ray burst maximum). Since, however, we have no means to determine n_0, the actual total number of electrons cannot be computed. Also the spectrum deduced from the X-ray burst no longer represents the spectrum resulting from the acceleration mechanism. Since the low energy electrons decay faster, the energy spectrum of the electrons which participate in the X-ray production becomes harder than the spectrum of electrons injected from the acceleration process (Brown, 1971; 1973a).

We speak about a *thin-target approximation* in the bremsstrahlung theory when

$$\tau_e < \tau, \tag{91}$$

i.e. electrons escape from the X-ray emitting region before they lose a significant fraction of their energy. Since in practice this can happen only when also condition (89) is fulfilled (low density of the ambient plasma), one can identify the thin-target and impulsive models (Brown, 1971). As long as the escape is not too fast, the correct N_e

can be obtained again from the maximum X-ray flux, and the electron energy spectrum of the emitting and injected electrons is the same, unless the escape depends strongly on energy (which however is to be expected).

Finally, we speak about a *thick-target approximation* in the bremsstrahlung theory when

$$\tau_e \gg \tau, \tag{92}$$

i.e. essentially all electrons lose all of their energy through collisions in the X-ray emitting region. The most likely situation fulfilling condition (92) occurs when (90) is valid, since both of them require high density of the ambient gas. Hence one can identify the thick-target and continuous injection models; the spectrum of the X-ray emitting electrons is then formed by equilibrium between the injection of newly accelerated electrons and the loss of energy through collisions.

Let us summarize the observations that favour one or the other of these two different interpretations of the hard X-ray bursts:

(a) The softening of the X-ray spectrum during the burst decay phase is often presented as an argument for the thick-target model (e.g. Van Beek et al., 1974; McKenzie, 1975). According to Van Beek et al. the elementary flare bursts should occur at $\lesssim 1500$ km height above the photosphere in order to explain the fact that their decay times are apparently independent of the photon energy. On the other hand, one can also explain the softening as due to penetration of high-energy electrons to deeper layers in the atmosphere (Brown, 1972a) or to preferential escape of high-energy electrons from the X-ray region (McKenzie et al., 1972).

(b) Tindo et al. (1972b) claim that their measurements of polarization of flare X-rays are in good agreement with continuous injection and radial movement of the electrons which emit hard X-rays. However, as we mentioned in Section 3.2.1.B, Korchak (1974) has shown that continuous acceleration cannot produce the polarization observed, unless it is composed of many repeated injections; Brown (1972b) and Petrosian (1973) have pointed out that radial streams of electrons should produce a brightening of the hard X-ray sources from solar center to the limb which has not been confirmed by observation (Pizzichini et al., 1974; Datlowe et al., 1974).

(c) There is still another reason why radial electron streams offer an attractive explanation for the hard X-ray emission and that is the model of the low-temperature flare described in Section 2.3.1. As soon as we accept the interpretation that the Hα kernels and the white-light flares are produced by accelerated particles streaming radially from the corona into the chromosphere, the same streams should also produce the EUV bursts and hard X-rays (cf. Kane, 1973b). The fact that we do not see any limb brightening of the hard X-rays speaks however against it.

(d) According to Datlowe et al. (1974) spectra of hard X-ray bursts near the limb are characteristically steeper. Behind the limb this spectral difference might be explained as due to an occultation of the low part of the X-ray source, emitting harder spectrum (Roy and Datlowe, 1975); however, this explanation fails, because the same effect of softening is observed also in X-ray bursts originating on the disc between

$\sim 60°$ to $90°$. A small variation of spectral index with longitude is predicted by thick-target streaming models of Brown (1972b) and Petrosian (1973), but these models, as we mentioned above, also predict a significant limb brightening that is not observed. Softening towards the limb should be also observed if Compton backscattering contributes significantly to the observed flux (Tomblin, 1972). However, the actual effect of such an admixture in the X-ray radiation should be quite small, according to the calculations of Santangelo *et al.* (1973). Thus no presently known theoretical model is able to explain the limb softening of the X-ray spectrum.

(e) The thick-target model assumes that the X-ray emitting region is in the chromosphere. Hence we would expect that the hard X-ray source would be occulted in flares occurring close behind the solar limb. This supposition has been checked by several authors. Kane and Donnelly (1971) estimated that the hard X-ray emission in a large burst of 1969, March 30, extended above 25 000 km over the photosphere. Roy and Datlowe (1975) found 25 over-the-limb events with a non-thermal component and concluded from it that at least some of the hard X-ray emission must come from considerable heights (of the order of 10^4 km) in the corona. This fraction should be significant, because the emission measure and the hard X-ray flux observed from over-the-limb bursts is not substantially reduced. On the other hand, McKenzie's (1975) data for 8 over-the-limb events are consistent with most of the hard X-rays being produced at a low altitude. Even in this case, however, one or two events indicated that a small fraction of hard X-rays probably arises at high altitudes.

Roy and Datlowe offered two explanations for the relatively large altitudes at which the hard X-rays are still observed: (1) The X-ray emission actually extends to heights in excess of 10^4 km. In this case the density is sufficiently low that escape loss may dominate collision loss, i.e. according to (91) we encounter thin-target emission. (2) A second possibility is that electrons accelerated in the flare travel over long distances along magnetic arches, reentering the high density regions far from their acceleration site. There are no observations that would support this idea. The same, however, is true for the interpretation (1), since observations in Hα and EUV support the thick-target model (cf. item (c)).

(f) If Equation (79) fits the observed X-ray spectrum, one can expect also that the energy spectrum of the injected electrons will be a power law (Equation (85)), with exponent $\delta \neq \gamma$. According to Brown (1971) $\delta = \gamma + \frac{1}{2}$ for the thick-target model and according to Equation (85) $\delta = \gamma - 1$ for the thin-target model. Under the assumption that the spectrum of escaping electrons recorded near the Earth is representative of the spectrum of accelerated electrons at the Sun, Datlowe and Lin (1973) checked upon the value of δ and found that it agrees with the thin-target model. This approach, however, is open to doubt. Datlowe and Lin set the escape probability at $<0.4\%$, thus indicating that the interplanetary electrons are very exceptional and scarcely representative of the whole assembly.

(g) The original Acton's (1968) and Arnoldy *et al.*'s (1968a) argument in favour of the continuous acceleration was the great similarity between the hard X-ray and the impulsive microwave bursts. As we shall see in Section 3.2.3, there is little doubt that both these phenomena have a common origin, i.e., the same acceleration process produces both of them. However, they do not seem to be produced by the same electrons in an identical volume of the solar atmosphere and energy losses in them might be of different kind: collisional in the X-rays, but radiative in the microwaves. Therefore, in particular, as far as rapid variations are concerned, it is easier to understand the similarity when the burst shape mirrors the acceleration process itself, and $\tau < \tau_a$ in both cases.

Thus, after summarizing the items (a)–(g), we arrive at the contradictory conclusion that several kinds of X-ray observations indicate a thin-target model for the hard X-ray bursts, while the burst fine structure, comparison with other types of data, as well as commonly accepted flare models, point to a thick-target emission. A plausible explanation may be that all hard X-ray bursts exhibit both kinds of emission: thick target for the electrons propagating along the magnetic field lines into the chromosphere, and thin target for electrons diffusing in the corona (Kane, 1973; Vorpahl and Takakura, 1974; Roy and Datlowe, 1975; McKenzie, 1975). The intensity of the X-ray emission then depends on the local ion density n_i and the instantaneous density of energetic electrons at that altitude. One also has to be aware of the fact that the thin and thick target models are extreme cases; probably they are never fully realized. Some electrons deposit their energy completely in the emitting region, while others escape before losing all their energy; thus we encounter in fact an intermediate case, which might be nearer to the thick- or thin-target model, respectively, in individual events, according to the magnetic configuration in which the acceleration process has been accomplished.

3.2.2. EUV Bursts

For many years our knowledge of flare spectra in the ultraviolet region was much worse than in the visible and X-ray parts of the spectrum. Only in the last few years satellite observations have been extended into the ultraviolet spectral range and give us now information about the transition layer between the chromosphere and corona in flares.

The ultraviolet spectrum is often divided in three parts:

UV: for $\lambda > 1500$ Å, which

is predominantly photospheric spectrum;

EUV: at ~ 300 to 1500 Å, which

is predominantly spectrum of the chromosphere and corona and the transition between these two layers;

and XUV: ~ 100 to 300 Å, which

contains lines of similar character; the separation has an instrumental reason: ~ 300 Å is the lower limit for normal-incidence optics.

Below ~ 100 Å we usually speak about soft X-rays. In this book we shall use the

term EUV for the whole region between the X-rays and the optical spectrum, since
the separation of XUV has no actual physical meaning and flare observations at
wavelengths much higher than 1500 Å are not available.

The first detected EUV burst ($\sim$1225 to 1350 Å (Figure 91)) was associated with
the proton flare of August 28, 1966 (McClinton, 1968; Friedman, 1969). Later on
many EUV bursts were observed either directly by satellites (Hall and Hinteregger,
1969; Hinteregger and Hall, 1969; Bruns *et al.*, 1970; Hall, 1971; Wood *et al.*, 1972;
Wood and Noyes, 1972; Kelly and Remse, 1972), or through their ionospheric
effects (Sudden Frequency Deviations (SFD), cf. Section 3.1.2) (Donnelly, 1969, 1970,
1971; Kane and Donnelly, 1971; Donnelly *et al.*, 1973).

Generally, the EUV bursts show very good time coincidence with the non-thermal
hard X-ray and the impulsive microwave bursts (Figure 58). If the hard component is
not present in the X-ray enhancement, an EUV burst may still be observed, but such
a burst is usually weaker and more slowly rising to the maximum than the EUV bursts
associated with impulsive X-ray events. Therefore, as in the X-ray region, the bursts
appear to have two components; one being of non-thermal and impulsive nature,
and the other one of quasi-thermal and gradual character (Castelli and Richards, 1971;
Wood and Noyes, 1972; Kelly and Rense 1972; Noyes, 1973; Donnelly *et al.*, 1973).
Since very little is known about the relatively weak gradual EUV bursts that accompany the soft X-ray bursts and microwave postburst increases, we shall confine our
discussion only to the typical impulsive EUV enhancements.

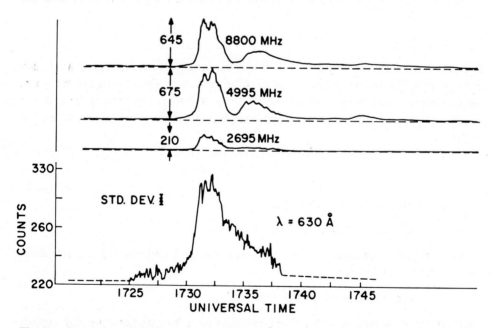

Fig. 58.   Correlation of EUV enhancement at λ 630 Å (O v line) with microwave burst, associated
with the 2b flare in 1971, August 1. (After Castelli and Richards, 1971; *Journal of Geophysical
Research*, Copyright by the American Geophysical Union.)

TABLE XIII

Important lines in the EUV region of the flare spectrum

Wavelength (Å)	Identification	I/I_0 [a]	I/I_0 [b]	k [c]	$h (\times 10^3 \text{km})$ [e]
1394	Si IV	(1.5)			
1335	C II	2.0			10.9 ± 1.3
1305	O I	2.0/1.5			
1243	N V	(1.7)			
1238.8	N V	2.5	1.3		
1215.7	H I (Lyα)		1.3	0.3	
1206.7	Si III		1.2	2.4 var	
1176	C III	7.2/2.0	3.0		
1085	N II	7.8/1.6	1.6		
1031.9	O VI	6.0/1.7	1.4	2.8	
1025.7	H I (Lyβ)	(1.3)	1.1	0.5	3.6 ± 2.4
991	N III	3.2/2.6	2.0		3.5 ± 2.1
977.0	C III	(3.5)	1.7	0.9	3.1 ± 2.0
972.5	H I (Lyγ)	(1.3)	1.1	0.7	
949.7	H I (Lyδ)	(1.2)	1.1		
834	O II, III		1.9		
791	O IV	2.7/1.6	1.8		5.1 ± 1.0
770.4	Ne VII	2.3/1.5	1.1 [d]	1.0	8.8 ± 1.4
765.1	N IV	2.2	1.4		
758	O V	(1.4)	1.8		
718	O II	(1.5)			
703	O III	1.4/1.4			
629.7	O V	1.9/1.7	1.1	1.2	
625.3	Mg X	3.2/1.3	1.1 [d]	0.3	10.6 ± 1.6
584.3	He I	1.4/1.2	1.1	0.3	3.0 ± 1.6
553	O IV	(1.6)	2.2		
537	He I	(4.7)			
521.0	Si XII	(1.2)			
499.3	Si XII	3.3/1.2	1.2 [d]	0.5	9.4 ± 1.7
465.2	Ne VII	2.9/2.1	1.2		
417	Fe XV	2.6/1.8			12.8 ± 2.2
368.1	Mg IX	(1.7)	1.1	0.5	
361	Fe XVI	2.1/1.6			15.2 ± 2.3
335.4	Fe XVI	1.4	1.2 [d]	0.4	
303.8	He II (Lyα)	2.7/1.8	1.1 [f]	1.4	
284.1	Fe XV		1.1 [d]	0.4	

[a] Relative line enhancement in flares, after Wood *et al.* (1972). The first number gives the mean of the three highest enhancements measured, the second one the median of all measurements. If only two flares were observed, no second number is given. If only one measurement was obtained, the value is in brackets.

[b] Relative enhancement of the integrated light in the line, in the spectrum of the imp. 2+ flare of 1967, March 22. I/I_0 at long wavelengths represents the flux close to the flare maximum, whilst shorter wavelengths were recorded at progressively later times. The only exception is the λ 553 O IV line, which was observed in the second order at the same time as the wavelength of 1106 Å. (Hall and Hinteregger, 1969).

[c] If A is the corrected area of a flare, the relative enhancement is proportional to $kA^{3/2}$ (Hall, 1971).

[d] Lines excited in later state of development of the flare.

[e] Heights of formation of the lines in active regions (no flare data), after Simon and Noyes (1972).

[f] In the flare of 1973, June 15, Tousey *et al.* (1973) observed 12 other lines of the helium Lyman series. Five Lyman lines were observed by Linsky *et al.* (1975) in the flare of 1972, August 7.

The main characteristics of these bursts can be summarized as follows (Donnelly, 1970, 1971; Hall, 1971; Wood *et al.*, 1972; Wood and Noyes, 1972; Noyes, 1973; Donnelly *et al.*, 1973):

The EUV burst occurs simultaneously at all temperatures from 10^4 to 10^6 K and coincides in time with the hard X-ray and impulsive microwave burst, often even in detailed time structure (also see Parks and Winckler, 1971). It also coincides with the white-light flare occurrence in the rare cases when this is observed (McIntosh and Donnelly, 1972). Similarly, it occurs during the flare flash phase, on the average two minutes before the Hα and soft X-ray maxima. The EUV burst life time is much shorter than the Hα brightening or the soft X-ray burst, but it is generally longer than the life-time of the accompanying non-thermal X-ray burst. It ranges from 1 to several tens of minutes, with the most common duration being 3 to 5 min according to Donnelly, 7 min according to Hall. The impulsive EUV flare radiation rises simultaneously with the net 10 to 1030 Å flash deduced from the SFD data which thus can be used as a substitute for direct EUV measurements if these are missing.

The EUV emission appears in many spectral lines ranging from neutral hydrogen and oxygen to highly ionized species such as Si XII and Fe XV and XVI (Table XIII), as well as in the recombination continua of neutral hydrogen and helium. Probably, at least in strong events, continuous emission other than radiative recombination continua also contributes to the EUV burst. The amount of energy radiated in the EUV resonance lines of C II, O VI, and Mg X is typically 10^{-1} to 10^{-3} times the energy radiated in Hα (and may be more, because the Hα energy is often overestimated, according to Zirin and Tanaka (1973)).

According to Hall and Hinteregger (1969; Hall, 1971) and Donnelly *et al.* (1973) the lines formed in the chromosphere and in the transition region zone dominate during the impulsive phase. The coronal emission lines, on the other hand, tend to dominate during the gradual, quasi-thermal phase which peaks later.

Kane and Donnelly (1971) and Donnelly (1971) have found some indication that the relative strength of impulsive EUV bursts, when compared both with flares and hard X-ray bursts, decreases towards the solar limb. This might indicate absorption of the EUV emission in the solar atmosphere.

Whilst Wood *et al.* (1972) find the EUV flares to be co-spatial with the associated Hα brightening to within the observational resolution of 35″, according to Donnelly (1970, 1971) the EUV burst appears to be associated with small impulsive portions of the Hα flare that are very bright and usually located near the edge of sunspots. The same conclusion can be drawn from the close relation of the EUV bursts (and, of course, hard X-ray and impulsive microwave bursts as well) to the white-light flares which also occur as small areas closely adjacent to sunspot penumbrae (Švestka, 1970a; McIntosh and Donnelly, 1972). Thomas and Neupert (1975) have found that multiple spikes of EUV emission during the impulsive phase occur at different locations within the flare and that these locations are not in all cases co-spatial with the Hα flare.

Widing and Cheng (1974; Cheng and Widing, 1975) used Skylab EUV observations

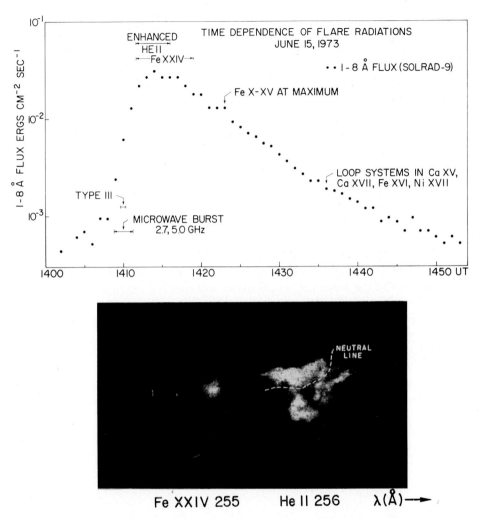

Fig. 59.   Upper part shows the soft X-ray emission of the flare of 1973, June 15 and a description of the emissions observed. He II and Fe XXIV are bright at flare maximum, but in different spatial locations, as one can see in the lower part of the figure. This part shows monochromatic images of the flare in the Fe XXIV (λ 255 Å) and He II (λ 256 Å) lines, obtained with the NRL slitless objective-grating spectrograph on Skylab. While the hot Fe XXIV emits on the top of the 'flare loop' above the $H_{\parallel} = 0$ line, the 'cool' He II is seen in the loop feet. (After Widing and Cheng, 1974; *Astrophysical Journal*, University of Chicago Press, Copyright by the American Astronomical Society.)

for a comparison of the EUV flare shapes and locations in different lines. As a rule they find in the early flare phase a hot core (represented by the Fe XXIV line at $\sim 1.6 \times 10^7$ K) centered over the zero line of the longitudinal magnetic field, while the cooler material (represented by the He II, Ne VII, and Fe XIV–Fe XVI lines) shows the familiar two-ribbon structure (cf. Section 2.1.3.E) co-spatial with the Hα flare (Figure 59). A similar observation was made by Neupert *et al.* (1974) when they compared the flare image in the 1.9 Å iron line with EUV lines. As the flare cools the hot cloud disappears and the region between the ribbons is filled gradually with emissions from ions of lower ionization temperature, exhibiting loop structures.

Cheng and Widing (1975) conclude from comparisons of the EUV spectrohelio-grams on Skylab and the magnetograms that the basic magnetic field configuration associated with flares is an arcade of loops with altitude ranging from 8000 to 20 000 km. In the case studied by Neupert *et al.* (1974) the height of the 1.9 Å top of the loop was estimated at 35 000 km.

As we mentioned above, lines in the EUV spectrum originate within a wide range of temperatures. The lowest temperature is deduced from the Lyman series of hydrogen: colour temperature of the hydrogen Lyman continuum is only 7500 K (Noyes, 1973). In the Lyman series of helium, Linsky *et al.* (1975) found a colour temperature of about 24 000 K in the Lyman continuum of a flare. Brightness tem-perature at the head of the Lyman continuum was about 41 000 K. Both these values are some 10 000 K higher than temperatures deduced from the quiet Sun spectrum. Other lines trace the transition region as one can see, e.g., from Figure 60.

Munro *et al.* (1971) have shown that the ratio of intensities from certain pairs of lines from ions of the beryllium iso-electronic sequence (e.g., C III, O V) are sensitive to the electron density through the influence of a metastable state (cf. similar consid-erations for X-ray lines in Section 3.1.5). From O V lines, 6 min after a flare maximum, Noyes (1973) thus found $n_e = 5 \times 10^{11}$ cm^{-3}, which is 5 times the preflare value,

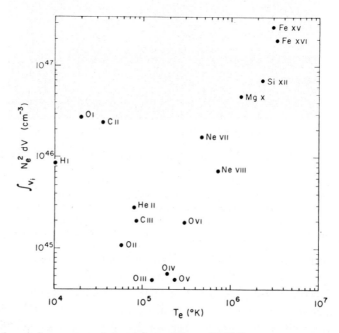

Fig. 60. Peak emission measure deduced from different EUV lines, versus electron temperature. The minimum at about 2×10^5 K and the increase in emission measure both toward lower and higher T values indicate the existence of a narrow transition layer in flares. (After Wood and Noyes, 1972.)

30 times that in a typical nonflaring active region, and 150 times that in the quiet Sun. Generally, Noyes reports increases by factors of 5 to 10 in the transition zone for small flares, also confirmed by increases in the emission measure. In a flare of Hα importance 1b, Donnelly and Hall (1973) found $n_e \gtrsim 10^{12}$ cm^{-3} from the C III lines.

Increased densities in the flare transition layer are not surprising. We have mentioned in Section 2.3 that the transition region in flares is probably shifted lower in the solar atmosphere, thus accounting for the increased density. Also densities deduced from the optical flare spectrum above the limb show an increase (cf. Figure 29). Nevertheless, Munro et al.'s (1971) method for Be-like ions may be misleading, similarly to the Gabriel and Jordan's (1969a, b) method for He-like ions discussed in Section 3.1.5, since n_e is not the only factor that determines the line intensity ratio, and the flare is far from the assumed stationary equilibrium. Indeed, in one flare Donnelly and Hall (1973) found a C III line intensity ratio three times larger than its maximum theoretical value. This points to unreliability of the method.

Purcell and Widing (1972) estimated the density from the ratio of the Fe XIV lines at 211 and 219 Å, applying Blaha's (1971) theoretical computations of the dependence of the ratio value on electron density. Four minutes after the maximum of a 2b importance flare they found $n_e = 3 \times 10^{10}$ cm^{-3} in the hot region where Fe XIV was emitted. Similar results were also found from Fe XV lines by Cowan and Widing (1973). Of course, as one can see from Figure 60, these densities correspond to flare regions with significantly higher temperatures of $\sim 3 \times 10^6$ K, as compared with $\sim 3 \times 10^5$ K in O V and $\sim 8 \times 10^4$ K in C III lines.

Brueckner (1974) has reported non-thermal Doppler broadening of EUV lines in flares with characteristic velocities of ~ 70 km s^{-1}. Non-thermal motions can be observed even before the flare onset as one can see in Figure 61; but as one also can see in the figure, the lines were broadened as soon as one and half hour prior to the flare. Thus the broadening may characterize active regions prolific in flares, but not the immediate pre-flare phase (Brueckner, 1975). Doschek et al. (1975) observed random velocities of 60 to 70 km s^{-1} in EUV lines of Fe XIX and Fe XXI (i.e., in hotter and thus higher regions) in the early phase of the 15 June 1973 flare, decreasing to some 30 to 40 km s^{-1} 20 min later. Some Fe XXI profiles are irregular, implying motions that are not strictly random. We mentioned similar effects in Section 2.2.9, when we talked about the asymmetry of flare lines (cf. Figure 32).

Most authors agree that the EUV burst is produced thermally in a region of high (chromospheric) density, which is being heated by collisional losses of the accelerated particles. These may be the same non-thermal electrons which are responsible for the impulsive X-ray and microwave bursts (Kane and Donnelly, 1971; Wood and Noyes, 1972; Hudson, 1972; Donnelly et al., 1973; Kane, 1973a), eventually with an admixture of protons accelerated at the same time (Švestka, 1970a; Najita and Orrall, 1970). These particle streams heat the transition layer and the chromosphere directly, thus producing the impulsive EUV burst ('non-thermal' by origin, but 'quasi-thermal' in its nature). The gradual (or 'thermal') EUV component may represent heating of nearby plasma by conduction.

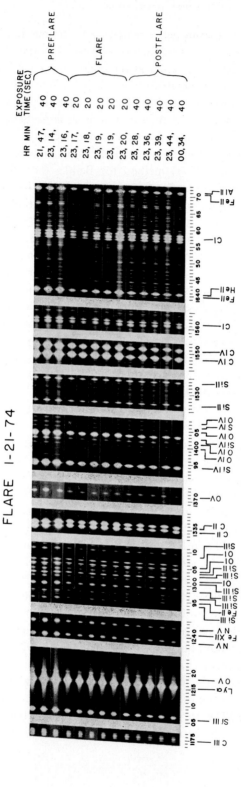

Fig. 61. Ultraviolet spectrum of the flare of 1974, January 21, photographed on Skylab. Already prior to the flare onset spectral lines were significantly broadened (cf. the 40 s exposures at 23^h14^m and 23^h28^m, e.g.) which indicates the presence of non-thermal motions in the pre-flare active region. (Courtesy of G. E. Brueckner, NRL Washington.)

On the assumption that EUV bursts are quasi-thermal in nature one can compute the emission measure from the line-intensities observed. The emergent intensity in the normal direction for an optically thin line is proportional to

$$\int_V n_e^2 \, dV \,,$$

where V is the volume over which the particular ionization stage is formed and n_e is the local electron density. The proportionality factor can be computed, after making assumptions about the temperature, the element abundance, and the processes that participate in the formation of the ion and in the level population. This has been done by Wood and Noyes (1972) who have found $\int n_e^2 \, dV$ to be between 5×10^{44} and $3 \times 10^{47} \, \mathrm{cm}^{-3}$, with the minimum values corresponding to a temperature range between 10^5 to 5×10^5 K, and growing both with increasing and decreasing ionisation, i.e., temperature (Figure 60).

The result is very interesting from the point of view that it closely resembles the results obtained for the quiet Sun (Athay, 1966) and active regions (Noyes *et al.*, 1970). It shows that the flare region has a temperature structure similar to that of an active region, in which a narrow transition zone (characterized by small $\int n_e^2 \, dV$ in lines produced in it) separates a region of very high temperature (where Ne VII and more stripped ions are formed) from the cooler ambient chromospheric plasma (characterized by neutral and singly ionized atoms). The data used had a low time resolution so that they may not refer to the earliest impulsive component. More probably, they indicate that a steep transition zone develops rapidly during the flare life. This is in agreement with the theoretical considerations discussed in Section 2.3.1.

There is a significant number of weak events which are easily detectable as apparent EUV flares but are not accompanied by any detectable flux in Hα, X-ray, or radio radiation (Wood *et al.*, 1972). On the average, these events are about a factor two less energetic than the average EUV burst associated with the smallest flare phenomenon, i.e., an *Sf* subflare. Out of 211 EUV bursts observed aboard OSO 6, 104 were clearly associated with Hα flares or subflares. Another 43 were still correlated with an observed X-ray or radio burst. But for 63 (30%) EUV bursts no associated enhancement in any other spectral region could be discovered. (The actual percentage may be lower, due to imperfect X-ray, Hα, and radio coverage.) According to Wood *et al.* this suggests that EUV observations may be a more sensitive indicator of instabilities in the solar atmosphere than are observations in the X-ray, Hα, or radio regions. And it may also indicate that besides flares and subflares there are still 'microflares', reflecting flare-like instabilities so small that the only way to detect them is with the EUV observations. Of course, one must not forget that there are also other kinds of bursts (in soft X-rays, or metric radio waves) which are not associated with enhancements in any other spectral region. Only careful statistics of these various isolated events can show whether there is any actual prevalence in the occurrence of EUV 'microflares'.

3.2.3. IMPULSIVE MICROWAVE BURSTS

A. *Characteristic Properties*

Impulsive microwave bursts, i.e., fast radio enhancements on frequencies above ~ 1000 MHz, occur simultaneously with hard X-ray and EUV bursts and have similar characteristics. In the vast majority of cases the impulsive burst is short, with a duration ranging between 1 and 10 min (Figure 62). According to Spangler and Shawhan (1974) who studied 259 microwave bursts at 15.4 GHz, burst duration at this frequency is most often between 1 and 2 min, similar to the hard X-ray bursts (see Section 3.2.1.A). Very short bursts are usually simple, but the longer ones often show complex fluctuations which are more pronounced at lower frequencies. Quite often the burst has a long-lived tail (Figure 62) which is called the post-burst increase and is probably of thermal origin (cf. Section 3.1.6). In some 10% of events, however, the impulsive burst itself is longer, with duration up to one hour (Takakura, 1967),

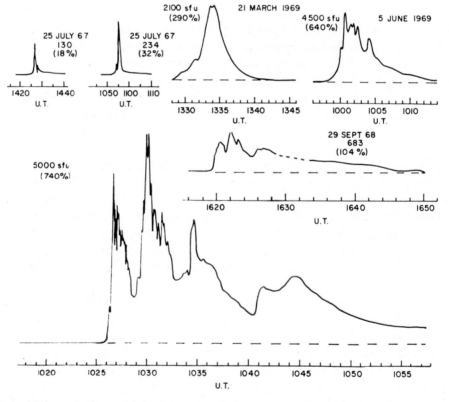

Fig. 62.   Examples of microwave bursts. Frequency 19 GHz (1.58 cm), peak flux intensities are given in sfu and in percent of the pre-flare flux. The July events show simple bursts with short duration; the March event is a simple burst lasting for more than 10 min; the June event illustrates a short-lived burst with complex structure; post-burst increases can be seen in the June and September events. The large event at the bottom occurred in 1968, November 18, and it represents an extremely strong microwave (IVμ) burst with very complex structure. (After Croom and Powell, 1971.)

and usually with very complex structures. The duration and complexity increase with increasing frequency (Figures 62 and 63). In such complex and long-lived cases only the onset part of the burst shows a good time coincidence and similarity to the associated hard X-ray burst, whilst the later flux variations have no counterpart in the hard X-ray region (Figure 63).

These long-lived microwave bursts (sometimes called microwave outbursts) are mostly associated with prolonged emission at lower frequencies up to the metric range, and form a part of a type IV burst (cf. Section 3.2.4). Therefore, they are called IVμ bursts. However, some on the type IV-associated microwave bursts, i.e., IVμ bursts by definition, may also have durations shorter than 10 min, sometimes even without any complex structure (Švestka and Olmr, 1966). The remaining short-lived microwave bursts, without any low-frequency counterpart, are usually called M-type bursts, but there is much inconsistency in this nomenclature in the literature. As a matter of fact there appears to be no principal difference between the M-type and

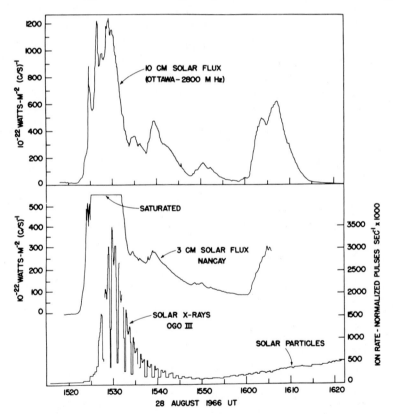

Fig. 63. Great microwave burst associated with the proton flare of 1966, August 28. Microwave records at 2.8 and 10 GHz are compared to >20 keV hard X-ray burst measured by OSO 3. The periodic decreases in the X-ray record are of instrumental origin (roll modulation). Note that only the initial microwave enhancement is reflected in the X-ray profile, while the secondary maxima have no counterpart in the X-ray flux. (After Arnoldy *et al.*, 1968a; *Astrophysical Journal*, University of Chicago Press, Copyright by the American Astronomical Society.)

IVμ bursts, apart from the fact that in IVμ bursts probably more particles are accelerated and/or more particles reach higher layers of the solar atmosphere where the low-frequency bursts are produced (Takakura, 1967).

Examples of M-type and IVμ spectra are shown in Figure 64. The spectrum of the microwave bursts is a broad-band continuum with maximum intensity generally at about 3 to 10 GHz (Takakura, 1967), the maximum tending to shift to higher frequencies for very strong bursts (cf. Figure 74); in particular, for strong IVμ bursts associated with particle emission in space, the maximum is found between 5 to 70 GHz, with the mean value at about 17 GHz (Croom, 1971). Exceptionally, in some active regions microwave bursts show unusual spectra – e.g., emission is limited to only short-centimeter (and possibly millimeter) wavelengths, with maximum shifted above 10 MHz even in rather weak bursts (Figure 65; Castelli $et\ al.$, 1974; Švestka $et\ al.$, 1974). The fact that repeated bursts usually show similar features indicates that the origin of such anomalies should be looked for in a particular magnetic configuration of the active region (probably causing an anomalously large depth in the atmosphere where the bursts are formed).

The intensity of a radio burst is commonly given in standard flux units (sfu) of 10^{-22} W m^{-2} Hz^{-1}. Whilst the quasi-thermal microwave bursts, discussed in Section 3.1.6, usually have their maximum flux below 40 sfu, the maximum flux of impulsive bursts is often much higher and it can exceed 10^4 sfu in outstanding events. For example, out of 148 microwave bursts recorded during $2\frac{1}{2}$ yr at 19 GHz (1.58 cm), 65 had maximum fluxes higher than 100 sfu, 15 higher than 10^3 sfu, and 2 higher than 10^4 sfu (Croom and Powell, 1971). Similarly, during two years of observations at

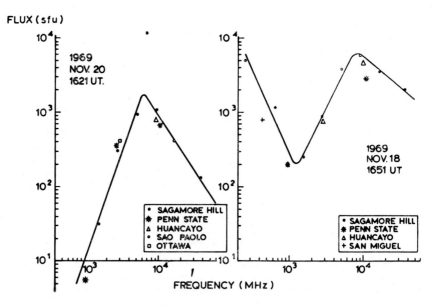

Fig. 64.   Radio spectrum of M-type (left) and IVμ-type (right) microwave bursts. (After Wood and Noyes, 1972.)

2800 MHz (10.7 cm), 116 bursts had maxima higher than 100 sfu and 6 higher than 10^3 sfu (Kundu, 1965). According to Tanaka (1969) the greatest burst ever recorded at 10 cm reached 32 000 sfu (on March 30, 1969).

All strong microwave bursts are associated with flares. Bursts of lower intensity ($\lesssim 100$ sfu) sometimes appear without reported Hα flares, and for weak bursts ($\lesssim 20$ sfu) the correlation with flares is only 65% (Kundu, 1965). However, the association with Hα brightening is usually very much improved if one looks at the Hα film records instead of using the published lists of flares. For instance, most bursts shown in Figure 65 have no Hα counterpart in the flare lists. Inspection of the Cal Tech Big Bear Observatory Hα film records for that period, however, confirmed the existence of Hα brightenings for all but one of the events (Castelli et al., 1973).

The correlation with reported flares appears to be much better when bursts at millimeter wavelengths are considered, where essentially all bursts are associated with flares (Shimabukuro, 1968; Croom and Powell, 1971). The reverse association is weaker: only about 60% of flares of importance 1 + (old classification) are accom-

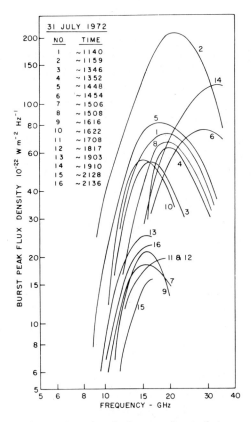

Fig. 65. Peak flux density spectra of a series of microwave bursts that occurred on 1972, July 31, in the active region that later produced the big 1972 August flares. Note the anomalous cut-off at ~10 MHz. (Castelli et al., 1974; *Journal of Geophysical Research*, Copyright by the American Geophysical Union.)

panied by bursts at 10 GHz (Kundu, 1955), and for smaller flares and subflares the correlation percentage is very small. This association, however, is different in different active regions. In some of them microwave bursts do not occur at all. In others, in particular where a magnetically complex configuration forms, microwave bursts accompany almost all flares, and even many subflares (Fokker and Roosen, 1961; Švestka and Simon, 1969; Švestka, 1971a; Matsuura and Nave, 1974; Krüger, 1973). This fact is demonstrated by Figure 79b in Section 3.2.6 and Figure 94 in Section 5.1.1.B.

Many authors have tried to determine the center-limb variation of the microwave burst intensity (Thompson, 1962; Kundu, 1965; Kakinuma et al., 1969; Scalise, 1970; Croom and Powell, 1971), but the results are fairly confused and contradictory. Obviously, the relatively small number of events used for the statistical study cannot smooth the extremely large differences between individual events. Theoretically, according to Takakura and Scalise (1970), one should expect a limb decrease at 1 GHz and a limb increase at higher frequencies, this being quite pronounced around 20 GHz.

It was believed a few years ago that the sources of microwave bursts were larger than the flare dimensions in the Hα light, of 1′ to 3′ in diameter (Wild, 1964; Kundu, 1965). However, observations at 9.4 GHz with a 24″ fan beam have shown that the angular size of the source of an impulsive burst does not exceed 0.5′ (2.3×10^4 km) (Enomé et al., 1969). Quite recently Kundu et al. (1974) observed a microwave burst at 3.7 cm, with 40% of the energy being released within a spherical volume of only 2″ diameter. This size increased to $\sim 5''$ in a later phase of the flare. At 11 cm the dimensions were about 5 times larger. Ogawa and Kawabata (1975), using the 35 GHz interferometer at Nagoya, found that the position of the microwave radio source in the flare of 1972, August 2 coincided with an Hα kernel.

In several sources Tanaka et al. (1967; Tanaka and Enomé, 1970) clearly distinguished a double structure roughly coinciding with sunspots of opposite magnetic polarity (Figure 66). This implies that the microwave burst may be produced inside the feet of a loop-like magnetic structure which stretches between the opposite polarities. This loop is probably also the seat of the hard X-ray burst, of the EUV burst and, in particularly strong events, of the white-light emission as well.

Janssens et al. (1973) observed quasi-periodic pulsations typically with periods of 10 to 20 s in 7 microwave bursts at ~ 3000 MHz. Their observations are similar to those in hard X-rays (cf. Section 3.2.1.A). Quasiperiodic fluctuations of intensity and polarization have also been reported by Erjushev and Tsvetkov (1973) at 3.15 cm, with periods from 6 to 320 s. In fact, the small number of pulses (five in a typical event) does not prove any statistically significant 'periodicity'. Also, as we showed in Section 3.2.1.A, there is evidence that, at least in some cases, the 'pulsations' actually reflect subsequent occurrence of bursts at different locations in the flare region. Thus, these results only give evidence that the microwave bursts are composed from many short-lived 'elementary bursts' similarly to the hard X-rays when observed with high time resolution (cf. Van Beek et al., 1974).

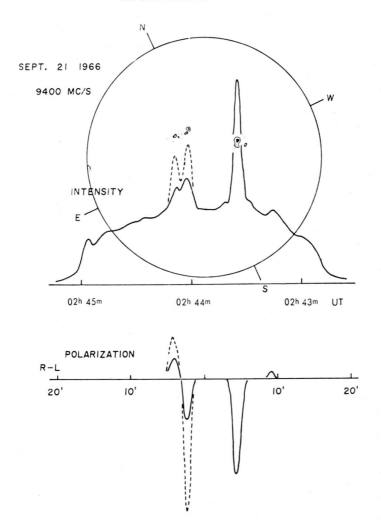

Fig. 66. Traces across the solar disc with a 1.1′ fan beam at 9.4 GHz showing flux enhancements above two active regions (solid curve), and an increased enhancement in the eastern region during a flare-associated microwave burst (dashed curve). The double structure of the microwave burst coinciding in position with the two big spots of opposite polarity can be clearly distinguished. The lower part of the figure shows the distribution of the circularly polarized component (R–L) across the disc. (After Tanaka *et al.*, 1967.)

Generally, the microwave bursts are partially circularly polarized and the sense of polarization reverses between 2 and 4 GHz (Kakinuma, 1958; Tanaka and Kakinuma, 1959; Tanaka and Enomé, 1970). In central parts of the disc one observes both senses of polarization, and when two peaks are locally separated, each of them has a different sense of circular polarization (Tanaka *et al.*, 1967; see Figure 65). When the central meridian distance exceeds some 40°, however, only one sense of polarization can be observed. At frequencies above 4 GHz its sign depends on the position of the radio source on the solar disc: the sense is different on the east and on the west side of the solar disc, as well as in the northern and the southern hemispheres. In the last solar cycle, for sources in the northern hemisphere, left-handed polarization was observed over the leading spot and right-handed over the following one, and vice versa in the southern hemisphere. The remaining polarization for regions close to the limb was the one corresponding to the leading spot in the eastern hemisphere and the opposite one in the western hemisphere (Takakura, 1967).

This polarization behavior of the microwave bursts can be understood if one assumes that all sources have a double structure and the radio emission is produced by gyro-synchrotron radiation. Then the sense of circular polarization from electrons with an isotropic distribution of pitch angles is generally that of the extraordinary mode, hence the polarization is different above sunspots with different magnetic polarity. The polarities of bipolar sunspot groups are reversed in the northern and southern hemispheres, hence the senses of polarization must be reversed in the same way. Since one can expect that the spectra of the two sources are generally different, due to differences in magnetic field and optical thickness, the spectra intersect at some frequency thus producing a change in the sense of the resulting polarization of the integrated source. However, this fairly simple explanation is open to some doubt since Tanaka and Enomé (1970) found that even the sense of polarization of each single component of a double source reverses at some frequency.

Another explanation has been given by Holt and Ramaty (1969; Ramaty and Holt, 1970) who suppose that the reversal occurs near the frequency where the microwave radio source ceases to be optically thin and becomes optically thick to its own radiation. However, as has been pointed out by Enomé (1971), the reversal frequency does not depend on the intensity of the bursts (Kakinuma, 1958) so that the optical thickness should also be independent of the intensity of the bursts. In such a case, the intensity must be determined by the source size, which leads to diameters in excess of 5' in some cases, in contradiction to interferometric measurements. Therefore, it seems that the frequency reversal must be due at least to some extent to propagation effects, which are also to be responsible for the polarization changes from the center to the limb, in a similar way as is the case with the slowly varying radio component (Cohen, 1960; Takakura, 1961).

B. *Interpretation of the Impulsive Microwave Bursts*

It is now generally believed that impulsive microwave bursts are produced by gyro-synchrotron radiation, as was first suggested by Takakura (1959, 1960). It is true that some bursts with fairly flat spectra can also be explained by bremsstrahlung (Hachenberg, 1959; Hachenberg and Wallis, 1961) from a hot sporadic condensation with $T \approx 10^7$ K and $n_0 \approx 10^{10}$ cm^{-3}. However, this process cannot explain the majority of burst spectra, in which the flux decreases with increasing frequency below 10 GHz, which should not be the case if only free-free emission were active (Takakura, 1967). Apart from this, it is also difficult to account for the time variation of the spectrum by free-free emission (Wild, 1964). Therefore, though bremsstrahlung can explain the post-burst increase (Section 3.1.6, Figure 53) and it might perhaps even contribute to weak impulsive bursts (≤ 50 sfu after Takakura, 1967), the prevailing part of the impulsive microwave bursts has to be attributed to gyro-synchrotron emission. An explanation of microwave bursts by coherent plasma emission (Bhatia and Tandon, 1970) requires unreasonably high magnetic fields, in excess of 10^4 G.

We do not intend to go into the details of the theory of synchrotron radiation. They can be found in the literature (e.g., Ginzburg and Syrovatsky, 1965; Kundu, 1965; Takakura, 1967; Krüger, 1972a). Generally we should speak about magnetic bremsstrahlung, which may be either gyro-emission if the electrons are non-relativistic, or synchrotron emission if the electrons are extremely relativistic. However, as with the X-ray production, in flares we meet with an intermediate case, when electrons are neither non-relativistic nor extremely relativistic. Most authors, therefore, speak about gyro-synchrotron radiation of the radio bursts.

Assuming the magnetic field and the energetic electrons to be homogeneously distributed throughout the emitting region, Takakura (1960, 1967) has shown that the radiation is almost entirely absorbed in the frequency range $f/f_H \leq 2$, where f_H is the gyro-frequency, and that in the frequency range $f/f_H = 2$ to 3 the extraordinary component of emission suffers selective absorption. Therefore, the radio spectrum should have a maximum at about 3 to 4 f_H, with a steeply decreasing slope toward lower frequencies. The gyro-frequency f_H depends (for electrons of mass m and electric charge e) only on the magnetic field strength H and on the electron velocity v,

$$2\pi f_H = (eH/mc)(1 - (v/c)^2)^{1/2}. \tag{93}$$

Thus the frequency of the maximum emission is determined by the magnetic field strength and the velocity distribution of the accelerated electrons, and as long as the electrons are non-relativistic or only mildly relativistic, the maximum is determined almost entirely by the magnetic field in the region where the burst is produced. The decreasing slope at higher frequencies depends predominantly on the energy distribution of the energetic electrons.

Takakura and Kai (1966) could fit the time variation of microwave spectra by using the theory of gyro-synchrotron emission under the assumption that the initial energy distribution of the accelerated electrons follows a power law. The same conclusion has been drawn from the hard X-ray bursts (cf. Section 3.2.1.C), and the power-law energy distribution is given by Equation (85). Takakura *et al.* (1968) have adopted this energy distribution and tried to find the value of δ in Equation (85) from the slope of the spectrum at very high frequencies where, for the extreme relativistic case, the flux should be nearly proportional to $f^{(1-\delta)/2}$. Then, after making the assumption that the burst decay is due solely to radiation losses (which may be true at very high frequencies) they have estimated the magnetic field strength, H, from the rate of loss. In this way they have found $\delta \simeq 3$ and $H \simeq 1000$ G. In type IVμ bursts Kai (1967) has found $\delta \simeq 2$ and H between 500 and 1000 G. An average spectrum for strong microwave bursts (22 events of the IVμ type, Figure 74) was published by Croom (1971). If one applies Takakura *et al.*'s procedure to this average spectrum in the frequency range between 40 and 70 GHz, where their assumptions should be correct, one gets $\delta = 2.4$, which is between the values found by Kai and by Takakura *et al.*

If a power law with constant δ is valid from $E = 5$ keV to 20 MeV, then 80% of radiation, for $\delta = 3$, is emitted by electrons between $E = 15$ keV and 1.5 MeV (Takakura, 1967); for $H = 1000$ G, non-thermal electrons between 50 and 500 keV emit 10 GHz radiation most efficiently (Takakura, 1969a). For the major flare of 1966, July 7, Takakura (1972) gives $H = 150$ to 700 G for a power-law spectrum with $\delta = 3$ in an environment with density of 3×10^9 cm^{-3}; Holt and Ramaty (1969) give for the same flare $H \lesssim 400$ G, $\delta = 3.5$, and density of 10^9 to 10^{10} cm^{-3}. Older determinations, without any specification of the electron energy spectrum, led to electron energies between 250 and 500 keV and magnetic fields between 700 and 2300 G (Takakura, 1972).

C. *Controversy with the Theory of Hard X-Ray Bursts*

Thus, it seems pretty well established that one can interpret both the hard X-ray and impulsive microwave bursts as being due to an electron population with a power-law energy spectrum, the X-rays being produced by bremsstrahlung from electrons with energies exceeding 10 keV, and the microwaves by gyro-synchrotron emission of electrons with slightly higher energies. However, if the absolute number of these electrons (i.e., the constant K in Equation (85)) is determined from the X-rays and from the microwave flux, respectively, we meet with a serious difficulty. In order to produce a peak flux of 100 sfu at 10 GHz, as in a typical microwave burst, one needs about 10^{29} electrons with energies in excess of 50 keV (Takakura, 1969a). In Section 3.2.1.D, however, we have seen that the lowest estimate of the total number of electrons needed for the production of a hard X-ray burst was $N_e(>20 \text{ keV}) = 3 \times 10^{34}$. With $\delta = 3$ this leads to $N_e(>50 \text{ keV}) = 5 \times 10^{33}$. Hence $N_e(\text{X-ray}) \simeq 5 \times 10^4$ N_e (radio) and estimates made for selected individual events where both the X-ray and microwave bursts have been simultaneously observed fully confirm this discrepancy (Peterson and Winckler, 1959; Holt and Cline, 1968; Takakura, 1969a). In all cases we find that 10^3 to 10^5 more electrons are needed for the production of the X-ray burst than for the observed emission on the microwave frequencies. This discrepancy may become still greater if thick-target emission (i.e., continuous acceleration) is assumed, since then the number of electrons deduced from the hard X-rays is the minimum value (cf. Section 3.2.1.E).

Several proposals have been advanced to remove this discrepancy. First of all, one may suppose that the volumes are not identical (Takakura, 1969a). The volume where the microwaves are produced may be smaller than that in which X-rays are emitted, since microwave radiation is formed only in the regions of strong magnetic field (i.e., close to sunspots), while X-rays can be produced wherever the density is high enough. However, unless the microwave sources are extremely small, the X-ray volume should then become comparable to the volume of the whole sporadic condensation, and many observations indicate that this is not the case (cf. Section 3.2.1.A). In vertical extension the microwave bursts certainly are not smaller than the hard

X-ray sources; observations from behind the limb have shown that the microwave sources extend above 2×10^4 km (Covington and Harvey, 1961; Bruzek, 1964a), and interferometric measurements near the limb lead to an altitude of $\sim 3.5 \times 10^4$ km (Tanaka and Kakinuma, 1961). Thus, in general, this explanation cannot solve the problem.

However, one can get differences in the volume of the microwave and hard X-ray sources if the magnetic field is not uniform (Takakura and Scalise, 1970; Takakura, 1972, 1973). For example, 80% of emissions at 9.4 GHz originate from a region in which magnetic field is ranging from 900 to 150 G, while it is 300 to 100 G at 3.75 GHz. Therefore, the volume of the effective source at any frequency is only 10% or less of the whole source in which non-thermal electrons are trapped. Though this fact cannot explain fully the above-mentioned discrepancy, it may diminish it by about one order of magnitude.

Another explanation may be that not only the volume, but also the electron densities in the two volumes are different so that the energetic electrons emitting hard X-ray bursts and those emitting associated microwave bursts are not the same. Takakura and Kai (1966) have proposed a flare model in which most of the energetic electrons are trapped in a region where the magnetic field is weak. The electron density in this region is fairly high, the electrons produce observable X-rays through bremsstrahlung, but the gyro-synchrotron emission produced by them cannot escape toward the observer, since it is absorbed in the assembly of thermal electrons of high density. If, for example, the electron density is 5×10^{10} cm^3, the plasma frequency $f_p \simeq$ 2000 MHz; on the other hand, if $H \simeq 100$ G, the gyro-frequency $f_H \simeq 280$ MHz; hence the maximum emission is produced below 1000 MHz, i.e. at frequencies much lower than f_p. Thus only a small fraction of the energetic electrons, those outside this magnetic trap, can contribute to the microwave burst that is observed. However, the close similarity between the X-ray and radio enhancements, often quite conspicuous even in details, seems to favor the opinion that not only the origin of the accelerated particles, but also the sources of these two bursts should be identical.

Therefore, efforts have been made to find another explanation for the differences in N_e which would preserve an identical source for both these kinds of radiation. Holt and Ramaty (1969) have emphasized that while the ambient solar atmosphere will hardly affect the production and propagation of X-rays, it may seriously modify both the generation and propagation of radio waves through a variety of processes, and this is now believed to be the main reason for the discrepancy in the N_e values deduced from the hard X-rays and the microwaves. Holt and Cline (1968) have supposed that most of the radio power from weakly relativistic electrons is in the low harmonics ($f < 3f_H$) which are heavily suppressed through gyro-resonance absorption by thermal electrons. However, at least in strong bursts, the major part of the radio emission is probably produced at higher harmonics. Therefore, Holt and Ramaty (1969) have proposed that the suppression of the radio emission is due to gyro-synchrotron self-absorption, i.e., to absorption by the non-thermal electrons themselves, which is effective at all frequencies and might make the radio source optically thick to its own radiation below the frequency of a few GHz. At higher frequencies the deficit may simply be due to the fact that Equation (85) is no longer valid for energies much in excess of 100 keV (which in fact has been deduced from the spectra of X-ray bursts, cf. Section 3.2.1.A), that is δ is equal to about 3 below 100 keV, but its value increases for energies above this limit.

In his more recent papers Takakura (1972, 1973) accepts Holt and Ramaty's explanation if it is improved by the assumption of a non-uniform magnetic field. After taking into account the gyro-synchrotron self-absorption, absorption by the ambient plasma and the Razin effect (cf. Section 3.2.4.D), Takakura succeeded in diminishing the three-orders-of-magnitude discrepancy to only a factor of 3 to 5, similarly to Holt and Ramaty's result. Nevertheless, the difference in N_e as deduced from the X-rays and from the microwaves, respectively, is still slightly bothersome, in particular when combined with the fact that the thick-target model for X-rays may give still higher values of N_e (Section 3.2.1.E), and with the troubles we have in the interpretation of the close similarity of these two kinds of bursts. One wonders whether this might perhaps be an indication that some basic item in the X-ray or microwave burst theories is wrong.

3.2.4. TYPE IV BURSTS

As we mentioned in the preceding section, the microwave bursts of Type IVμ are accompanied by long-lived enhancements at lower frequencies. Such an increase in intensity over practically the whole frequency spectrum is called the type IV burst.

Originally, this name was given to an enhancement of continuum on meter waves by Boischot (1957), who thus extended Wild and McCready's (1950) earlier classification of type I, II, and III metric bursts (cf. Sections 3.2.5 and 3.2.6). Type I denotes a 'noise storm', i.e., a disturbed period of hours or days duration which consists of a long series of bursts called 'storm bursts' superimposed on a weak background continuum. These type I enhancements are not associated with flares, but they form above particular active regions.

A type IV burst, when recorded by a radio spectrograph in the meter range, differs from type I mainly in four aspects: only continuum is present, smooth, steady and free from the individual storm bursts; its intensity is higher than in type I storms; its duration is from some 10 min to several hours; and it is associated with flares, usually large ones. However, the classification of type IV bursts by dynamic spectrographs need not be fully identical with the definition of type IV we have adopted, i.e., continuum increase throughout the whole spectrum. Sometimes there are periods during type I storms when continuum prevails over the storm bursts, and such periods may erroneously be classified as type IV. Sometimes there appear increases in the meter wave continuum which are without any response at higher frequencies, and they may even be associated with flares. It is to be emphasized that these are not type IV bursts according to the definition we have adopted. On the other hand, there may be actual type IV bursts which are not observed in meter waves, either due to the proximity of the flare to the limb, when meter waves are heavily absorbed in the solar atmosphere,

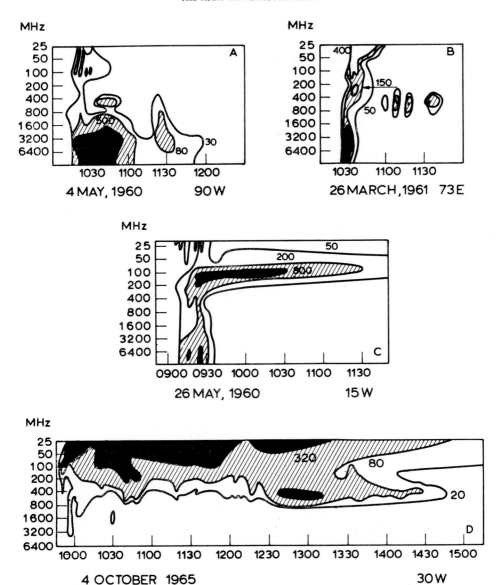

Fig. 67.   Examples of spectral diagrams of type IV bursts after Fokker *et al.* (1967). The diagrams have been constructed from single-frequency records and frequency is here plotted against time. Numbers inside the graph give relative intensities.

(A) and (B) represent bursts associated with flares close to the limb. (C) is a typical moderate type IV. (D) represents an event with a very weak microwave component. Flares (A), (C), and (D) were associated with particle events in space, but in case (D) no electrons were recorded. Flare (B) was far on the eastern hemisphere so that the particle stream might have missed the Earth.

or due to the Razin effect which cuts off the low frequencies if the ratio n_e/H is very high (Section 3.2.4.D).

Figure 67 summarizes the different cases that usually occur. The events (A) and (B) show spectral diagrams of flare-bursts close to the limb, in which the low-frequency emission is suppressed. In cases similar to (A) the dynamic spectrographs do not discover any type IV burst at all. The event (C) shows a fairly typical spectrum of a moderate type IV burst produced by a flare in central parts of the disc. Finally, event (D) shows a very strong type IV burst at meter wavelengths with a very weak micro-wave component. This case can still be classified as type IV (and, of course, it is classified as a very strong type IV in the meter range), due to the extension of emission below 1000 MHz. Nevertheless, this flare was anomalous from the point of view of particle production (cf. Section 5.3.1.B), and there are other cases when the microwave emission is completely missing.

We have devoted some time to this apparently formal problem because it is of basic importance when correlations with type IV bursts, e.g., of particle emission, are made. The high-frequency part of the type IV is of fundamental importance; when it is missing, the radio burst and the associated flare are quite flaccid from many points of view. Consequently, the statistics based only on dynamic spectra in the meter range may be misleading; many 'dynamic type IV's are not type IV at all.

Figure 68 shows the scheme of a fully developed type IV burst, after Wild (1962). It distinguishes five different burst components, and each of them can be more or less developed in individual bursts actually observed. Type IVμ has been discussed in the preceding section; the 'microwave early burst' corresponds to that part of the IVμ burst which correlates closely with hard X-rays; IVmA and IVmB are two components

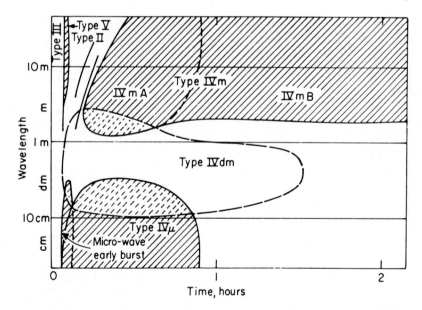

Fig. 68.   Schematic structure of a fully developed type IV burst, after Wild (1962).

of the meter type IV, of which IVmA is sometimes called 'moving type IV,' and IVmB 'stationary type IV'. Finally, IVdm is the burst component between the metric range and microwaves, and our knowledge of it is fairly poor.

A. *Type IVmA Bursts*

The moving type IV component is observed almost exclusively on frequencies below 200 MHz (Kundu and Firor, 1961; Fokker, 1963b; however, a IVmA observation at 408 MHz was made by Boischot and Clavelier, 1967), and it is not present in all type IV bursts (Weiss, 1963b). Usually it starts a few minutes after the onset of type IVμ and the onset delay time increases with decreasing frequency. A typical type IVmA burst lasts several tens of minutes and its source moves outward (not necessarily radially) with a speed of the order of 100 to 1000 km s^{-1}. The source ascends to some top height in the corona, which may amount to 2, but sometimes more than 5 or even 10 solar radii. Sometimes it ceases to be visible while still rising; in other cases it stops there or moves slowly downward with decreasing intensity (Boischot, 1958; Wild *et al.*, 1959a; Kundu and Firor, 1961; Thompson and Maxwell, 1962; Philip, 1964; J. Warwick, 1965; Malitson and Erickson, 1966; Boischot and Clavelier, 1967a,b; Wild, 1969; Kai, 1969, 1970; Smerd and Dulk, 1971; Wild and Smerd, 1972; Stewart *et al.*, 1974; Riddle *et al.*, 1974; Gergely and Kundu, 1974; McLean, 1974).

The size of the source is usually large, about 10' or more in diameter. However, in the limb event studied by Boischot *et al.* (1967) the source dimensions at 169 MHz were only 2', increasing to $\sim 7'$ only toward the end of the observation. The radiation is emitted within a wide cone of angles (unlike type IVmB, it is quasi-isotropic) and it may be weakly circularly polarized.

In the last few years excellent studies of moving type IVmA bursts have been performed with the 80 MHz Culgoora radioheliograph. Figure 69 shows the development of one of the bursts of this type which was recorded in 1968, November 22 (Wild, 1969). One can see clearly that the burst region had the form of an expanding loop or arch, with opposite polarization in the two feet, which is, on a large scale, the same effect that is observed in microwave bursts (Section 3.2.3.A). After some 20 min of development bright condensations *A* and *B* formed in the arch feet, one with left-handed and the other with right-handed polarization. A few minutes later a third, unpolarized condensation *C* formed on the top of the arch, which became the most intense component of the burst for a period of several minutes. All of these components were rising with speeds which might have been as high as 1000 km s^{-1} in the initial phase, and decayed after 01 45 UT. At about the same time two other sources, *D* and *E*, became visible, again with opposite polarization, with *E* decreasing downward and *D* appearing stationary. This stationary source then stayed in position above the flare for more than two and a half hours.

From the burst record (top part of Figure 69) one can see that the moving sources were responsible for the emission from ~ 01 00 to 01 40 UT, while the source *D* obviously produced a distinctly delayed emission from ~ 01 50 to > 04 00 UT. Such a behavior is quite typical for many type IV bursts on meter waves and, particularly

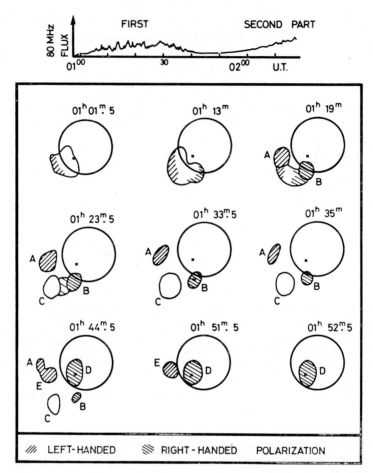

Fig. 69.   Sources of the type IV burst in 1968, November 22, as recorded by the 80 MHz radioheliograph at Culgoora, from $01^h01.5^m$ to $01^h52.5^m$ UT. A, B, and C are sources of the moving type IVmA; C was unpolarized and located near the top of the expanding magnetic arch, A and B were circularly polarized in the opposite sense and located in the arch legs. D (and possibly E) is the stationary type IVmB. Position of the associated flare is marked with a cross.

   On the top: Record of the integrated 80 MHz flux showing the 'two parts' of the type IV, corresponding to the moving and stationary sources, respectively. (After Wild, 1969.)

in the older literature, authors often speak about the first and second parts of a metric type IV burst, following the classification proposed by Pick (1961). Wild's observations demonstrate in a very illustrative way the physical origin of these two parts.

   The flare which gave rise to this burst (marked by a cross in Figure 69) started at about 00 50 UT and was classified as of importance 2. Wild believes that this flare was the source of the stationary burst *D*, but that the moving burst was associated with a prominence that was activated by the flare on the eastern limb close to the center of the expanding arch. Since similar cases have been observed several times (Wild *et al.*, 1968; Labrum and Smerd, 1968), Wild suggests that moving type IV bursts are commonly caused by eruptive prominences activated by the IVmB-associated flare. If this were true, it would imply that IVmA is present only if a nearby prominence (or filament) is available to be disturbed. If such a prominence is missing, only the IVmB type develops. However, as Křivský (1969) has pointed out, the apparent association with a prominence may just be a chance coincidence, and one can easily construct a plausible geometrical model in which the flare is the source of both the components of the metric type IV burst (and it is quite easy to do so for the event demonstrated in Figure 69). Since to explain the metric burst one needs electrons with at least the same energy as is required for the IVμ component (see below), Křivský's interpretation seems the more likely.

The two-dimensional observations with the Culgoora radioheliograph revealed that the term 'moving type IV burst' involves a spectacular variety of configurations that might signify several different physical phenomena (Wild and Smerd, 1972; McLean, 1974). The following three types have been recognized:

(a) *Magnetic arch.* An example of this is the event we described above (Figure 69). Of the 25 moving type IV sources observed at Culgoora before 1973, 6 cases might belong to this type.

(b) *Advancing front.* This type of burst, first described by Kai (1970), also has an arch shape, but it differs from the 'magnetic arch' in three aspects: it seems to be closely related to a type II burst; it has a large extent perpendicular to the direction of motion; and it is only weakly polarized. One can suppose that we observe in this case an advancing shock front which first produced a type II burst. Three or four bursts observed at Culgoora might be of this type.

(c) *Isolated source.* This term was introduced by Smerd and Dulk (1971) for a uniform radial motion of a 'blob' source to heights of many solar radii, and it also covers many cases when the source is observed to break up into a number of sources moving with similar velocities. Most of the moving type IV bursts belong to this category (e.g., Sheridan, 1970; Riddle, 1970; Dulk and Altschuler, 1971) which sometimes is also called 'ejected plasma blob'. An example is shown in Figure 70. As the 'blob' moves, the degree of polarization is mostly increasing to very high values of 80 to 90%. This type of moving type IV usually continues to move without any obvious deceleration until it fades.

There are also some special events that cannot be easily identified with any of these categories. One example is the burst observed by Boischot and Clavelier (1967a, b), mentioned earlier, which had very small dimensions and sharp cuts-off on both the high and low frequency edges of the spectrum. Another case was described by Riddle and Sheridan (1971) who observed a row of bright patches in a moving type IV burst, forming a jet.

The moving type IV bursts have been ascribed to gyro-synchrotron emission from energetic electrons trapped in a magnetic cloud ejected from the flare region. Boischot and Denisse (1957) were the first to suggest that synchrotron radiation is the source cf the type IV burst emission; assuming a magnetic field of 1 G, they showed that 8×10^{32} monoenergetic electrons with energy of 3 MeV were required to account for

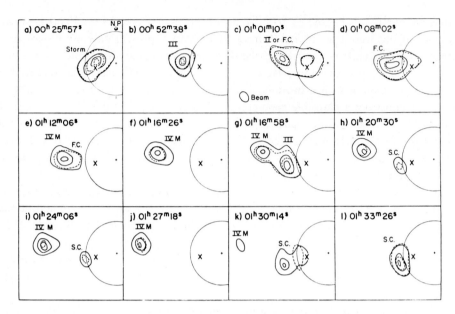

Fig. 70. Contours of radio brightness from the Culgoora radioheliograms for the event of 1970, April 29, demonstrating a 'blob-type' moving type IV burst. The flare position is shown by a cross. Dashed and full lines refer to the left-handed and right-handed circular polarization, respectively. (After Dulk and Altschuler, 1971.)

the observed intensity at 169 MHz. In a more recent study (cf. Section 3.2.4.D) Boischot and Clavelier (1967a, b) have found 3000 monoenergetic 3 MeV electrons cm^{-3} in a field of 0.5 G. However, with our present knowledge it is essentially impossible to determine uniquely both the energy of the electrons and the magnetic field strength. If one assumes that fairly strong magnetic fields frozen in the magnetic cloud can be carried up to significant heights above the solar surface, one can also fit the observations with electrons of much lower energy, e.g., 50 to 500 keV, assuming a power-law distribution with $\delta = 4$ and $H = 20$ G (Takakura, 1969b). According to Dulk (1973) this might indeed be the case, because only $\gtrsim 10$ G fields carried out within the source can explain the high degree of circular polarization observed in many moving type IV bursts.

The great variety of the phenomena that are called 'moving type IV bursts' indicates how difficult it may be to explain this part of the type IV emission. Let us take Wild's (1969) observation (Figure 69) as an example. If the moving burst is associated with the active prominence, as Wild suggests, sources A and B can hardly be of synchrotron origin, since their radiation corresponds then to the ordinary mode (with respect to the magnetic polarities near the prominence). On the other hand, if the burst is associated with the flare itself, the radiation appears to be the extraordinary mode, in agreement with the theory of synchrotron radiation. In that case we probably observe in A and B the radiation from electrons confined in the narrowing feet of the arch, but we can see only that part of them which is above the 80 MHz plasma level. An alternative explanation (Wild, 1969) is that the emission is due to radiation from plasma oscillations set up by electron streams travelling along the arch, in the two regions where the magnetic arch intersects the 80 MHz plasma level. The seemingly radial ascent may be attributed to tangential motion due to the arch expansion so that only the geometry changes the height. On the other hand, there seems to be little doubt that the unpolarized source C was due to synchrotron radiation. Its location near the top of the arch may possibly be due to the greater emissivity of synchrotron radiation at greater heights where the refractive index approaches unity (Boischot and Clavelier, 1967a, b; Wild, 1969; cf. Section 3.2.4.D).

Interferometric observations of the moving type IV bursts have opened the possibility of a comparison of radio and optical measurements for events observed close to the limb. In 1972, August 12, Riddle *et al.* (1974) observed an event occurring $\sim 17°$ behind the western limb in the region McMath 11 976, which produced the well-known series of big August 1972 flares. Figure 71 illustrates the observations: first, there was a surge (crosses) which rose to 1.3 solar radii at 20^h41^m and returned afterward. At 20^h55^m a spray became visible (dots) in the form of a twisted flux loop, speeding up to 510 ± 60 km s^{-1}. Starting at 20^h52^m a moving type IV burst was observed, propagating up to 5.5 solar radii (circles). As one can see in Figure 71, the radio material was moving well ahead of the Hα material, probably representing a shock front emitting synchrotron emission.

A similar set of data was obtained also for two possibly homologous flares in 1973, January 11, near the western limb by Stewart *et al.* (1974). It appears that again a

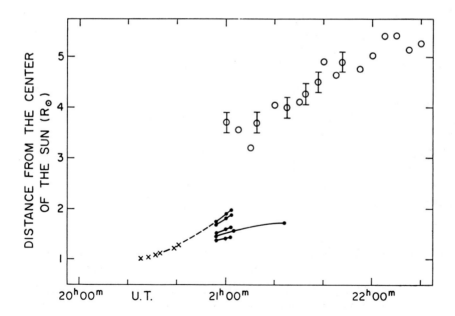

Fig. 71.   Outward movement of the coronal disturbance of 1972, August 12. Crosses refer to the surge, dots to the spray, and circles to the moving type IV radio burst. (After Riddle *et al.*, 1974.)

common disturbance initiated a spray, a moving type IV burst and density change in the K-corona. All these disturbances travelled with projected speeds between 300 to 600 km s^{-1}, behind a shock moving with higher velocity in front of them. Stewart *et al.* also estimated the energies contained in the moving type IV burst. While $\sim 10^{27}$ erg in mildly relativistic particles can account for the radio emission, the energy in the magnetic field of the plasmoid may have been 10^{30} to 10^{31} erg.

Gergely and Kundu (1974) observed four moving type IV bursts with a swept-frequency interferometer and found that all four bursts were associated with depletions in the electron content of the white-light corona. The homology that is sometimes observed in the occurrence of coronal transients (also Stewart *et al.*'s events were probably homologous) indicates that the regeneration of magnetic configurations in the corona should take place over a time scale of several hours.

Robinson and MacQueen (1975) have described a moving type IV burst in 1973, September 14–15, associated with a mass ejection event observed with the white-light coronagraph on Skylab. The radio emission occurred in a region occupied by an expanding loop system seen in white light. According to Gosling and Dulk (1975) at each height the radio emission began and peaked before the arrival of the most densely populated loops at that height. The radio emission continued, but was considerably diminished, during the passage of the dense loops, and migrated toward the foot-points of the outwardly expanding arch.

All these observations show that moving type IV bursts are associated with ejections of matter from the Sun, which may manifest themselves as sprays in Hα light and expanding loops, arches, or bottles in the white-light corona. The IVmA radio emission always precedes the moving optical source; as a rule, however, in addition a stationary type IV continuum source (IVmB, described in the next section) appears at lower altitudes during the time when the moving type IV burst decays (Gergely and Kundu, 1974).

However, the association of moving type IV bursts with optically visible ejections, well documented in a few individual cases, is not valid generally. Among 38 coronal transients, observed with the white-light coronagraph on Skylab, only 5 were associated with a type IV burst (Gosling *et al.*, 1974). In some cases, when the source was far in the invisible hemisphere, the radio emission might have been absorbed in the overlying corona. However, there obviously still remain many coronal transients close to the limb and on the visible hemisphere that are not accompanied by any radio emission of type IV or type II.

B. *Type IVmB Bursts*

The stationary type IV burst (for example, source D in Figure 69) may follow type IVmA, but it also may occur without any preceding moving burst. Its duration is usually one or several hours, and if the burst lasts for a long time it often transforms to a 'continuum storm' after one or two hours of its duration. Sometimes this continuum storm may further develop into an ordinary type I noise storm, with a duration of many hours or even days (Pick, 1961; Boischot and Pick, 1962). The size of the

IVmB source is smaller than that of type IVmA, usually 3′–5′, and its emission is directive. Due to this, the IVmB component is missing in flares that occur very close to the solar limb (cf. Figure 67). For example, in the set of 174 type IV bursts listed by Švestka and Olmr (1966) for the period 1956–1963, the percentage of type IV bursts in which the IVmB component was observed was 77% for flares within $\simeq 30°$ distance from central meridian, 57% between 30° and 60°, and only 30% for distances exceeding 60° (this percentage may actually be still lower, since certainly many limb type IV bursts are not classified as type IV at all due to their very short duration).

The radio source of IVmB appears to be fixed close to the critical plasma layers at given frequencies, and it is usually strongly circularly polarized, according to Weiss (1963b) and Kai (1965a) in the ordinary mode. This makes it difficult to interpret these bursts by gyro-synchrotron emission, and therefore plasma waves have been proposed as the source of their origin (Denisse, 1960; Kai, 1965b). It is supposed that these plasma waves are excited by energetic electrons accelerated earlier during the flare flash phase, which are trapped in the corona and slowly diffuse downward. However, according to Takakura (1962, 1967), the gyro-synchrotron mechanism cannot be completely discarded, since the mode of polarization might be established erroneously, due to a polarization sense reversal during the propagation, or due to the assignment of the radio source to a wrong magnetic polarity. In Figure 69, for example, the polarization may be extraordinary mode, if the sources E and D are considered coupled.

C. *Type IVdm Bursts and Radio Pulsations*

At decimeter wavelengths the continuum enhancement is usually characterized by large variability (cf. Figure 67a, b, d) and complexity. It is often difficult to decide whether we actually meet with a distinct burst component (as seems to be the case in Figure 67b), or whether the dm part is only an extension of the IVm (Figure 67d) or IVμ (Figure 67a, c) bursts. Also, the duration is very variable, from 10 min up to a few hours. In events with great variability the individual impulsive enhancements are strongly circularly polarized as in the ordinary mode.

At the low decimetric and upper metric frequencies series of peculiar and often pulsating fine structures are observed in some type IV bursts (Young *et al.*, 1961; Abrami, 1970; De Groot, 1970; Rosenberg, 1970; Slottje, 1972; Gotwols, 1972; Wild, 1973; Achong, 1974; Maxwell and Rinehart, 1974). Examples of these pulsations are shown in Figure 72. Their periods are usually between 1 and 3 s and there is some indication of increase of the period with decreasing frequency (Achong, 1974). The pulsations occur most often between 100 and 300 MHz, simultaneously over a frequency range typically of 2:1 (Wild, 1973). However, pulsating emission was reported also above 700 MHz (Young *et al.*, 1961; Gotwols, 1972) and below 28 MHz (Achong, 1974). The pulsations characteristically occur during the late part of a type II burst preceding the type IV.

The origin of the pulsations has been attributed by Rosenberg (1970) to synchrotron radiation from electrons trapped in a flux tube undergoing radial oscillations from

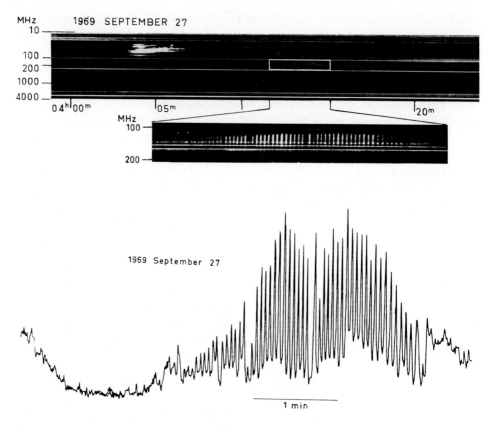

Fig. 72.   Radio pulsations recorded during the type IV burst in 1969, September 27. These pulsations, of periodicity ~3 s, were visible between 100 and 200 MHz for about 3 min. A type II burst set in about 10 min before the pulsations. The record below traces the radio pulsations at 150 MHz. (After Wild, 1973.)

standing MHD waves. Rosenberg's mechanism was incorporated in a model proposed by McLean *et al.* (1971) to account for the association with the type II bursts. The model is indicated in Figure 73. An ascending MHD shock front (which also produces the type II burst) intercepts a discrete flux tube that loops high in the corona, compresses the magnetic field, and generates large amplitude Alfvén wavetrains which then travel ahead of the shock, around the top part of the loop in opposite directions. As we shall see in Section 6.6.2 (Figure 128) this is a configuration suitable to Fermi acceleration of electrons and protons. When the shock reaches the top of the loop its front becomes essentially aligned with the flux tube over a length which is large compared to the tube diameter. Thus the tube is excited into radial oscillations as Rosenberg proposed it.

This model is very attractive. As we shall see in Sections 3.2.5 and 5.1.1.A (Figure 92), type II bursts appear to be associated with the second-step acceleration of energetic particles in flares, and this second-step acceleration may be due to the Fermi process. Thus McLean *et al.* succeed in explaining in one simple model two important observed properties of type II bursts: how they produce the radio pulsations and how they accelerate particles. Nevertheless, as we shall see in Section 6.6, there are limitations in this and other Fermi-like models that are difficult to overcome.

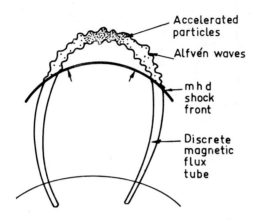

Fig. 73.   McLean *et al.*'s (1971) model of the radio pulsations; the pulsations occur when a type II burst propagating upward excites oscillations in a closed magnetic loop structure.

D. *Type IV Burst Spectrum*

If one plots the peak flux versus frequency, one obtains radio spectra of type IV bursts, examples of which are shown in Figure 74. Characteristically there is high intensity in the meter range, a minimum at 500 to 4000 MHz, a peak near 10 GHz, and a decline to higher frequencies. Such spectra were first constructed by Castelli *et al.* (1967) who have shown that this 'U-type' shape of the spectrum (i.e., U-shaped below ~ 5 GHz) is a characteristic feature of all flares producing strong proton fluxes in space. Since, however, a well-developed type IV burst is a typical characteristic of all 'proton flares' (cf. Section 5.1.1.A), this spectrum shape is in fact characteristic for all strong and well-developed type IV bursts. Later on Croom (1971) extended the spectrum to higher frequencies and arrived at the 'mean spectrum' shown at the bottom of Figure 74.

Spectra of this shape are constructed from the fluxes recorded at the earliest maximum of the radio burst, and it is important to note that the maxima need not necessarily be simultaneous at all frequencies. The first peak usually coincides with, or is very close to, the hard X-ray burst maximum. Later in the burst development secondary maxima that are even higher than the primary one can occur at some frequencies

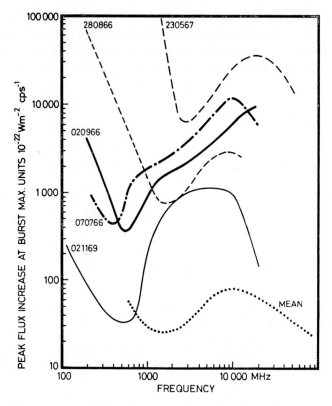

Fig. 74.   Typical spectra of type IV bursts constructed from single-frequency observations by plotting the maximum flux versus frequency. (After Castelli and Guidice, 1971, and Croom, 1971). The dotted curve shows the 'mean' spectrum as obtained by Croom from an analysis of 22 strong type IV bursts (all accompanied by protons in space). This curve gives intensity relative to intensity at spectral peak, hence the absolute flux value is arbitrary.

(cf. Section 3.2.3.A and Figure 63). The spectrum at these maxima looks entirely different (Figure 75), and these maxima are not associated with any increase in the X-ray emission. Thus, obviously, the secondary flux increase does not reflect any new acceleration process, but only modulations in the radio flux production by the magnetically trapped and stored electrons, for reasons which we do not yet know. Yurovsky and Babin (1970) found a time correlation between a secondary 200 MHz flux enhancement and high-speed ejecta from a limb flare; Anastassiadis *et al.* (1964) and McKenna (1965) found a correlation between the secondary maxima and the covering of sunspot umbrae by the Hα flare. But these may be chance coincidences.

Figure 75 shows the type IV burst spectrum at three successive maxima during the 1966, July 7 event. The first spectrum is identical with that shown in Figure 52. The other two spectra are different, showing a maximum instead of a minimum in the decimeter range. It is of interest to notice that spectra similar to the second and third ones in Figure 75 have also been found by Castelli *et al.* (1967) in the earliest maxima of bursts associated with flares that did not produce protons in space. Such bursts

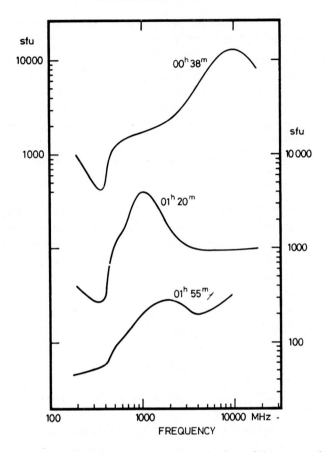

Fig. 75. Spectrum of the 1966, July 7, type IV burst at the time of three successive flare maxima. (After Enomé, 1969.)

should be considered as type IV's according to our definition, since their emission is extended from centimeter to meter wavelengths, but their intensity is generally very low: some 100 sfu in comparison to 1000 sfu or more in 'typical' type IV bursts.

In Section 3.2.3.B we discussed the microwave part of the spectrum; the peak frequency is mainly determined by the magnetic field strength in the region where the IVμ burst is produced, and the decline toward increasing frequency depends on the energy distribution of the accelerated electrons. The flux decrease toward lower frequencies can be explained by absorption of the radio emission by ambient electrons at the gyro-frequency and the low harmonics (Takakura, 1960), and by self-absorption (Ramaty, 1969). The decimeter part of the spectrum is not well understood, since in fact we are not sure how the radiation is formed. In the meter range, where with the IVmA bursts we again meet with synchrotron radiation, the interpretation is seriously complicated by effects of the coronal plasma upon the emission and propagation of the radio waves. In addition to the influence of the ionized medium on the propagation (waves with frequency lower than the plasma frequency, $f_p = 9 \times 10^3 \, n_e^{1/2}$, cannot propagate), plasma also influences the generation of the synchrotron emission since it has a frequency-dependent index of refraction. The index of refraction is usually less than unity. In consequence, the phase velocity of light exceeds c at frequencies of the order of magnitude of f_p, which results in the suppression of synchrotron emission at these wavelengths (Ramaty and Lingenfelter, 1967; Boischot and Clavelier, 1967; Ramaty, 1968). This suppression which causes a low-frequency cut-off in the synchrotron radiation is often called the Razin effect (Razin, 1960; Ginzburg and Syrovatsky, 1965; M. Simon, 1969).

The Razin effect, for the relativistic case, becomes important for frequencies below

$$f_R \simeq 20 n_e / H_\perp, \tag{94}$$

where n_e is the density of ambient electrons and $H_\perp$ the magnetic-field component perpendicular to the line of sight in the emitting region. One can expect that the maximum in the meter range should be close to $f \simeq f_R$. The Razin frequency f_R equals 200 MHz for $H_\perp = 10$ G and $n_e = 10^8$ cm^{-3}, or for $H_\perp = 1$ G and $n_e = 10^7$ cm^{-3}, hence synchrotron emission in type IVm bursts might be suppressed at frequencies near to and below a few hundred MHz. Boischot and Clavelier (1967) have studied a distinct rising source associated with the limb flare of 1966, September 14, and under the assumption that the low-frequency cut-off observed was due to the Razin effect, they could interpret the observation as emission from 3 MeV electrons in a region with an ambient density $n_e = 10^7$ cm^{-3} and a magnetic field of 0.5 G. Unfortunately, this event was fairly atypical: as we mentioned in Section 3.2.4.A, the size of the moving source was small (2′) and it emitted at 408 MHz. These are not common characteristics of type IVmA bursts.

One can generally say that if frozen-in magnetic field of a significant strength (> 10 G) is carried up in the magnetic arch into the corona, the Razin effect should not be important; if, on the other hand, the magnetic field in top parts of the magnetic arch is weak, the Razin effect may plan an important role. Eventually, the Razin effect might also explain the triple structure of the IVmA burst in Figure 69. In regions A and B the magnetic field is still strong so that 80 MHz are emitted; in higher layers H decreases faster than n_e, hence 80 MHz emission is suppressed; only in the top part, where n_e is also very small, 80 MHz again becomes visible. Generally (Dulk, 1973), in the moving type IV bursts the Razin effect should not be important in isolated sources, where rather high magnetic field must be carried out in order to explain the high degree of circular polarization. But it may be important in advancing fronts, where synchrotron radiation presumably comes from energetic particles accelerated in a shock front with lower magnetic field strength.

In the case where the gyro-frequency f_H is less than about one-third of the plasma frequency f_p, the Razin effect can influence even the microwaves (Holt and Ramaty, 1969). Since

$$f_p = e(\pi m)^{-1/2} n_e^{1/2}, \tag{95}$$

while f_H is given by Equation (93), the Razin effect becomes important for microwaves as soon as

$$H \lesssim \tfrac{2}{3} c(\pi m n_e)^{1/2} = 1.1 \times 10^{-3} \, n_e^{1/2} \,. \tag{96}$$

Thus, for example, the Razin effect should be taken into account when $H \lesssim 35$ G at plasma density 10^9 cm^{-3}, or $\lesssim 350$ G at 10^{11} cm^{-3}.

3.2.5. TYPE II BURSTS

Payne-Scott *et al.* (1947) were the first to detect fairly intense radio bursts on meter waves, with duration of a few minutes, which became progressively delayed as the frequency was decreased. When, a few years later, the technique of dynamic radio spectrography was introduced by Wild and McCready (1950), a number of similar slowly drifting bursts were recorded, and they have been designated as type II metric radio bursts.

A typical example of a type II burst is shown in Figure 76. It is characterized by an abrupt cut-off in frequency; above the cut-off no radiation is detected, while below the cut-off the radio event starts with a fairly high intensity. It usually has the form of two relatively narrow bands (of several MHz width) with a frequency ratio of approximately 2:1, one being the fundamental band and the other (present in about 60% of cases) the second harmonic. The starting cut-off frequency of the fundamental is commonly close to or below 100 MHz, but in some events the burst sets in higher, up to 250 MHz (Maxwell and Thompson, 1962; Kundu, 1965). The bands drift to lower frequencies at a rate of some 100 kHz up to 1 MHz s^{-1}, and become invisible between 20 and 30 MHz, which corresponds to heights of 2 to 3 solar radii (Takakura, 1969b). The total duration of a type II burst is typically between 5 and 10 min.

Warwick (reference in Dulk *et al.*, 1971) observed several type II bursts drifting down to 10 MHz and one can suppose that many more bursts drift below 10 MHz but cannot be observed because of ionospherically reflected interference. At these lower frequencies type II bursts could be detected by spacecraft radio receivers. So far only two such events have been recorded: in 1971, June 30 and in 1972, August 7–8 by Malitson *et al.* (1973, 1974; also in Stone and Fainberg, 1973; Fainberg and Stone, 1974). The latter burst, for which more convincing data are available, originated in a flare at 15^h on August 7; it was recorded at 155 kHz at 2^h on August 8 (at a distance of $\sim 60 \, R_\odot$ from the Sun) and could be followed down to 55 kHz at 8^h on August 8 (at a distance of ~ 130 solar radii); the shock travelling with an average speed of 1270 km s^{-1} arrived at the Earth in the evening hours on August 8 causing a sudden commencement at $23^h 54^m$.

Palmer (1974) has tried to relate the first type II-like burst observed in space on 30 June 1971 to a type II burst recorded on the Sun. The association, however, is doubtful since the travel time appears to be too long. Palmer explains this discrepancy as due to the fact that the shock wave interacted with a streamer and only then became visible on the low-frequency radio waves.

There are also a few indirect measurements indicating continuing propagation of the type II burst-producing shock through interplanetary space, after it has ceased

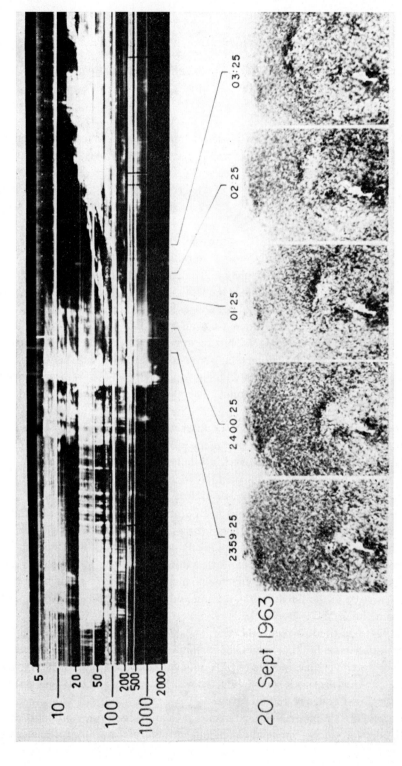

Fig. 76. Example of a type II burst as recorded by a dynamic radio spectrograph (onset at 23ʰ59ᵐ, end toward 24ʰ05ᵐ UT). A series of type III bursts precedes the type II event. The bottom part shows the flare-associated travelling coronal disturbance as a dark marking propagating upward on successive photographs. It seems to be well correlated in time with the type II burst occurrence. (After Wild, 1968.)

to be visible on radio waves in the corona. Perhaps the most convincing of these has been presented by Gosling *et al.* (1975) for the type II burst that was observed at $12^h 20^m$ in 1973, September 7. Forty-seven minutes later (when the next photograph was taken) the white-light coronagraph on Skylab showed a disturbance moving with a speed of 960–1300 km s^{-1} through the corona, higher than the speed of most other ejections detected by the Skylab coronagraph; two days later the plasma detector on Pioneer 9 detected this disturbance in the solar wind near 1 AU. A similar event has been reported by Brueckner (1974).

If one assumes that the emission occurs near the plasma frequency, i.e., near the level where the coronal electron density reduces the refractive index to zero (an idea put forward by Martyn (1947) and confirmed by Wild (1950; Wild *et al.*, 1953)), the frequency drift corresponds to ascending velocities between 500 and 5000 km s^{-1} with the most common values between 1000 and 1500 km s^{-1} (Maxwell *et al.*, 1964). Sometimes there appears to be an indication of deceleration (Maxwell and Thompson, 1962) or acceleration (Weiss, 1965) with height; however, both the deduced velocity and its variations depend strongly upon the coronal density model adopted.

The motion also need not be strictly radial. A significant fraction of type II bursts seems to show evidence for tangential motions in the solar atmosphere with speeds of 1000 to 2000 km s^{-1} (Weiss, 1963a). The spectra of these presumably tangential bursts are characterized by little or no drift in frequency with time, and by broader bandwidth. Sometimes multiple type II bursts are observed, composed of bands with different drift rates.

Usually the drifting bands are not smooth and continuous but fluctuate in intensity over periods of seconds. In some cases the bands consist of a series of short-duration, narrow-band bursts, which themselves show the type II frequency drift, while some other type II bursts exhibit the so-called herring-bone structure (Roberts, 1959), characterized by a rapid succession of short-lived, broad-band elements with fast upward and downward frequency drifts. The peak intensities of type II bursts commonly exceed 1000 sfu and the bursts appear to be unpolarized (Komesaroff, 1958; Akabane and Cohen, 1961); the only exception has been found for the herring-bone structure elements, which at times exhibit strong circular polarization (Stewart, 1966).

The angular dimensions of type II burst sources are quite large. According to Weiss and Sheridan (1962) the 'core' of the source typically has a half-power diameter of about 6', but this core is surrounded by a 'halo' with a diameter of about 40'. Radio-heliograms of type II bursts made at Culgoora (Kai and McLean, 1968; Kai, 1970) confirm that the size measured between 1/8 peak-intensity contours is of the order of that of the optical disc. This shows that the emission cone of the shock wave may be very wide, in extreme cases up to $\sim 180°$ (Wild and Smerd, 1972). Within this cone, however, the burst intensity may show several peaks (according to Uchida (1974) in preferable directions where the Alfvén velocity is low).

More than 90% of type II bursts are clearly associated with flares (Swarup *et al.*, 1960; Maxwell and Thompson, 1962). The missing $< 10\%$ can be ascribed to flares behind the limb so that probably all type II bursts are flare-associated. On the other hand, only a very small fraction of flares are accompanied by type II bursts, the associated flares being mostly of large or medium size. The percentage association increases with the flare importance and amounts to about 30% for flares of importance 3. As we shall see later, essentially all flares accompanied by type II bursts are sources of protons in space (Section 5.1.1.A).

As we mentioned when describing the IVmA bursts in Section 3.2.4.A, some type II bursts appear to be associated with fast ejections of material from flares. One example is the above-mentioned event of 1973, September 7, described by Gosling et al. (1975). Another case was observed by Stewart et al. (1974) in 1973, January 11. In the latter case the type II moved with speeds of 800 to 1200 km s^{-1} in front of a white-light cloud and Hα spray moving with velocities of 300 to 600 km s^{-1}. J. Warwick (1965) has shown that in several cases the time interval between the start of a flare-associated prominence on the limb and the appearance of a type II burst at a given frequency was in accordance with the ascending speed of the prominence.

On the other hand, there is essentially no association between type II bursts and surge prominences (Swarup et al., 1960), and the really effective ejections are probably events with supersonic speeds (Giovanelli and Roberts, 1958), either sprays or more compact ascending arches (J. Warwick, 1965; Munro, 1975). However, as E. Smith (1968) has pointed out, there are also many limb events of similar characteristics which occur without any associated type II burst. This has been confirmed by Skylab observations, as we mentioned in Section 3.2.4.A: out of 38 coronal mass ejections observed with the white-light coronagraph on Skylab, only 12 were associated with type II and/or type IVmA bursts (Gosling et al., 1974).

Type II bursts appear to be associated with emission or absorption clouds in Hα light or with expanding arc features originating in the flare and travelling paths of 200 000 km or more with lateral velocities between 500 and 2500 km s^{-1} (Athay and Moreton, 1961; Moreton, 1964; Harvey et al., 1974; see Section 4.3). The time association between the blast wave travelling laterally on the Sun and the type II burst travelling upward has been found to be extremely good in some cases (Dodson and Hedeman, 1968a; Wild, 1968; Harvey et al., 1974; Figure 76).

The type II burst often occurs in association with a type IV burst. According to Maxwell and Thompson (1962) about 30% of all type II bursts are associated with type IV, and the reverse association is still stronger: In Švestka and Olmr's (1966) catalogue of type IV bursts, 41% of type IV's were clearly associated with type II, in another 40% this association was possible (spectral data were missing), and only 19% of type IV's were clearly without any accompanying type II event. Two-dimensional observations of combined type II + IV bursts by the Culgoora radioheliograph described by Kai (1969, 1970), Smerd (1970), Stewart and Sheridan (1970), and Dulk and Altschuler (1971) indicate that both type II and type IVmA bursts, when observed together, are generated by a common disturbance. The reader is referred to Section 3.2.4.C for a discussion of the possible interference of the type II burst with the type IV source leading to pulsations in the type IV continuum (Figures 72 and 73).

Nevertheless, statistical comparisons of the occurrence of type II + type IV bursts

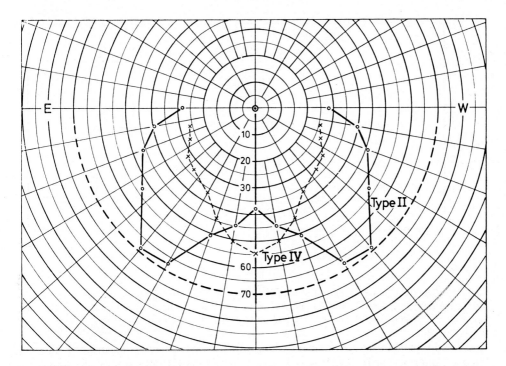

Fig. 77. Polar diagrams of the type II and type IV occurrence. Both curves have been smoothed and the type IV curve (corresponding to smaller number of cases) has been multiplied by factor 2. The vertical line points to the Earth. (After Švestka and Fritzová, 1974.)

are influenced by the difference in directivity in these two types of radiation. As one can see from Figure 77, the number of observed type IV bursts decreases much faster toward the limb than that of type II bursts. (The decrease in the number of type II bursts near central meridian, if real, remains unexplained.) Thus the number of type II + IV bursts may be underestimated, in particular close to the limb. Since type II bursts manifest an acceleration process giving rise to relativistic electrons (see below and Section 5.1.1.A) and type IV bursts manifest a storage of mildly relativistic electrons in the solar corona, their common occurrence is to be expected as a rule.

Typically, the type II burst begins near the maximum of the associated flare which implies that it originates before the Hα maximum during the flash phase. Figure 78 shows the distribution of time intervals between the peak of the impulsive microwave burst (generally coincident in time with the peak of the hard X-ray burst) and the observed onset of a type II burst on metric waves (Švestka and Fritzová-Švestková, 1974). The median time interval is 2.0 min. This corresponds to a shock wave speed of about 1500 km s^{-1} if we assume that the shock originated at the time of microwave burst maximum. Thus Figure 78 confirms that the type II burst originates during the time of the non-thermal acceleration process that gives rise to the hard X-ray and impulsive microwave burst. This is also confirmed in individually studied cases (e.g., Sakurai, 1971). On lower frequencies, of course, the delay between the microwave

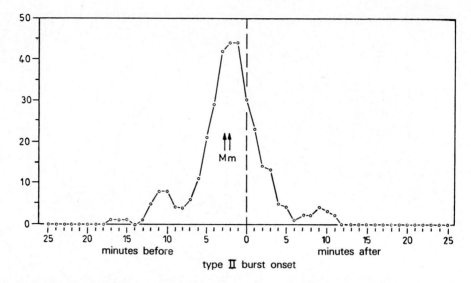

Fig. 78.   The frequency distribution of the occurrence of the microwave burst maximum in relation to the metric type II burst onset. On an average, the onset of the metric type II is delayed by 2 min relative to the microwave burst. (After Švestka and Fritzová, 1974.)

peak and type II onset is significantly larger. What appears to be important is the fact that among 166 type II bursts studied Švestka and Fritzová did not find any single case when type II would become visible before the onset of the microwave burst.

As we shall discuss more in detail in Section 5.1.1.A (Figure 92) Švestka and Fritzová have presented evidence that the type II bursts are intimately associated with the second-step acceleration of particles to energies above ∼ 10 MeV. When assuming that the impulsive hard X-ray and microwave bursts are the manifestations of the first acceleration step (to energies ≲ 1 MeV), the short-time interval in Figure 78 implies that the secondary acceleration (when present) follows immediately upon the primary one. The fact that in some 20% of cases the type II is observed even prior to the micro-wave-burst maximum indicates that the two steps can be accomplished simultaneously: while some particles are still being pre-accelerated, other particles, pre-accelerated earlier, already enter into the second acceleration step. This conforms to the fact that the white-light flare emission (Section 2.2.11.B) and γ-rays from flares (Section 5.4.3) are observed simultaneously with the hard X-ray and impulsive microwave bursts, though they need ≳ 30 MeV protons for their production.

This time schedule is to be changed, however, if we adopt McLean et al.'s (1971) point of view that the second step acceleration is accomplished when the type IV pulsations are observed (Section 3.2.4.C; Figure 73); the pulsations usually occur somewhat later when the type II burst is seen at ∼ 15 to 30 MHz (Wild, 1973) and when it reaches an altitude of about 500 000 km above the photosphere (Maxwell and Rinehart, 1974).

As we shall see in the next section, fast drift bursts (type III) are explained as plasma waves excited by a stream of fast electrons. This explanation cannot be applied in the case of the type II bursts, since their velocity is in fact less than the mean speed of the thermal electrons ($\simeq 7000$ km s^{-1} at $T \simeq 10^6$ K). However, the speed is several times greater than both the sonic and the Alfvén velocity in the corona, and therefore the disturbance can be attributed to a shock wave. The very large dimensions of the type II burst source indicate a quasi-spherical shock expanding outward from the flare and exciting plasma oscillations of a given frequency from regions where the shock front crosses the appropriate plasma level (McLean, 1967; Wild, 1968). This explanation is also supported by the good time correlation of type II bursts with the blast waves observed in projection on the solar disc and with ascending arches and sprays on the limb.

However, the shock itself cannot directly excite plasma waves because its velocity is smaller than that of the thermal electrons. Therefore, one has to assume that electrons are accelerated to much higher energies in front of or behind the shock wave. Several different mechanisms have been proposed: Westfold (1957) and Welly (1963) have suggested a hydrodynamic shock wave, where the accelerated electrons are the thermal run-away electrons ejected from high-temperature plasma behind the shock; in that case the shock must be quite strong, with a Mach number ($=$ shock velocity/ sound velocity) $M > 6.5$ and its maintenance would need an energy of at least 10^{32} erg (Takakura, 1966), which is equal to the total energy release in the largest flares. On the other hand, Uchida (1960), Sagdeev (1962), and Pikelner and Ginzburg (1963) have proposed a hydromagnetic shock wave, with fast-drifting electrons in the hydromagnetic shock front as the source of plasma waves. Finally, Tidman and Dupree (1965; Tidman, 1965) have suggested a turbulent collisionless bow shock wave analogous to that of the Earth, behind which suprathermal electrons originate in a turbulent region with a disordered magnetic field.

Uchida's hypothesis of collisionless hydromagnetic shock has been the most widely accepted one. At a shock front propagating across the magnetic field, electrons drift perpendicularly to both the magnetic field and the shock front. The drift velocity is proportional to $\Delta H / \Delta l$, where both the thickness Δl of the shock front and the increment of the magnetic field ΔH at the shock front are functions of the Mach number M_A (Shock velocity/Alfvén velocity). According to Sagdeev (1962) the drift velocity exceeds the thermal one as soon as the Mach number is greater than

$$M_{Ac} \approx 1 + \tfrac{3}{8}\left[nkT \left/ \frac{H^2}{8\pi} \right]^{1/3}, \right. \tag{97}$$

making the assumptions that the shock is weak and $H^2/8\pi \gg nkT$. Condition (97) may determine the height range in the corona where the type II burst becomes visible (Takakura, 1966). One can suppose that the accelerated electrons give rise to plasma waves and thus to the type II emission, and in the case that small streams of electrons consecutively originate on or near the front, miniature type III-like bursts can repeatedly take place at intervals of a few seconds, giving rise to the 'herring-bone structure' sometimes observed (Sheridan et al., 1959; Wild, 1968). As to the theory of plasma waves and their coupling with the electromagnetic waves observed, the reader is best referred to the review paper by Takakura (1967) or to Kundu's (1965) book.

However, one has to admit that the interpretation of the type II disturbance still encounters some difficulties. From a comparison of the source position of type II's with the magnetic field line configuration, Dulk *et al.* (1971) have inclined toward the opinion that type II sources propagate parallel to magnetic field lines in the corona rather than perpendicular to them. They consider a compressional slow-mode gas-dynamic shock propagating along the open magnetic field lines (also discussed by Kopp (1972)) as the simplest explanation. On the other hand, Uchida *et al.* (1973), who have carried out detailed computations of the propagation of magnetohydro-dynamic fast-mode wavefronts expanding into the corona, suggest that a type II burst occurs when the shock front proceeds into a low-Alfvén-velocity region (i.e., region of higher density and/or lower magnetic field, cf. (112)). Uchida (1974) has shown that the propagating shock wave is refracted by magnetic and density structures of the corona toward the pre-existing low-Alfvén-velocity regions. The strength of the shock, which is otherwise weak, is then significantly enhanced due to the focussing effect by refraction and also due to the lowered Alfvén speed. Thus the shock strength builds up above a certain critical value and mechanisms for the acceleration of particles in the shock front begin to work. According to Uchida, both the blast waves observed on the disc (Section 4.3) and type II bursts are produced by the same shock propagating through the corona. Also the blast waves do not propagate into regions of high Alfvén velocity as Uchida *et al.* (1973) verified by comparing several of them with coronal magnetic fields computed by the method of Altschuler and Newkirk (1973).

Another puzzling problem is the fact that in some type II bursts the individual harmonic bands are themselves double: the splitting is about 10 MHz in the funda-mental band and twice as great at the second harmonic (Roberts, 1959). Various explanations have been suggested by Sturrock (1961), Takakura (1964), Zheleznyakov (1965), Zaitsev (1965), Tidman *et al.* (1966) and McLean (1967), but each of them meets with difficulties of some kind. More recently Smerd *et al.* (1974) proposed that the two bands of a split-band burst correspond to emission from in front of and behind the shock front. The upper-frequency band comes from behind and the lower-frequency band from in front. According to this interpretation the amount of splitting is a measure of the Mach number of the shock which would thus be in the range from 1.2 to 1.7. These rather small values are compatible with a laminar shock structure, but are below the values required for a turbulent type II shock which appears, e.g., in the theory of Smith (1972b).

More general discussions of the physical processes in the shock front can be found in Zaitsev (1965, 1968), Fomichev and Chertok (1965, 1967), and D. F. Smith (1972b).

3.2.6. Type III Bursts

A. *Type III Burst Occurrence*

Type III bursts are the most frequent discrete radio events on the Sun. They occur either singly or in groups, and there are days when more than 50 such bursts or groups

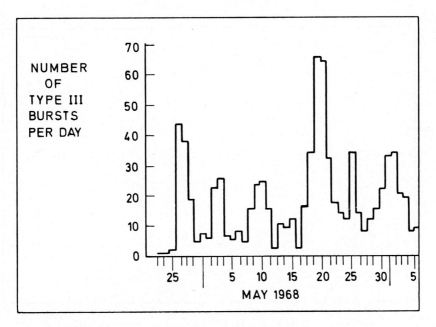

Fig. 79 a.   Illustration of the fast variations in the daily number of type III bursts. The very fast rise on
April 26 was due do a rise of the type III-active region on the eastern limb.

of bursts are recorded on the Sun; prolonged storms of weak bursts sometimes occur over periods of the order of one hour. On the other hand, there are many days without any type III burst activity at all, even when there are active regions and flares on the visible solar disc.

The daily number variations are often very large. On some occasions the sudden increases clearly coincide with the appearance of an active region on the eastern limb (Figure 79 a) but the type III activity never lasts for the whole time of the passage of the region over the disc. At other times the increase starts without the appearance of any new region on the Sun, and it must be due to some situational change in a region already existing on the disc. The period of strongly increased type III activity may last for several days, but on some occasions these periods are quite short, lasting only 10 to 20 h. All this shows that type III bursts originate during a particular phase of development of certain active regions, as was first noted by Loughead *et al.* (1957).

We have seen in Section 3.2.3.A that the microwave bursts occur most frequently in active regions which are magnetically complex. Such regions are characterized by high radio flux at 2800 MHz and therefore the daily number of microwave bursts shows a clear correlation with the daily 2800 MHz flux, as we see in Figure 79 b. The same figure, however, shows that no such correlation exists for the type III bursts. This means that type III burst occurrence is not restricted to magnetically complex active regions, and, as a matter of fact, we still do not know what are the characteristics of the active regions that produce them.

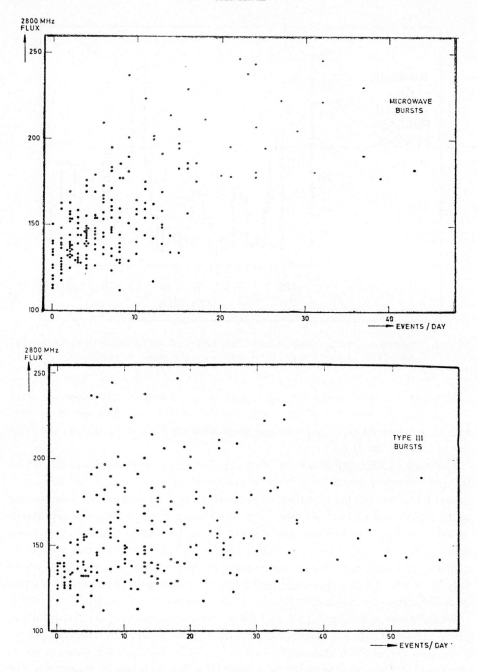

Fig. 79b. While the number of microwave burst occurrences correlates with the 2800 MHz flux
        (upper graph), this correlation is completely missing for the type III bursts (lower graph).

In contradistinction to the bursts of type II and IV, which are essentially always associated with flares, type III bursts appear to be predominantly active-region-associated rather than flare-associated phenomena. Type III bursts in an active region tend to occur when a flare is formed, but many of them also occur independently of flares. It is fairly difficult to establish the correct percentage of flare associations with type III bursts, since due to the very frequent occurrence of this type of bursts the number of chance coincidences may be very high (and so far we very rarely know the actual position of the burst on the solar disc). Some authors (Loughead *et al.*, 1957; Swarup *et al.*, 1960; Malville, 1961) have considered a type III burst as flare-associated when it occurs during the lifetime of a flare. In such a case, the association is fairly high: more than 60% of type III bursts occur during flares and about 20% of flares (or subflares) are associated with them. However, the type III burst occurrence shows a very high peak during the flare flash phase (Swarup *et al.*, 1960; Malville, 1961). If one makes the assumption (apparently justified when the chance coincidences are statistically estimated) that only type III bursts occurring during the period between the flare onset and maximum are physically connected with the flare, the percentage association is significantly smaller. Statistics made by the author for selected parts of the years 1968 and 1969 have shown that at least 70% of type III bursts occur without any event classified as a flare or subflare in the optical region. According to Kuiper (1973) about 35% of all type III bursts are associated with flares or subflares.

As a rule intense type III bursts are more likely to be flare-associated than weak bursts. The same is true for the type III association with X-ray bursts (Kahler, 1972;

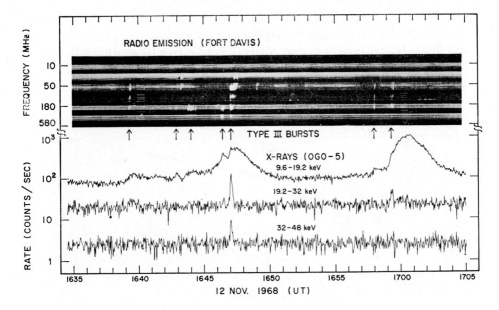

Fig. 80.   An example of time-correlation between impulsive hard solar X-rays and type III solar radio bursts in a dynamic spectrum covering 10 to 580 MHz frequency range. Note the event at 16^{h}47^m UT in particular. (After Kane, 1972.)

Kane *et al.*, 1974) and microwave radio bursts. Kane (1972) has found that $\sim 32\%$ of the impulsive hard X-ray bursts (which closely correlate with the impulsive microwave bursts) are associated with type III bursts; on the other hand, the converse correlation is only $\sim 25\%$ for the most intense type III bursts of intensity class 3 and very considerably less for weaker type III bursts so that the overwhelming majority of type III bursts are not associated with any detectable impulsive X-ray enhancement. However, in the rare cases when such an association occurs, the time correlation between the two emissions is invariably very good (Figure 80).

Kane's (1975) most recent study indicates that the probability of simultaneous occurrence of impulsive X-ray and type III radio bursts increases with the increase in the hardness of the X-ray spectrum. One cannot exclude the possibility that type III bursts include two different classes of events: strong bursts, visible low in the corona which are mostly (or always) flare- and X-ray-associated, and weak bursts (often 'storms' of type III bursts), which usually appear only at high levels in the corona and show no association with flares. (See, e.g., Kai and Sheridan, 1974, and the discussion in their paper.)

There appears to be no dependence between type III burst occurrence and the size of the flare (Giovanelli, 1959), and the vast majority of the associated optical events are subflares (Malville, 1961; Chin *et al.*, 1971; Kane, 1972). In the case that a type III (or mostly a group of type III bursts) is associated with a major radio event, the type III bursts usually occur at the very beginning of the event (Figures 68, 76 and 81). Some of them may be coincident with the impulsive microwave burst, but this is not a general rule (see, e.g., Švestka and Fritzová-Švestková, 1974).

Recently several authors have tried to detect the optical chromospheric counterpart of the type III bursts by studying high-resolution Hα photographs. Kuiper and Pasachoff (1973) have investigated six hours of intense type III activity in the region McMath 10 743. Seventeen type III groups occurred during that time and, as the authors say, 'eleven were associated with the start of some activity in Hα'. Similarly,

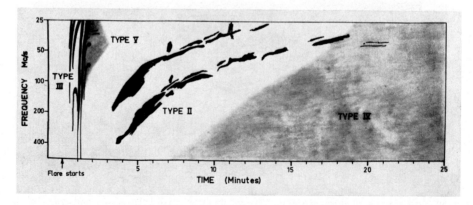

Fig. 81.   A schematic drawing of the most common succession of type III, V, II, and IV bursts in the dynamic spectrum of a large solar flare. (Courtesy of CSIRO Radiophysics Division, Sydney.)

Vorpahl and Zirin (1973), who studied type III bursts associated with flare activity in 1969, May 19–20, conclude that 'almost every type III burst had some associated Hα activity'. Kuiper (1973) has made a similar statement about type III burst activity in March and April 1971. Zirin and Lazareff (1975) claim that many flares (15 out of 21) occurring in front of a rapidly moving spot produced type III bursts, while flares in other locations in the same active region were much less productive in type III activity (3 out of 13). This result is somewhat similar to Axisa's (1974) statement that flares located outside the general bipolar pattern of the active region are characterized by a higher type III burst association than those taking place inside of it. However, apart from Axisa all the authors are unable to define any typical form of the Hα activity associated with type III bursts, and hence the correlations they offer are open to some doubt.

On the other hand, Axisa *et al.* (1973a), who have studied 44 groups of type III bursts, claim to find a typical chromospheric activity associated with these radio events: absorbing features extending along the $H_{\parallel} = 0$ line at the border of an active region and moving both downward and upward in the atmosphere. According to these authors most of the type III-associated flares are also related to an absorbing feature of this type. Since type III bursts are often assumed to be due to electrons propagating along coronal streamers (cf. Section 3.2.6.C) and coronal streamers develop above $H_{\parallel} = 0$ lines, Axisa *et al.* (1973b) believe that the association discovered by them is real and easily understood. Mercier (1973, 1974) has extended this study to 76 bursts in 28 different active regions and confirmed relation with a nearby filament for 57 of them. Kane *et al.* (1974) discovered in a few cases increase in brightness along the absorption feature and soft X-ray emission roughly coinciding with a type III burst. Priest and Heyvaerts (1974) even offered a theoretical interpretation for the 'neutral line absorbing features associated with type III radio bursts', in terms of newly emerging magnetic flux, penetrating an area of opposite polarities; a current sheet is formed between the new and old fields which can give rise to a thermal instability as soon as its length is large enough. Thus a dense and cool plasma is produced (the absorbing feature) and particles, accelerated through a tearing mode instability in the compressed current sheet, emit the type III radiation (cf. Section 6.5.2.A).

All these studies seem to give a rather consistent picture of what is happening when a type III burst is observed. Nevertheless, some doubts about the reality of this association between type III bursts and dark Hα features still remain. We compare here two phenomena that occur very frequently on the Sun so that the probability of a chance coincidence is very high. This is even more pronounced in that the authors take as positive evidence for the association not only the filaments existing in the active region at the time of the burst, but also filaments that existed there even two or three days before (cf. Mercier, 1973).

Summarizing, our present knowledge indicates that type III bursts might occur mostly in the peripheral parts of active regions, often where a dark filament with distinct motions extends (or extended before) from the active region to the surroundings along the $H_{\parallel} = 0$ line. But we still do not know what makes some active regions extremely prolific in type III bursts, while other regions (looking very similar) do not produce them at all.

B. *Properties of Type III Bursts*

The type III burst is characterized by a short duration and very fast frequency drift from high to low frequencies (Figures 76 and 80). The bursts are identified over a great range of frequencies from several hundred MHz, which corresponds to the base of the corona, down to tens of kHz, corresponding to heights of about 200 solar radii, i.e., the distance of the Earth.

Slysh (1967), Hartz (1969) and Alexander *et al.* (1969) were the first to observe type III bursts propagating deep into interplanetary space. In later years the satellites OGO 3 (Dunckel *et al.*, 1972), OGO 5 (Haddock and Alvarez, 1973) and IMP 6 (Frank and Gurnett, 1972) extended the data up to the distance of the Earth from the Sun. The lowest frequency to date at which a type III burst has been detected is 10 kHz (Kellogg *et al.*, 1973).

Figures 82a, b show the propagation of a type III burst between 1 MHz and 30 kHz according to Fainberg *et al.* (1972; Fainberg, 1974). As we can see from it, the burst duration increases with decreasing frequency. It is a second or less at

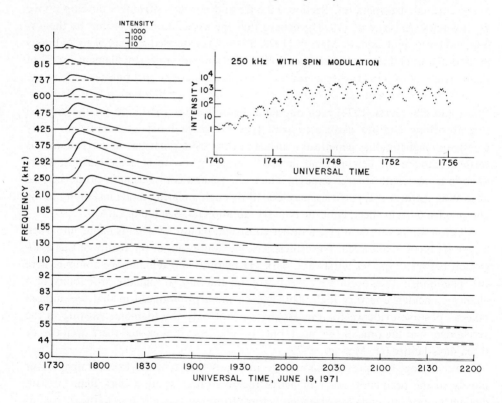

Fig. 82a.   A type III burst observed between 1 MHz and 30 kHz on IMP 6.
(After Fainberg *et al.*, 1972.)

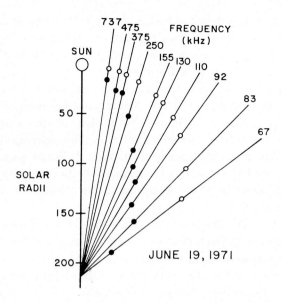

Fig. 82 b. Direction of arrival of the type III emission (in the ecliptic plane) for the records in Figure 82 a. The open circles mark the closest approach of the burst arrival directions to the Sun. The black circles are the intersections of these lines with spherical emission levels for each frequency. Thus the black circles track the probable trajectory of the stream of electrons through interplanetary space.
(After Fainberg *et al.*, 1972.)

frequencies above 200 MHz (according to Elgaroy and Lyngstad (1972) the minimum duration is 0.2 s), tens of seconds near 10 MHz (Kundu, 1965), minutes near 1 MHz (Alexander *et al.*, 1969; see also Figure 82 a), tens to 100 min in the 100 kHz range (Dunckel *et al.*, 1972), up to a few hours close to the Earth (Frank and Gurnett, 1972; cf. Figure 82 a).

Alvarez *et al.* (1972) have found an almost one-to-one correlation between intense type III bursts at very low frequencies and ≳ 20 keV electron events detected at 1 AU from flares located in the western solar hemisphere. On the other hand, not one of the type III bursts associated with flares located at longitudes east of 9° W showed any correlation with an electron event (cf. Section 5.3.1 and Figure 108). This can be easily understood if one assumes that the type III bursts are produced by streams of non-relativistic electrons which propagate from the Sun to the Earth along Archimedean spiral field structure in interplanetary space.

This actually has been confirmed by direct observations. Frank and Gurnett (1972) and Lin *et al.* (1973) succeeded in correlating a type III burst, propagating through space, to direct measurement of onset of a stream of non-relativistic electrons near the Earth (cf. Section 5.3.1.A). According to Lin *et al.*, the electron spectrum peaked at ≳ 100 keV at the type III burst onset and at ∼ 15 keV at the burst maximum.

Essentially no type III bursts from flares located more than 30° E are observed below 100 kHz (Alvarez *et al.*, 1973), while at higher frequencies this limit shifts toward the eastern limb. A deficit of hectametric type III bursts near the eastern limb was also confirmed by Sakurai (1974). This can be considered as more evidence that the trajectories of the burst exciters have curvatures similar to that of the interplanetary magnetic field lines (as in Figure 82b).

The number of bursts decreases with decreasing frequency in the hectometer and kilometer range (Alexander et al., 1969; Alvarez and Haddock, 1973). This is certainly partly due to sensitivity limitations, but it also indicates a decrease in density of the stream as it propagates outward; most bursts are observed only in the higher frequency range. The angular diameter of the type III bursts sources amounts to a few minutes of arc at 80 MHz and it increases sharply with decreasing frequency (Wild, 1968; Kundu and Erickson, 1974). At the Earth the diameter of the electron cloud is about 10^8 km (Lin, 1974).

When the bursts extend far from the Sun, several authors found evidence for a rather constant velocity through most or all of the bursts' travel in space (Alexander et al., 1969; Fainberg and Stone, 1970; Haddock and Alvarez, 1973). The derived velocities are typically 0.2 to 0.6 times the velocity of light, which correspond to energies of 10 to 130 keV for electrons. The velocity mean is between 0.3 and 0.4c (Fainberg and Stone, 1970). Haddock and Alvarez (1973) give as the mean $0.32c \pm 0.02c$. However, an analysis of IMP-6 data (Fainberg et al., 1972) indicates a slowdown at greater distances. It can be explained either by an increase in velocity dispersion inside the cloud of the propagating electrons or by a decrease in the exciter velocity with increasing distance from the Sun (Evans et al., 1973).

Thirty percent or more of type III bursts show a harmonic structure, but it is not easy to recognize, because the rapid frequency drift, broad bandwidth, and grouping of bursts tend to merge the fundamental and the second harmonic. As a matter of fact, there are differences in opinion regarding which component actually is observed. According to Haddock and Alvarez (1973) the fundamental often disappears below a certain frequency range as the burst drifts while the second harmonic remains. Thus, below 1 MHz the second harmonic predominates. This agrees with Fainberg's (1974) conclusion that the densities derived from the emission levels favor the harmonic interpretation. Also observations of the type II burst in space (Malitson et al., 1973; see Section 3.2.5) gave for the type II fundamental component the expected velocity only when the scale was based on the assumption that the associated type III bursts were observed at the harmonic. The direct comparison of the type III bursts with electrons recorded in space by Frank and Gurnett (1972) also shows a better fit if the second harmonic is predominant in the type III burst emission near the Earth (Lin et al., 1973; Lin, 1974).

However, in some cases the change from the fundamental to the second harmonic apparently occurs at very low frequencies. For example, Alvarez et al. (1974) could explain the emission as at the fundamental of the local plasma frequency down to 230 kHz. It was only below ~130 kHz that the observations could not be reconciled with this assumption and a change to the second harmonic had to be supposed.

About one-half of the bursts have been found to show a certain amount of circular polarization at frequencies below 200 MHz (Komesaroff, 1958; Rao, 1965), apparently in the ordinary mode (Enomé, 1964). In order to detect linear polarization, one has to use a very narrow band receiver, because any linear polarization of the radio source may be distorted by differential Faraday rotation in the solar corona (Akabane and Cohen, 1961). The narrow band-width measurements give polarization close to linear rather than circular (Akabane and Cohen, 1961; Bhonsle and McNarry, 1964; Chin et al., 1971; Dodge, 1972), but it is still not clear whether this reflects the polarization intrinsic to the generating mechanism or whether the polarization is modified through propagation effects (Fokker, 1971 a).

De la Noë and Boischot (1972) have drawn attention to the fact that in the decametric range one observes a large number of bursts which have a strong tendency to appear in chains and drift in frequency in a manner similar to the type III bursts. The polarization differs from one burst to another but sometimes reaches 100%. De la Noë and Boischot call them type III b bursts; it appears that they often are precursors of a normal type III burst. According to Baselyan et al. (1974) the type IIIb chains are excited by faster electrons than the ordinary type III.

From simultaneous observations of hard X-ray and type III bursts, Kane (1972) has estimated that the total number of $\gtrsim 20$ keV electrons responsible for a group of intense type III bursts is $\lesssim 10^{34}$. Lin et al. (1973) have compared simultaneous observations of type III emission and solar electrons in the interplanetary space and have concluded that the total number of $\gtrsim 20$ keV electrons is $\sim 10^{33}$. The reader is referred to Section 5.3.1 for more detailed discussion of the electron observations in space.

C. Interpretation of Type III Bursts

The generally accepted explanation of type III bursts is that proposed long ago by Wild et al. (1954): each individual burst is radiation from an agent that moves outward through the corona and excites plasma oscillations of progressively diminishing frequency as the stream passes through coronal plasma of diminishing electron density. Observations in interplanetary space and near the Earth have confirmed beyond any doubt that the exciting agent is a stream of accelerated electrons as was proposed by Wild et al. (1959 b) and by de Jager (1960, 1962) long ago. However, we meet here with the difficult problem of the stability of such an electron stream. As Sturrock (1964) has pointed out, such a stream exciting plasma oscillations should be damped out in a fraction of a second, while we know that the type III burst stream can travel through space up to the distance of the Earth.

The reason why the electron stream should decay very fast can be briefly summarized in the following way. A group of fast electrons produces the type III burst through generation of longitudinal electron plasma waves at frequencies near the local plasma frequency as they pass through the coronal (and later interplanetary) plasma. These plasma waves then scatter off ion density fluctuations to transform into electromagnetic radiation near the plasma frequency (fundamental) and off other plasma waves to produce the second harmonic. As the stream passes upward through the corona, faster electrons outpace slower ones and low energy electrons are lost from the stream by collisions. Thus there appears a 'hump' in velocity space: within a range of phase velocities the velocity distribution function $f(v)$ is an increasing function of v. When this happens, the longitudinal electron plasma waves grow; the developed two-stream instability decelerates the stream and stops it over a short distance.

Several proposals have been made regarding how to stabilize the stream. Sturrock (1964) suggested that the two-stream instability might be stabilized by non-linear effects. Kaplan and Tsytowitch (1967) suggested stabilization by non-linear scattering of the plasma waves, which has been found inadequate by several other authors (D.F. Smith, 1970a; D. F. Smith and Fung, 1971; Zheleznyakov and Zaitsev, 1970a; Melrose, 1970a). Stabilization processes involving ion sound waves have been suggested by Sturrock (1965), Melrose (1970a, b) and D. F. Smith (1972a), but none of the suggested sources of the ion sound waves appears to be acceptable for streams deep in space. Zheleznyakov and Zaitsev (1970a, b; Zaitsev *et al.*, 1972) suggest that the instability may be overcome if the waves generated by the faster particles (at the front of the stream) were reabsorbed by the slower particles behind them. This process might indeed stabilize the stream provided that the reabsorption is as complete as Zaitsev *et al.*'s model requires. In another model proposed by Papadopoulos *et al.* (1974) the stabilization mechanism is the oscillating two-stream instability which depends quite sensitively on the effective width of the electron stream. According to D. F. Smith (1974) the stream is likely to have a broad distribution which makes this mechanism inactive. Melrose (1974b) has suggested an idea of 'individual electrons': Each electron propagates along the magnetic field lines independently of the others; the stream remains identifiable because of an initial collimation and small velocity dispersion of the individual electrons. The 'collective effects' braking the stream would then not be effective since the time required for development of the instability is comparable to the time available for growth of the waves.

While it is difficult to decide now which of these mechanisms is the correct one, the great number of ideas shows that there are ways to stabilize the stream of electrons. It would be much easier to stabilize a stream of protons (Tsytowitch, 1966, 1967; Kaplan and Tsytowitch, 1967; Friedman and Hamberger, 1969b; D. F. Smith, 1970b), but the correlation of the low-frequency type III bursts with electron events near the Earth (and the direct observation of the electrons by Frank and Gurnett (1972) and Lin *et al.* (1973)) has proved definitely that protons are not involved in the type III production. (Their energy would be 20–100 MeV for the velocities observed.)

Another problem is the way the streams propagate through the solar corona and interplanetary space without losing their identity as a stream. The most reasonable explanation appears to be that they travel along a current sheet (i.e., coronal streamer) which acts as a channeling agent and helps to keep the transverse coherence of the type III burst sources (McLean, 1970; Wild and Smerd, 1972; Sturrock, 1972). While observations by Fainberg and Stone (1971) confirm this supposition by deducing from type III bursts densities a factor 2 to 3 times higher than the densities of normal coronal streamers measured with a K-coronameter, other authors (Kuiper, 1973:

Kai and Sheridan, 1974) have presented some evidence that not all type III exciters travel in dense parts of the coronal streamers. Leblanc *et al.* (1974) even claim that the centers of type III activity avoid the regions of high density in the corona, in contradistinction to Axisa *et al.* (1973 b) who associate type III bursts with dark filaments, above which the coronal streamers develop. (As a matter of fact, however, these are not the typical dark filaments that underlie streamers.) From a theoretical point of view D. F. Smith and Pneuman (1972) concluded that type III bursts cannot propagate along the dense parts of coronal streamers because the transverse magnetic field there is so strong that it inhibits particle motion along the sheet. However, this conclusion was re-examined by Priest and D. F. Smith (1972) who found the transverse magnetic field significantly smaller than assumed earlier. Thus type III bursts probably travel along current sheets, but observational evidence on it is still inconclusive.

Many other features remain unexplained. For example, emission originating near the plasma level should be strongly focussed, with the result that the bursts would be observed only near the center of the solar disc. However, the distribution of type III bursts appears to be fairly uniform across the disc (Fokker, 1970). Another problem is the high-frequency starting point of the burst. It may be determined by the formation of the above-mentioned hump (Takakura, 1966), but also the ratio of gyrofrequency and plasma frequency may play an important role in it (Fokker, 1969). Some type III bursts are also observed in the decimetric range, and according to Malville (1967) the starting frequency varies, on an average, from 300 to 500 MHz during the solar minimum to 150 to 300 MHz during the maximum years.

D. *Type U Bursts*

The height in the corona to which a type III burst penetrates does not depend solely on the efficiency of the electron stream, but also on the magnetic configuration in the active region where the burst is formed. When the magnetic field lines are open, the electron stream can propagate very far from the Sun; in some cases, as we have seen, out to the Earth and even beyond it. On the other hand, some streams are stopped in the corona, giving rise to type III bursts with a sharp low-frequency cut-off and some are curved downward, giving the so-called inverted U-bursts (first discovered by Maxwell and Swarup (1958) and Haddock (1958)).

These U-bursts, much rarer phenomena than the typical type III's, are usually bent at heights lower than one solar radius (Fokker, 1970), but on some occasions the reversing frequency can be very low, thus indicating a turning point at several or even tens of solar radii (Figure 83). Stone and Fainberg (1971) suggest that such a 'high' U-burst follows a magnetic bottle, which is to be understood as the remnant of an earlier flare that gave rise to heated coronal plasma expanded high into space and confined by a magnetic field. Evidence for the existence of such magnetic bottles has been derived (Schatten, 1970) from Levy *et al.*'s (1969) measurements of Faraday rotation during the occultation of a radio source (Pioneer 6) by the solar corona, and from expanding radio arches detected by the Culgoora spectroheliograph (Smerd and Dulk, 1971).

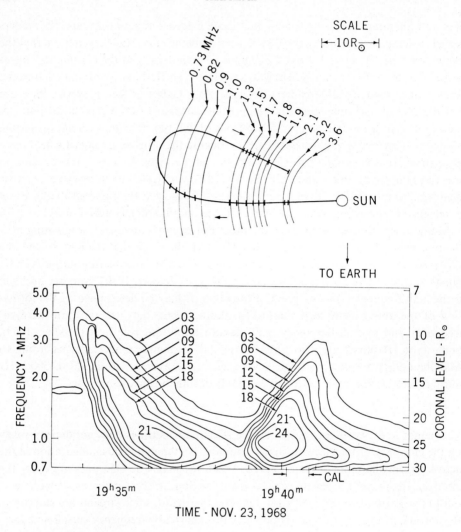

Fig. 83.   Dynamic spectrum of a *U* burst which reversed at an anomalously high altitude of ∼35 solar radii. The inferred trajectory of the electron stream producing this burst is shown at the top. (After Stone and Fainberg, 1971, and Fainberg and Stone, 1974.)

By using the Culgoora radioheliograph, Labrum and Stewart (1970) have found that the two branches of a U-burst may be interpreted as being produced in different active regions. This indicates that the U-bursts might appear when there is a magnetic connection between two different centers of activity (Fokker, 1971b). However, Caroubalos *et al.* (1973) who made very detailed observations of another U-burst with the Nançay radioheliograph, found their observations consistent with a burst trajectory along a magnetic arch connecting opposite polarities in the same active region (McMath 11 433). Maybe both these cases are possible.

E. *Type V Bursts*

Some type III bursts are followed at metric wavelengths by a short-lived (1–3 min) broadband continuum, which is called a type V burst (Figure 81). The III + V bursts appear more likely to be associated with flares (Švestka and Fritzová-Švestková, 1974) and with electrons directly recorded in space (Švestka, 1969) than simple type III's. Therefore, a type V burst occurrence is probably the manifestation of a rich stream of electrons, of which a fraction is trapped in the corona and becomes visible either through synchrotron radiation (Wild *et al.*, 1959b) or through plasma waves (Weiss and Stewart, 1965).

Type III burst characteristics appear to be strongly dependent on the magnetic field configuration in the source region. Thus in some regions the majority of type III bursts are combined with type V, in other regions type V bursts do not occur at all. The same is true for the U-bursts (Fokker, 1971b). Vorpahl (1973) is probably right when guessing that the magnetic field structure, rather than (or at least as well as) the efficiency of the acceleration process, determines the strength and kind of many type III bursts observed.

CHAPTER IV

FLARE-ASSOCIATED OPTICAL PHENOMENA

Besides the optical flare proper a large number of various optical phenomena are observed in the solar chromosphere and corona that obviously are related in some way or the other to the flare event. We may divide them into four different types:

(a) Activation of filaments (or prominences when seen on the limb) *preceding* the optical flare.

(b) Blast waves originating in the flare and traveling within large solid angles both into the corona as well as along the solar surface, giving rise to *winking* or *disparitions brusques* of distant quiescent filaments.

(c) Prominence ejections during the flare (*surges* and *sprays*).

(d) *Loop prominences* which start developing during the flare and proceed in the corona for hours after the chromospheric phenomenon has faded.

4.1. Pre-Flare Filament Activation

As was mentioned in Section 2.1.2.B, flares originate close to the line of zero longitudinal magnetic field $H_\parallel = 0$, dividing the magnetic polarities. This zero line is also identical to the position of quiescent dark filaments, which begin to form in active regions during a later phase of their development. The filaments then stay on the disc for many weeks after the sunspots in the active region have disappeared and the surrounding plage has faded. Therefore, it is not surprising that these filaments, when present, are greatly disturbed if a flare occurs in close vicinity to the $H_\parallel = 0$ line along which they extend.

What is somewhat surprising, however, is the fact that such a filament disturbance often starts prior to the flare, preceding it by between a few minutes and one hour. About 45% of major flares are preceded by such filament activations (Smith and Ramsey, 1964; Bruzek, 1969b) and flares that occur in old regions without sunspots are quite generally preceded by filament motions (Dodson and Hedeman, 1970). Thus, considering that quiescent dark filaments do not exist in relatively young active regions, in which many flares occur, one is inclined to suspect that the magnetic field changes which lead to the pre-flare activation essentially precede all flares.

Filament activation frequently takes the form of expansion, enhanced internal motions, and a slow rise into the corona. Sometimes, for some period of the disturbance, the filament oscillates and the spectral lines in the filament spectrum may be shifted alternatively to the red and violet; thus when observed through a narrow-band Hα filter, the image of the filament is shifted alternately out of and into the filter passband and gives a 'winking' impression. Toward the end of the activation, when the flare proper starts to develop, the activated filament is quite often suddenly accelerated and it usually disappears completely during the flash phase of the flare.

This disappearance may take several different forms: in some cases the filament as a whole begins to rise and expand very fast until it eventually fades high in the corona; in other cases the filament is disrupted and ejected with velocities of several hundred km s^{-1} upward (see, for example, Dodson and Hedeman, 1968b). There also seem to exist cases of a sudden downflow of the filament into the chromosphere (Kiepenheuer, 1964). The whole process is very similar to the phenomenon of 'disparition brusque' which will be discussed in Section 4.3.2. Several interesting individual events have been described in detail by Bruzek (1958), Smith and Ramsey (1964), Ramsey and Smith (1965, 1966a), Hyder (1967b), Dodson and Hedeman (1968a), Dodson et al. (1972), Švestka and Simon (1969), Michalitsanos and Kupferman (1974), and Rust et al. (1975). These descriptions show that while all the events have some disturbance of the pre-existing filament in common, they differ very much in the details; we meet with winking filaments as well as with continuously dissolving ones, with slow motions as well as with fast ejections, some filaments bend, untwist and change their positions while others preserve their shapes during the whole disintegration, and some filaments, even when rather close to the flare position, appear to be completely unaffected. Ramsey and Smith (1966b) have also described a case of a slowly expanding succession of chromospheric arc structures preceding the flare.

One has to be aware of the fact that our means of observation of these events are very unsatisfactory. Observations with narrow passband filters in the Hα line center may easily miss many fast-moving phenomena and sometimes they can give us quite a misleading picture of what is happening. Skylab observations missed many pre-flare events, since observations usually started only after the flare set in. The only information we have so far obtained from Skylab shows that EUV lines may be broadened prior to the flare (Figure 61), indicating the presence of large non-thermal motions in higher atmospheric layers (Brueckner, 1974b) and that soft X-rays begin to be slightly enhanced for minutes or tens of minutes prior to the flare onset, thus indicating a pre-heating in the corona (Petrasso et al., 1975; Vorpahl et al., 1975). Data from OGO 5 and OSO 7 show that some of these pre-flare enhancements in soft X-rays are directly associated with the filament activation (Roy and Tang, 1975). Covington (1973) observed a 'pre-flare absorption' on 2800 MHz radio flux which coincided with the filament activation. This, however, also might have been the decay phase of a preceding 'rise and fall' event so that the observation is inconclusive.

In a typical event observed in the Hα line the flare emission appears in the form of two bright ribbons on both sides of and parallel to the dissolved filament (Figures 19 and 20). This again is similar to the flare-like brightenings observed in association with the disparition brusque (Section 4.3.2). One can suppose that similarly to the disparition brusque (Figure 87) the disrupting filament produces soft X-rays. Roy and Tang (1975) indeed have observed that rapid increase in soft X-rays accompanies the phase of fastest expansion of the filament, and they suggest compression of the coronal gas by the expanding prominence as an explanation for it.

Hyder (1967a, b) has proposed that the chromospheric flare brightening is due to the prominence material which falls along the arched field lines into the chromosphere where the kinetic energy of the fall is dissipated. While this explanation might be acceptable in the case of a disparition brusque and it perhaps can also be applied to some local brightenings in flares which occur in old regions without sunspots, it fails to explain the flare emission generally, for several reasons:

(a) Some two-ribbon flares in sunspot groups develop when dark filaments in them have not yet formed (Švestka and Simon, 1969). Hence, if the brightening is caused by material falling down, this cannot be the material of a disrupted prominence.

(b) Hyder's mechanism is unable to produce other kinds of radiation observed in flares. In particular, there is no obvious way to produce the explosive and impulsive phase by a raining-down of material (Zirin and Russo Lackner, 1969).

(c) Finally, direct comparisons of falling material and flare brightenings show only a poor spatial correspondence (cf. Section 2.1.3.G).

Therefore, the appearance of a major flare does not seem to be a product of the filament activation as Hyder (1967b) has suggested. Nevertheless, the filament disintegration is closely related to the flare and demonstrates that changes in the magnetic field close to the zero line start to occur in many (or all) cases several minutes or tens of minutes earlier than the actual flare onset is observed.

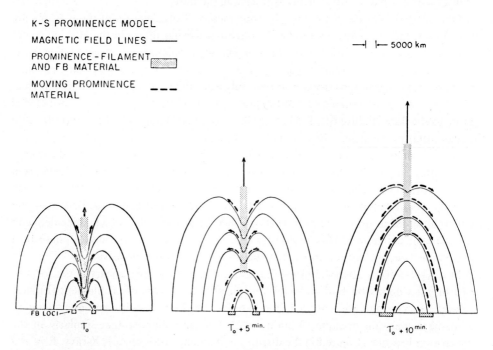

Fig. 84.   The left drawing shows schematically Kippenhahn and Schlüter's model of the magnetic configuration supporting a quiescent prominence (i.e., dark quiescent filament on the disc). Arrows and the subsequent two drawings show the field-line changes and gas flows expected during a pre-flare filament activation or a disparition brusque. (After Hyder, 1969a.)

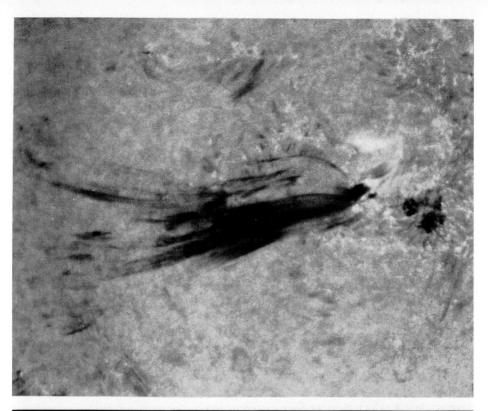

Fig. 85a.   Two big surges, one observed in projection on the solar disc (above) and the other one on the solar limb (below). (Courtesy of D. M. Rust, Sacramento Peak Observatory.)

Figure 84 shows the model of a quiescent filament as postulated by Kippenhahn and Schlüter (1957). The filament is supported by the magnetic field and its matter is collected in a dip which coincides with the zero line when projected onto the solar disc. The magnetic dip on the top of the field lines preserves the stability of the quiescent prominence (Pikelner, 1971). Therefore, the filament disintegration can be explained by a straightening of the field lines on the top of the magnetic arches, thus removing the dip. If the magnetic field is frozen into the gas, this straightening leads to a rise of the filament gas, which is eventually carried up to an unstable configuration on the top of convex magnetic arches. In consequence the filament loses its magnetic support, it expands, parts of it escape into the corona (heating it) and other parts flow along the arched magnetic field lines into the chromosphere (Hyder, 1967a).

Our example does not imply that the old Kippenhahn and Schlüter concept is the only model that can explain the filament activation. Dungey (1958) and Raadu and Kuperus (1973) have presented different models which can equally well explain the phenomenon. The only purpose of Figure 84 is to demonstrate that the filament activation obviously implies a rearrangement of the magnetic field structure; since it precedes the flare it gives important evidence that the magnetic field is (often or always) rearranged before the flare-producing instability develops. As we shall see in Chapter VI, Syrovatsky (Section 6.5.3.A) and Pustilnik (Section 6.5.2.D) have proposed flare models in which this pre-flare restructuring of the magnetic field is a necessary condition for building the flare-producing current sheet.

Thus the activation of a filament clearly demonstrates that the stabilizing agent, i.e., the magnetic field in the active region, becomes unstable, and that this destabilization starts long before the onset of the optical flare. However, quite a number of the activated filaments do not disappear completely. Many of them begin to reappear again during the decay phase of the flare, or they resume their former stable shape within several hours (Bruzek, 1969b). This shows that the gross configuration of the magnetic field does not undergo any strong permanent change.

Using movies taken in the green coronal line, Bruzek and DeMastus (1970) have found a case of coronal arch expansion preceding a flare. This case showed a very close similarity to what is seen in $H\alpha$ quiescent filaments. About $1\frac{1}{2}$ h before the flare a bright hill appeared at the base of a system of stable coronal arches, and it expanded with a relatively small velocity of ≤ 5 km s^{-1} for more than 1 h. The expansion velocity began to increase some 15 min before the flare onset, at a rate indicative of an explosion. Shortly before the arch disappeared high in the corona (6 min before the flare onset), the expansion velocity increased to about 100 km s^{-1}.

4.2. Surges and Sprays

When a flare starts, it is quite often accompanied by gas ejections which manifest themselves as moving prominences on the limb or Doppler-shifted dark filaments on the disc (Figures 85a, b). We meet here with a great variety of events but generally can divide them into two groups: surges, in which the gas goes up and down again (sometimes along the same magnetic lines of force), and sprays, in which most of the ejected gas does not return to the active region of its origin.

4.2.1. SURGES

A typical surge has a shape of a straight or slightly curvilinear spike which grows upward from the chromosphere with velocities typically between 50 and 200 km s^{-1} (Figure 85a). After reaching the maximum height (from 20 000 up to 200 000 km) the material is usually seen to descend, as it seems along the original trajectory. Lifetimes of these spikes (Westin, 1969) are mostly between 10 and 30 min. Surges have a strong recurrence tendency and the surge ejections often repeat themselves at intervals of the order of 1 h at apparently exactly the same place (Rosseland and Tandberg-Hanssen, 1957; Lodén, 1958; Giovanelli and McCabe, 1958; Kiepenheuer, 1960). Giovanelli and McCabe (1958) have found that many surges on the disc are preceded by a diffuse expansion of a part of the flare, or by the formation of a small bright area, which is short-lived and as it becomes transparent and fades (typically after ~ 6 min) the surge issues from it. Giovanelli and McCabe have called this phenomenon a 'flare puff' and one cannot exclude the possibility that the puff-surge sequence would be found for all surges if the observations were carried out with adequate time and angular resolution. Roy (1973) has found these pre-surge brightenings for all 30 surges he studied, and he identifies them with the moustaches (cf. Section 2.2.11). However, as Bruzek (1972) emphasizes, there are many moustaches which do not give rise to any detectable surge, hence a moustache appearance does not necessarily imply the surge occurrence. What Bruzek (1969b) finds curious is the fact that no chromospheric brightening is observed toward the end of the surge when the gas material is obviously coming back and falling down onto the chromosphere. According to the Hyder's infall-impact theory (Sections 2.1.3.G and 4.1) such a brightening should be expected.

The surge sources are usually situated close to the edges of sunspot penumbrae (Giovanelli and McCabe, 1958), according to Rust (1968) above small, changing, satellite sunspots which represent small 'islands' of polarity reversals near a large spot. This is also confirmed by Roy (1973), but only for a fraction of surges. According to his results, a more general characteristic is the presence of evolving magnetic features, and surges appear to be related to rising flux of one polarity into a region of stronger opposite flux. Recently Rust (1975), by using He λ 10 830 spectroheliograms, has found that the surge bases (i.e., the flare-puffs after Giovanelli and McCabe (1958), satellite sunspots after Rust (1968) or evolving magnetic features after Roy (1973)) are closely related to the 'moving magnetic features' detected by Vrabec (1973) in sunspot surroundings.

Zirin and Severny (1961) and Ioshpa (1962, 1963) have found fairly strong magnetic fields in surges, of about 100 to 200 G. According to Tandberg-Hanssen and Malville (1974) this is the upper limit for the magnetic field strength, which usually exceeds 30 G; but Harvey (1969) actually did not find stronger fields than ~ 30 G in surges.

Surges are a very common phenomenon and many flares are accompanied by them. However, the actual percentage of flares associated with surges is not well known, since due to low contrast, high radial velocities, and short lifetime, many small surges can easily escape observation. On the other hand surges occur so frequently in growing active regions that many surge-flare associations must be chance coincidences. According to Gopasyuk and Ogir (1963) all flares are accompanied by surges. Other authors have found only a fraction of flares associated with this type of prominence ejection, e.g., 20% of all flares and subflares according to Giovanelli and McCabe (1958). Surges and even very large ones (Gopasyuk and Ogir, 1963) are often associated with quite small flares, even when statistically the percentage of association increases with increasing flare importance (Westin, 1969). But more often (according to Westin in 50% of large surges and 70% of small ones) one can see a surge without any chromospheric brightening large enough to be classified as a flare or subflare.

One can suppose that a surge ejection is due to some sudden change in the distribution of the magnetic field lines in the active region; this change may in some cases be initiated by the flare process occurring in the same region of activity, but in most cases it develops independently of flares. This recalls to some extent the generally loose, but in some individual cases very well-defined, association of type III bursts with flares. Indeed, some remote relation between type III bursts and surges cannot be excluded: Kahler (1973a) has found that 28% of flares with type III are accompanied by surges, whereas the association is only 17% for the assembly of all flares. Kahler has concluded from it (but he does not work it out) that streams of > 10 keV electrons should play some role in the production of surges.

In any case, the original magnetic configuration must be easily restored, since surges are more repetitive than any other type of solar activity. This is a problem that cannot be solved easily. Many years ago Schlüter (1957) suggested that a diamagnetic plasma cloud forms in an active region due to the existence of anti-parallel electric currents, and this cloud may eventually be expelled by the tensions of the magnetic field. This process is often referred to as the 'melon-seed effect'. More recently another mechanism has been proposed by Altschuler *et al.* (1968) who have considered the coupled effects of changing magnetic and velocity fields. Plasma in a relatively cool and dense region tends to descend, its downward motion across magnetic field lines induces an electric current, and this current distorts the dominant magnetic field. An 'island' with reversed polarity forms (somewhat analogous to the magnetic dip in Figure 84, but at a lower atmospheric level), and if a perturbation of this dynamic state occurs, a disturbance is generated which propagates toward weaker magnetic fields with the Alfvén velocity determined by the strong self-induced magnetic field in the cloud. This model places the surges exactly on the 'islands of polarity reversals' where Rust (1968) has observed them. Of course, in order to explain the repetitive surges, the initial situation must be restarted after the disturbance, like when a jumper exercises on a trampoline.

Sturrock (1972b) suggests that the 'satellite spots' occur where an 'errant' magnetic flux tube is 'hooked' through the photosphere to form an apparent field reversal. A field-line reconnection occurs at a cylindrical current sheet surrounding the errant flux tube. Heating in the current sheet or particles accelerated there cause an annular brightening at the base of the surge. The surge itself is seen as material is ejected by magnetic stresses from the field-line reconnection site. Rust (1975) finds support for Sturrock's model in the fact that the 'moving magnetic features' which seem to be related to the surges are often found in pairs of positive and negative fields.

4.2.2. SPRAYS

The term 'spray' was first used by Warwick (1957) for very high-velocity ejecta occurring in conjunction with flares. One can define them as flare-associated prominence ejections characterized by velocities that as a rule exceed 670 km s^{-1}, i.e., the velocity of escape in the chromosphere. These high velocities distinguish them from surges. While surges remain compact during all of their lifetime, sprays usually disintegrate into many fragments and only a very small fraction of them returns to the solar surface (Figure 85b). They also differ from eruptive prominences (E. Smith, 1968). While eruptives start ascending slowly and then accelerate to high velocities only at a late phase (like the pre-flare activated filaments, for example), sprays reach the characteristic very high velocities within a minute or two of the start of the phenomenon. One might also distinguish sprays from the high-speed ejecta of compact parts of flares, such as the example shown in Figure 18, but opinions on this point are

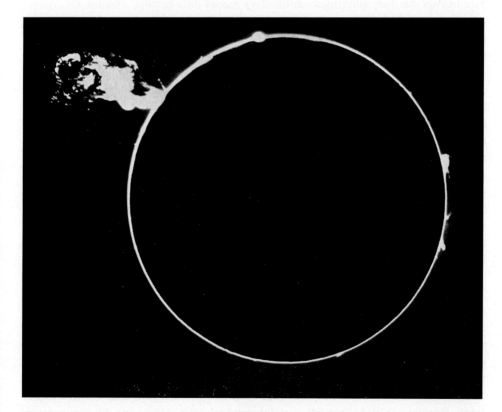

Fig. 85b.   Example of a spray on the solar limb. (Courtesy of R. T. Hansen, HAO – NCAR, Boulder.)

different; e.g., E. Smith (1968) just characterizes sprays as being parts of the flare itself. Observations of transients in the white-light corona (Koomen *et al.*, 1974; Wagner *et al.*, 1974; Riddle *et al.*, 1974; Stewart *et al.*, 1974a, b; Brueckner, 1974a; Gosling *et al.*, 1975; Munro, 1975) show still greater variety of coronal disturbances associated with Hα ejections that can be classified as sprays.

As a matter of fact, all these high-speed phenomena are so complex and variable that it appears almost impossible to classify them properly. Spray observations on the disc are particularly difficult, since large Doppler shifts carry them outside the narrow pass-band of most patrol instruments. Therefore, the normal disc patrol detects only sprays moving nearly transversely to the line of sight, and in consequence the velocities of disc sprays generally appear lower than the spray velocities on the limb (E. Smith, 1968). This makes the classification of a fast moving disc event as a spray fairly difficult.

Zirin (1966) is of the opinion that the ejections of material observed in explosive flares (Athay and Moreton, 1961; see example in Figure 76) would be always seen as sprays on the limb. E. Smith (1968), on the other hand, states that this need not be the same phenomenon, and certainly not all explosive flares are accompanied by sprays nor vice versa; according to E. Smith, sprays should generally appear bright and flare-like on the disc, differing from the flare proper only by their motion. Of course, the emission can change to absorption after the density and temperature in the spray have decreased and the spray should also be dark when observed far from the Hα line center.

A list of events which may be considered sprays has been compiled by E. Smith (1968). According to this survey sprays can occur with very minor flares. On the other hand, there is convincing evidence that no spray was associated with several major flare events close to the limb. Sprays themselves do not seem to be associated with any appreciable X-ray or radio emission. However, as far as the relatively small number of events makes a statistical comparison possible, spray-associated flares appear to show a higher association with type II bursts than non-spray flares. This reminds one of the association between type II bursts and explosive flare ejections (cf. Figure 76), and type II bursts and flare waves, as we shall see in Section 4.3.1. The similarity of the velocities of the flare-waves, disc flare high-speed ejecta, sprays, and type II bursts suggests that we may indeed see in these events different aspects of the same basic phenomenon.

Recent observations of spray-associated coronal transients (cf. Sections 3.2.4.A and 3.2.5) have given great support to the supposition that sprays are related to the type II as well as to the moving type IV bursts. In Figure 71 we showed the association between a moving type IV in 1972, August 12, and a spray, expanding with apparent velocity of about 510 km s^{-1} in the form of a twisted flux loop into the corona (Riddle *et al.*, 1974). Stewart *et al.* (1974a) have found that a common disturbance initiated the spray, moving type IV burst and density change in the corona in 1973, January 11, while a type II burst travelled in front of these disturbances, with a higher speed. Brueckner (1974a) reports coronal clouds in 1971, December 14, moving outward with projected velocities between 950 and 1100 km s^{-1}. Traced back, these clouds coincide with a type II radioburst moving with a velocity comparable to that of the clouds. Wagner *et al.* (1974) observed an opening of green corona structure in 1972, February 17, cospatial with the disappearance of a white-light streamer that followed a behind-the-limb flare associated with types II and IV. With the white-light coronagraph on Skylab Gosling *et al.* (1975) observed in 1973, September 7, material moving with a speed of 960 to 1300 km s^{-1} from the Sun, following again a flare associated with types II and IV. Many coronal transients are not associated with recognizable radio bursts or visible flares (Koomen *et al.*, 1974; Munro, 1975). However, essentially all flares during the Skylab period that ejected high-velocity material in Hα were associated with coronal transients. The observations of spray-associated coronal transients made it possible to estimate the mass and energy involved in a spray event. Stewart *et al.* (1974a) find the total number of particles participating in a spray-associated white-light coronal disturbance equal to 10^{40}, with total mass $\simeq 1.7 \times 16^{16}$ g and $\sim 10^{31}$ erg involved in the kinetic energy of the expelled gas. The energy in the flare spray itself was $\lesssim 10^{30}$ erg, i.e., much smaller than in the white-light cloud.

4.3. Flare Waves

Some flares initiate various kinds of disturbances in distant regions of the solar disc, after time intervals which correspond to a propagation velocity of the disturbing agent of the order of 1000 km s^{-1}. The disturbances reported in the past 40 yr were either of the form of an activation of a distant filament (first discussed in detail by Dodson (1949)), or of apparent initiation of another flare in a distant active region (Richardson, 1936, 1951; Becker, 1958; Valníček, 1961). Finally, in 1960 the existence of flare wave disturbances was proved by direct optical observations (Moreton and Ramsey, 1960; Athay and Moreton, 1961).

At the present time there are four different types of evidence for the existence of flare waves (S. Smith and K. Harvey, 1971):

(a) A fast-moving diffuse front is seen in emission in the Hα-line center and in absorption (or absent) in the Hα wings, propagating away from the flare (Athay and Moreton, 1961; Zirin and Werner, 1967; Hyder, 1967b; Dodson and Hedeman, 1968; Zirin and Russo Lackner, 1969).

(b) Progressive, short-lived brightenings of small points appear in the chromosphere in places passed by the wave.

(c) An abruptly initiated activation or oscillation of a filament which stood in the way of the passing wave is occasionally seen, in some cases eventually leading to a 'disparition brusque'.

(d) Onset of a sympathetic flare in another active region is observed at the time when the region is reached by the wave.

This last point, however, is still open to doubt. If some flares were actually instigated by flare activity in another active region, we should expect an excess in the statistical occurrence of small ($\lesssim 10$ to 20 min) intervals between successive flares. However, statistical studies of the frequency of the flare occurrence did not show any indication of such an excess, and the interval distribution is exactly what can be expected from purely random incidence (Waldmeier, 1938; Fritzová, 1959; H. Smith, 1962b; Smith and Smith, 1963). This contradicts the fact that individual direct observations seem to show the occurrence of sympathetic flares (Becker, 1958; Valníček, 1961; S. Smith and K. Harvey, 1971; but none of them saw the associated wave). One should be aware of the fact that statistically, when completely random incidence is assumed, one out of five time intervals between successive flares is shorter than 20 min during the years of maximum activity (Fritzová, 1959). When small subflares are considered, the frequency of short time intervals is still higher. In many such cases, then, the occurrence of a 'sympathetic' flare can easily be suspected, even when the flare appearance is purely random. We should be extremely careful about concluding from one isolated pair of observations that one flare instigated the other.

Simnett (1974b) believes to have observed sympathetic flares in regions $\sim 100°$ apart which, according to his assumption, were connected through a loop fed with energetic particles. Similarly, Gergely and Erickson (1975) believe to have found sympathetic flares in active regions that were common sources of radio storms. In both these studies, however, the authors underestimated by about a factor two the random occurrence of <20 min time intervals between flares so that their evidence is invalid (Fritzová and Chase, 1975). There is still some indication of excess number of <20 min intervals in the material studied by Gergely and Erickson, but its statistical significance is small. X-ray pictures of the Sun on Skylab revealed about 100 loops interconnecting active regions (Chase et al., 1975), but no homologous flares could be discovered in any of them (Fritzová and Chase, 1975).

A study of the time difference between the flare and the onset of activation of a distant filament yields propagation velocities between 400 and 2200 km s^{-1} (45 cases). with a mean value of 880 km s^{-1}. Measurements of the motion of 15 visible wave fronts yielded velocities between 440 and 1100 km s^{-1}, with a mean value of 600 km s^{-1} (S. Smith and K. Harvey, 1971). Thus we meet with supersonic velocities of the order of the Alfvén velocity in the solar corona. According to Ramsey and Smith (1965) 15% of major flares have detectable wave effects.

4.3.1. The Wave Fronts

The traveling wave fronts are usually semicircular, extending across a sector of 60 to 120° (S. Smith and Harvey, 1971), and they are sometimes called 'Moreton waves'. However, this term may also involve sprays projected on the disc. The sprays may have a common origin with the wave fronts but do not represent the same phenomenon: while sprays are material ejected from the flare region and emitting or absorbing in Hα when traveling through the corona, the wave front seems to be a pure wave effect which causes a depression and successive relaxation of the fine structure of the chromosphere (Dodson and Hedeman, 1964a; Bruzek, 1969). The spicules and fibrils are forced to move downward for about 40 s with a velocity amplitude of 6 to 10 km s^{-1} in the leading wave front and afterward they return more slowly to their former positions. In the well studied flare of 1966, August 28, specific portions of the solar disc exhibited an absorption feature for 2 to 3 min (Dodson and Hedeman, 1968). As a general picture one sees then at Hα +0.5 Å a traveling dark front some 30 000 km wide followed by a wider bright disturbance, while at Hα −0.5 Å the situation is opposite; the thin preceding front is bright and the wide front behind is dark. The leading dark Hα +0.5 Å front coincides with the leading bright Hα −0.5 Å front completely within the time and spatial resolution available (Moreton, 1964; Anderson, 1966; Hyder, 1967b; Dodson and Hedeman, 1968). In line center the wave is usually in emission.

The front usually becomes visible only at a distance of more than 100 000 km from the place of its origin. This makes it difficult to establish the exact place and time of its onset as it is necessary to extrapolate the velocity curve back along the time coordinate. Even when errors of some ±2 min cannot be excluded, this extrapolation clearly indicates that the waves originate during the flash phase of the flare (S. Smith and K. Harvey, 1971). However, any more precise determination is difficult. In the case of the flare of 1966, August 28 (Figure 86), Zirin and Russo Lackner (1969) identified the origin of the wave with a spray which was ejected from the western part of the flare. Dodson and Hedeman (1968) estimate a similar position for the source, but put the wave origin 2 min earlier, when the separation of the two bright flare ribbons first became clearly visible. The association between the wave and a spray appears very attractive as we emphasized in Section 4.2.2. However, in the second best observed case of a flare wave, on 1963, September 20 (Figure 76), when a spray also occurred, Zirin and Werner (1967) had to conclude that the wave origin and the spray most probably did not coincide in time.

28 Aug 66
15:31

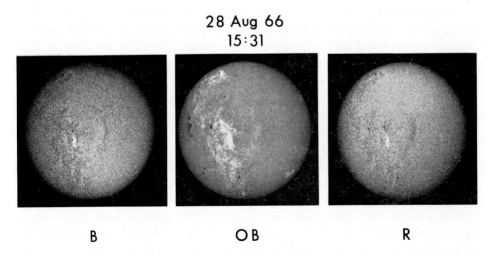

B   O B   R

Fig. 86. Direct photograph of the flare wave propagating from the flare of 28 August 1966 (B = blue, R = red wing of Hα). It can be seen as a semi-circle with center near the flare. (Courtesy of C. Hyder and J. W. Evans, Sacramento Peak Observatory.)

Some flare waves can be followed on the disc up to distances exceeding 500 000 km and they propagate with fairly constant speed (Moreton, 1964; Dodson and Hedeman, 1968). Some of them are restricted to rather narrow cone of propagation (much less than 90°) while others, like that in Figure 86, have well-developed semi-circular shapes. Different waves propagate in various directions on the Sun, perhaps with slight preference toward the west (S. Smith and K. Harvey, 1971). As we mentioned in Section 3.2.5 (see also below) they probably move toward the regions of low Alfvén speed (Uchida et al., 1973).

We have already shown in Figure 76 that some flare waves seem to be well corre-lated with type II bursts (Uchida, 1963; Lin and Anderson, 1967; Kai, 1970; K. Harvey et al., 1974). On the other hand, a statistical analysis of this correlation carried out by S. Smith and K. Harvey (1971) for 50 flare waves, leads to the result that only 36% of waves were associated with a type II burst. This percentage is higher after 1963 ($>50\%$), probably due to more complete and more sensitive radio spectrum observations. Even this rather loose correlation, however, indicates a common origin of the visible wave fronts and type II bursts, because visibility conditions of these two phenomena are different (Uchida, 1974; see below). Thus one can expect that some-times only one or the other phenomenon is observed even when they are both pro-duced by the same wave effect.

Theoretical considerations about the generation and propagation of flare waves have been published by Anderson (1966), Meyer (1968), Uchida (1968, 1974), Uchida et al. (1973) and Kassinsky and Krat (1973). First of all it is clear that the waves cannot be propagating through the chromosphere, even when the propagating agent strongly influences the chromospheric structures. The chromospheric sound velocity is of the order of 10 km s^{-1} and the chromospheric Alfvén velocity in a 5 G magnetic field is about 50 km s^{-1}. Thus the Mach number of the waves observed would be of the order of 10 and a shock wave of such strength would dissipate strongly. Its velocity would decrease with increasing distance from the source, which has not been observed, and it could not travel the very long distances reported. Therefore, the wave propaga-tion must occur in the corona, where both the sound and Alfvén speeds are about a factor 10 higher than in the chromosphere.

Anderson and Meyer have proposed that the coronal waves are reflected from a certain level in the corona back onto the chromosphere where they cause the observed chromospheric disturbances. Meyer has shown that the fast hydromagnetic mode suffers downward refraction in the coronal density gradient and upward reflection at the density jump to the chromosphere. Thus the wave can be trapped between these two reflecting levels and might be guided in this region to a very large distance. In Uchida's interpretation the hydromagnetic fast mode propagates quickly in the corona, and the skirt of this wave front sweeps over the chromosphere with a velocity much exceeding the fast mode velocity in the chromosphere itself. In Kassinsky and Krat's (1973) interpretation the flare wave might be an effect analogous to the very long gravity wave ('tsunami') after an ocean bottom earthquake. If the energy density of magnetic configurations is much smaller than the energy density of the tsunami wave, the wave can go through.

Uchida's model is the most widely accepted one. On a computer, Uchida *et al.* (1973) and Uchida (1974) simulated the weak MHD fast-mode shock propagation. They have shown that the shock wave is refracted toward the low Alfvén velocity regions in the corona, where the shock is amplified. Comparisons with realistic three-dimensional distributions for the coronal magnetic field and electron density give reasonably good results for observed wave fronts as well as for type II burst sources observed by the Culgoora radioheliograph. However, the chances of occurrence of these two phenomena are not necessarily equal. If the source point of the wave is surrounded by strong magnetic field everywhere, the wave may be reflected upward and little chance for any observable wave front in the chromosphere can be expected. This, e.g., might have been the case with the big August 1972 flare events. On the other hand, if such a focussing upward is missing, or even a refracting region exists in the corona, the radio burst may be too weak to be observed (also see Section 3.2.5).

Bruzek (1952b, 1958), Öhman (Öhman and Öhman, 1953; Öhman *et al.*, 1962) and Yajima (1971) suppose that large flares also generate a slow wave disturbance which propagates at about the normal velocity of sound. Its occurrence has been inferred from activation of distant filaments a long time after the flare onset. The corresponding velocities lie between 60 and 200 km s^{-1}. If such slow wave disturbances are real, the slow and fast waves might represent two modes of propagation of the same primary disturbance (Bruzek, 1969). Of course, the chance of accidental coincidence greatly increases as the time interval between the source phenomenon and its effect grows.

4.3.2. DISPARITIONS BRUSQUES

The fact that a traveling flare wave activates distant filaments has been established by many authors (Dodson, 1949; Becker, 1958; Bruzek, 1958; Dodson and Hedeman, 1964a; Ramsey and Smith, 1966a; S. Smith and Harvey, 1971; and others). Most common are filament oscillations, leading to 'winking' of filaments as described in Section 4.1. Such a winking filament (alternately disappearing and reappearing in the passband of the monochromator) can become stabilized again and resume its original quiescent shape. In some cases, however, the activation proceeds further, as in the pre-flare cases described earlier, and the filament erupts. For such phenomena very often the French term 'disparition brusque' (DB) is used. As a matter of fact, this term has been used by various authors for three different phenomena:

(A) Disappearance of a quiescent dark filament far from any active region giving rise to slight brightenings along the filament channel. Such brightenings usually are not classified as a flare (Figure 87).

(B) Disappearance of a dark quiescent filament in an old (spotless) active region giving rise to brightenings along the filament channel which can be classified as a 'spotless' two-ribbon flare (examples in Figures 19 and 20).

(C) Disappearance of a dark filament in a fully developed active region giving rise to a 'regular' two-ribbon flare (cf. Section 2.1.3.E).

Case (A) is the typical DB which possibly can be related to a propagating wave front. Case (B) is generally considered a DB, but its origin may be different (e.g., emergence of new magnetic flux below the filament as Rust (1967) has suggested). Finally, case (C) probably has no relation to propagating wave fronts at all, being caused by internal variations of magnetic field inside the active region (cf. Section 4.1). If some of the flares in (B) and (C) are instigated by propagating shock fronts we encounter the case of sympathetic flares. In the following we shall consider only the case (A) as a DB, may be with some admixture of events belonging to the category (B) since these two cases are sometimes difficult to separate.

The DB is a fairly common phenomenon, and according to Dodson *et al.* (1972) about 30% of all large filaments in the last two solar cycles disintegrated in the course of their transit across the solar disc. This percentage seems to be still higher for filaments in the equatorial zone ($\sim 50\%$ according to Waldmeier (1938)). However, according to M. and L. d'Azambuja (1948) only one in three cases of disappearance is permanent. Usually within hours or days the filament reappears at the same position, often with nearly the same shape. It is to be understood that many DB's are

SEPTEMBER 1, 1973

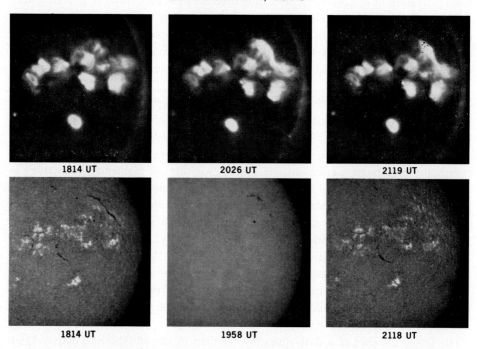

Fig. 87. 'Disparition brusque' of a quiescent filament as seen in the Hα line (below) and in soft X-rays (above). The Hα frame at $19^h 58^m$ is out of the line center and shows the moving material. Note that only very small brightening accompanied the Disparition Brusque in the chromosphere. In the corona, on the other hand, the DB produced a very strong enhancement in brightness; the filament, completely invisible prior to its activation, became an outstanding feature when rising into the corona.

not directly observed. They are mostly deduced only from the fact that a filament, present at the evening of one day, is missing next morning.

The fast flare waves discussed in the preceding section usually cause only the winking effect. According to Bruzek (1958, 1969) the DB is mostly produced by a slower wave which might be another mode of the disturbance. Some observations of individual events support this hypothesis (Dodson, 1949; Dodson and Hedeman, 1964a; Yajima, 1971), but more data are certainly needed to be sure, since chance plays a very significant and confusing role in correlations of this type. So far two facts seem to have been well established:

(a) Some DB's appear to be caused by flare waves, because – apart from direct observations of individual events – a comparison of the occurrences of major flares and DB's shows a clear peak on the day when a large flare occurred (Dodson et al., 1972).

(b) However, DB's can also be initiated by another type of disturbance. Many large filaments disintegrate without the occurrence of an obviously associated prior flare (for example, see McCabe (1970)), and if one wants the item (a) to be valid generally, one has to assume the existence of an unknown flare behind the solar limb in all of these cases. This certainly may be true in some of them (according to Yajima (1971) filaments can be affected as far as 700 000 km from the flare source), but the actual number of such cases appears to be too high to substantiate this point of view. As we mentioned above, Rust (1967) suggested that the eruption occurs when new fields emerge under the filament. According to Bruzek (1952a) the magnetic field in the surroundings of the filament may also become unstable if a new active region emerges in its vicinity, up to the distance of about 25°.

Summarizing, we can say that all the cases (A), (B), and (C) are due to an instability of the magnetic field in the surroundings of the filament. In case (C) this instability is an internal affair of the active region, according to Rust (also see Rust et al., 1975) caused by new magnetic flux emerging under the filament. This obviously happens quite often in an evolving active region. In old regions (case (B)) this happens rarely. Nevertheless, the approximately 7% of flares occurring there (Dodson and Hedeman 1970) still might have the same origin. In fact, old filaments represent a rather extensive coronal cavity the stability of which can be disturbed by new flux emerging quite far from the filament itself. This is also true in case (A) where, as we mentioned above, Bruzek (1952a) suspects even regions emerging 25° from the filament to be connected with the instability. The flare wave distortion of the magnetic field is probably a secondary effect, responsible for more DB's in the category (A) and probably also for a few ones in category (B).

Skylab observations (AS & E, 1975) have revealed that the DB process is associated with strong heating in the corona, visible in soft X-rays (Figure 87). It is probably a phenomenon similar to the X-ray enhancements observed by Roy and Tang (1975) during the preflare filament activation. As we mentioned in Section 4.1, Roy and Tang explain it by a compression of the coronal gas by the expanding prominence. Hyder (1973) has proposed shock waves as the source of the X-ray emission, since

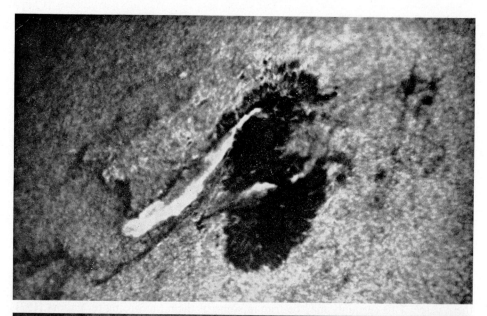

Fig. 88.   The loop prominence system observed in Hα in the late phase of development of the flare of 1974, September 10. The upper filtergram, made in the helium D₃ line, shows the brightest parts of the flare ribbons in which the loop system is anchored. (Courtesy of J. M. Beckers, Sacramento Peak Observatory.)

Mach-4 shocks are capable of heating the corona to $> 10^7$ K. He refers to four X-ray associated prominences with speeds possibly in excess of the speed of sound in the corona (> 150 km s^{-1}).

A detailed study of the DB process has been presented by Hyder (1967a, b). As a rule the DB leads to a successive brightening of the chromosphere in the form of two rows of brightened points distributed along both sides of the space formerly occupied by the filament (Bruzek, 1957). We already said in Section 4.1 that Hyder explains this effect by a flow of the prominence material into the chromosphere along the magnetic field lines, as is indicated in Figure 84. This might be a correct explanation for the events of category (A), for some brightened points in category (B), but certainly not for the large flares in categories (B) and (C). Pustilnik (1973) tries to explain the DB and the DB-associated flares by a balloon mode of a trough instability in a quiescent filament. His ideas, based on Kippenhahn-Schlüter's model of the filament and Syrovatsky's configuration of the magnetic field, are described in more detail in Section 6.5.2.D.

4.4. Loop Prominences

The two-ribbon flares that form in a well-developed active region are often accompanied in Hα light by a phenomenon called loop prominences (Figures 9, 15, 88 and 89a). On rare occasions this loop system is also indicated in spotless two-ribbon flares (Figure 20). Its development is shown in Figure 24, and Bruzek (1969b) describes it in the following way: When observed in Hα on the limb, the flare develops a bright mound which grows at a rate of less than 20 km s^{-1}. At about flare maximum the mound gradually divides into a number of bright loops with material streaming downward in both legs of the loop. This system of loop prominences continues to grow at a steadily reduced rate up to a maximum height of 60 000 km or more, and it may last for ten or more hours after the flare decay. When observed on the disc, the loops resemble bright or dark bridges between the two bright flare ribbons, obviously being rooted in them. Usually, in the Hα line, it is difficult to see the loops on the disc, but in some outstanding cases they form an arcade bridging the zero line and connecting regions of opposite magnetic polarities (Figure 88).

Figure 89a shows a well developed loop prominence system as it appeared 5 h after the onset of a major limb flare in 1968, November 18; also shown is the configuration of the magnetic field lines fitting the system and computed in the current-free approximation from a magnetogram taken when the active region was close to central meridian (Roy, 1972). The fit is surprisingly good. It confirms the arcade structure of the loops and, from the theoretical point of view, it indicates that the flare energy, if stored in the magnetic field before the flare occurrence, had to be almost completely released during the flare phenomenon; if this were not the case, the current-free approximation should not have worked. Of course, one has to keep in mind that, according to Levine and Altschuler (1974), only strong currents produce striking deviations from the current-free approximation in the computed results.

1520 UT

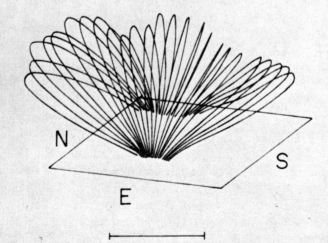

N

S

E

100 ARCSEC

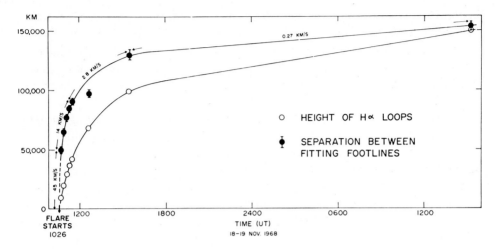

Fig. 89b.   Growth of the loops shown in Figure 89a and the rate of separation of the feet of the loops computed from it. Note the similarity with the observed rate of expansion of the bright flare ribbons shown in Figure 23. (After Roy, 1972.)

The next figure (Figure 89b) shows the measured heights of the Hα loops and the separation of the bright flare ribbons at their feet as deduced from the computations (Roy (1972); compare this with Figure 23). Similar results were obtained by Rust and Roy (1971), and also the big flares of August 1972 showed a relaxation of magnetic field in the post-flare phase manifested by a progressive decrease in the shear of the loops (Zirin and Tanaka, 1973; Tanaka and Nakagawa, 1973).

A superficial look at a loop prominence system might give the impression that the loops expand. This, however, is not the case. The growth of the loop prominence system is due to the generation of higher and higher loops while the smaller ones fade. The lifetime of a single loop is about 20 min (Bruzek, 1969b). Each individual loop starts as a rapid brightening of a knot above existing loops; the knot grows for several minutes and eventually it flows downward along the magnetic field lines.

Loops are visible also in a number of coronal lines (Waldmeier, 1945, 1947, 1960; Evans, 1957; Newkirk, 1957; Meyer, 1968; Rust and Roy, 1971), and, as OSO 7 (Chapman and Neupert, 1975) and Skylab experiments (AS & E, 1975; Robinson and MacQueen, 1975) have shown, also in white-light corona, EUV lines, and soft X-rays (Figure 20b). A correlation of soft X-ray enhancements with loop system occurrence was also reported by Brinkman and Shaw (1972), Machado et al. (1972), and Waldmeier (1973). This shows that besides the low-temperature portion which emits

←

Fig. 89a.   The loop prominence system accompanying the flare of 1968, November 18, as photo-graphed in the Hα line at the Sacramento Peak Observatory. The fitting computed magnetic field lines are presented from the same view angle (90° W) as the prominence on the limb, and also with a view from an angle of 75° W which clearly shows the 'loop tunnel' schematically drawn by Bruzek in Figure 24. (After Roy, 1972.)

in the Hα line there must be loop parts with temperatures in excess of 10^6 K. Pallavicini *et al.* (1975) interpret their observations of the flare of 15 June, 1973 on Skylab as X-ray loop systems formed at successively increasing heights. This indicates that one observes in X-rays the same phenomenon of successively excited loops as it is observed in the Hα line, but an improved spatial resolution is needed to be sure about it.

According to Fisher (1974) the loop structures in Hα and Fe XIV differ significantly. In contradistinction to this, McCabe (1973) found remarkable similarities in the structure of loop prominences when observed in Hα and coronal lines. The picture she obtains appears to be a system of cool loops underlying a corresponding hot loop system, but closely related insofar as the individual associated loops are not entirely separate from each other. Figure 20b rather supports McCabe's conclusion, since the (very weak, but still visible) Hα loops seem to extend to the same altitudes as the X-ray loops.

The Hα line profiles in the loops are often extremely wide, up to ± 60 Å, which corresponds to proton energies of 40 keV and indicates jet-like velocities of the order of 1000 km s^{-1} (Jefferies and Orrall, 1963, 1964, 1965a). According to Hyder (1964) the magnetic field strength inside the loops is about 50 G. This agrees with fields computed by Rust and Roy (1971) who showed that the field strength in the best fitting field lines varied from ~ 10 G at the top ($\sim 100\,000$ km high) to several hundred gauss near the footpoints.

Loop prominences were explained originally as condensation of coronal matter (e.g., Kleczek, 1957, 1958, Lüst and Zirin, 1960; Olson and Lykoudis, 1967). However, there does not seem to be enough material in the pre-flare corona to build the loops (which represent 10^{15} to 10^{16} g), and also the explanation of the high velocities observed meets with difficulties (Kleczek, 1964; Jefferies and Orrall, 1964, 1965b). Therefore, it seems much more probable that the material is injected into the magnetic loops from below, in the form of nonthermal particles accelerated during the flare process (Jefferies and Orrall, 1964, 1965b). Trapped particles with energy of the order of 10 keV and lower are then thermalized near the top of the loop and make the loop structure visible. The total energy loss of a typical loop system exceeds 10^{31} erg by a factor of 2 or 3; hence the loop system is actually a very energetic late phase of the flare event.

Jefferies and Orrall's interpretation is able to explain the broad line profiles, it is not limited in the material available, and it explains why loop prominences occur predominantly in association with the two-ribbon flares in sunspot groups, i.e., flares that produce energetic particles.

Bruzek (1964b) has concluded that loop prominence systems are associated exclusively with proton flares. Švestka (1968) confirmed that essentially all loop prominence systems occur in active regions that were productive of high-energy protons. Therefore, it seems reasonable to suppose that all loop prominence systems are fed by accelerated particles as Jefferies and Orrall assume. However, high-resolution Hα observations in recent years (Rust and Bar, 1973; Michalitsanos and

Kupferman, 1974) revealed clear indications of loops even in spotless active regions (Figure 20) that generally are not believed to be capable of producing high-energy protons (cf. Dodson and Hedeman, 1970). A simultaneous photograph in X-rays (Figure 20b) confirms that even in a spotless flare the hot-loop system can be fully developed. Thus, in order to give rise to the loops, one probably does not need protons accelerated to extremely high energies as in proton flares; a milder acceleration is obviously sufficient to produce them.

More recently, the condensation model has been revived by De (1973, 1975) who assumes that loop prominences are inductive filamentary current paths that begin to form following a flare. This model disagrees with Rust and Roy's (1971) and Roy's (1972) results which showed a good agreement of the loop structure with the current-free approximation. De also does not offer any solution for the old problem (Kleczek, 1964) that there is not enough material in the corona to condense. Sturrock (1973) suggests that this missing material is gas 'evaporated' from the chromosphere. Streams of energetic particles impinge upon the chromosphere, heat the upper layers to $\sim 10^7$ K and the heated gas expands rapidly into the corona. Some of this gas will cool very fast and becomes visible in Hα light. Cool gas at coronal heights will tend to stream down as one observes. It is questionable, however, whether the velocities may be as high as we observe. We shall speak more about the idea of gas evaporation when discussing Hirayama's (1974) model in Section 6.5.1.D.

CHAPTER V

PARTICLE EMISSION FROM SOLAR FLARES

Table XIV summarizes the different kinds of energetic particle emission from the Sun:

Permanent particle emission is observed in association with specific active regions during their transit over the visible hemisphere of the Sun. The particle flux sets in on about the second day after the appearance of the region on the eastern solar limb (cf. Figure 100) and continues until the region moves to about 40° behind the western limb (Fan *et al.*, 1968). Figure 90 shows an example of such a long-lived flux increase. We can see that the 'permanent' character of proton flux is recorded only at very low energies, close to 1 MeV. At energies above 10 MeV the proton flux variations show a clear impulsive nature which raises a doubt that the emission really is of a permanent character. It is more likely that an active region of this type is capable of producing many discrete particle events associated with flares, and it is storage in the solar corona

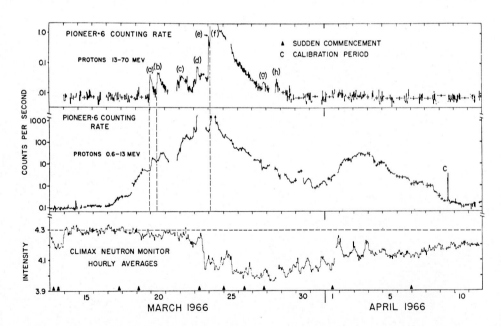

Fig. 90. The 'permanent' emission of protons as recorded at energies >13 MeV (above) and >0.6 MeV (below) during the transit of an active region across the solar disc. (After Fan *et al.*, 1968; *Journal of Geophysical Research*, Copyright by the American Geophysical Union.)

or in space that produces the rather smooth, permanent-like character of the flux variations at lower energies (Fan *et al.*, 1968; Anderson, 1969; McCracken and Rao, 1970). Thus the proton observations do not give convincing evidence that low-energy particles are continuously accelerated in an active region. But such a long-lived flux enhancement obviously characterizes active regions which are very efficient in producing particles, since essentially all large proton flares occur in regions which are also the source of this 'permanent' particle emission (Švestka, 1970b).

This leads us to the *discrete particle events* on the Sun (Table XIV). We can divide them into two classes:

(a) Those which are clearly associated with a flare on the solar disc and set in within a short time interval after the flare, more or less corresponding to the expected travel time of the particles from the Sun to the Earth; we call them *prompt events*; and

(b) Those where the flare association is not so clear, either because no obvious

TABLE XIV

Scheme of the energetic particle emission from the Sun

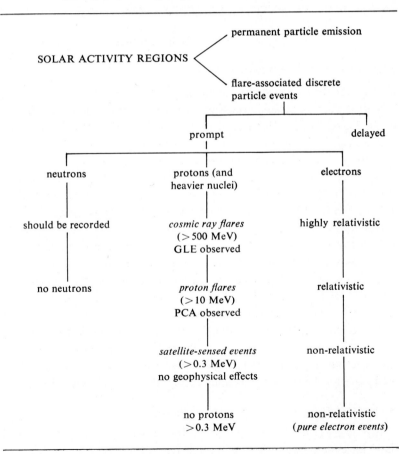

flare can be found as their source, or because of an unusually long interval between the flare and the particles' arrival at the Earth. In the first case we may still be dealing with a 'prompt' event if the flare is hidden behind the western solar limb; or, at lower energies, we may meet with a cloud of particles stored for a long time in space; there may also exist acceleration processes on the Sun or in space of which we yet know nothing. In the other case we meet with a *delayed event*, and this delay can be due to several causes. The flare source can be near the east limb, where the particles lack direct access to the Earth and can be brought here only through diffusion or by a corotation effect; or the particles can be trapped behind a shock in space and get to the Earth only when the shock arrives; the particles can also be delayed through some other kinds of magnetic inhomogeneities in space; and finally the delay may be due to anomalous particle storage in the vicinity of the Sun.

Since in this book we are not interested in the details of particle propagation in space, but only in the origin of the particle flux on the Sun, we shall confine our interest to the prompt events, those clearly associated with flares. The propagation effects will be mentioned only as far as it is necessary to understand the processes on the Sun. In our discussion we shall closely follow the scheme of Table XIV.

5.1. Solar Protons

5.1.1. Cosmic-ray and proton flares

Before 1956 the only evidence for energetic particle production in flares was the observation of flare-associated transient increases in the cosmic ray flux at the surface of the Earth. Such an increase (called 'ground level effect' (GLE)) requires protons with energies in excess of 500 MeV. Flares (called 'cosmic-ray flares') which produce enough particles of such a high energy are very rare. Even with the improved techniques of the neutron monitors after 1956 only twenty such events were recorded from 1942 (the first observation) to 1972. (See reviews by Carmichael (1962), Webber (1964), McCracken (1969), Hultqvist (1969) and Dodson and Hedeman (1969b).)

Our knowledge on the proton emission from flares improved enormously after 1956, when the first riometers at the Earth's polar caps were able to record cosmic noise absorption obviously caused by protons ejected from flares impinging into the polar ionosphere and causing anomalous ionization in the ionospheric *D* layer (for a review see Hultqvist, 1969). In comparison to the GLE's, the protons that cause these 'polar cap absorptions' (PCA) are of much lower energy: below 100 MeV, with maximum effect near 10 MeV. The number of PCA events is about ten times higher than that of the cosmic-ray flares, which demonstrates that the number of proton-producing flares greatly increases with decreasing peak energy of the accelerated particles. The flare events that produce PCA's are often called 'proton flares'.

A. *Characteristics of the Flare on the Sun*

Ellison *et al.* (1961) were the first to draw attention to the fact that all cosmic-ray flares have the typical two-ribbon shape described in Section 2.1.3.E, with the ribbons embedded in large sunspots (Figures 15 and 88). The same is true for the majority of proton flares, in particular for those which produce a strong PCA effect. The flash phase, during which the first acceleration step (and possibly the whole acceleration process) takes place, seems to be coincident with the initial fast expansion of the flare ribbons and the onset of its deceleration (Křivský, 1963; Švestka and Simon, 1969; Figure 91). Nevertheless, as we mentioned in Section 2.2.1, any attempts to find differences in the physical parameters (T_e, n_e, N_2) between proton and non-proton flares from their chromospheric spectra failed (Steshenko, 1971b).

In the radio range, a proton flare is characterized by the occurrence of a type II radio burst (Švestka and Fritzová, 1974) and by a strong type IV burst (Dodson *et al.*,

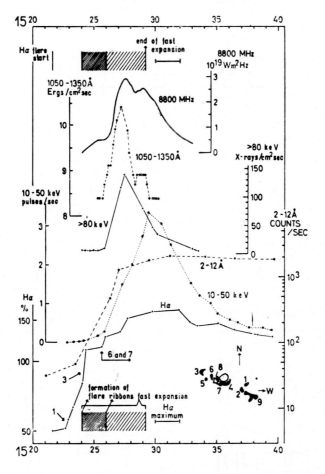

Fig. 91.   A synopsis of the time-development of different kinds of electromagnetic radiation from the proton flare of 1966, August 28. From below: Hα intensity; medium hard (10 to 50 keV), soft (2 to 12 Å) and hard (>80 keV) X-rays; UV-burst; and microwave burst. Note the coincidence of the impulsive phase with the latter part of the fast expansion phase of the flare ribbons. (After Švestka and Simon, 1969.)

1953; Hakura and Goh, 1959; Warwick and Haurwitz, 1962; Pick, 1966; Fritzová
and Švestka, 1966) with the typical spectrum shown in Figure 74 (Castelli *et al.*, 1967;
Castelli and Guidice, 1971), and with abnormally strong emission in the centimeter
and millimeter regions (Croom, 1971). Type III bursts are often present, but their
association with proton flares does not seem to be higher than with flares of other
types; therefore, the type III bursts appear not to be intimately associated with the
proton acceleration process.

The association of type IV bursts with proton flares can be easily understood;
some accelerated particles are trapped in coronal magnetic fields and when the density
of mildly relativistic electrons is large enough then electrons become visible through
synchrotron radiation in the radio continuum. This obviously is the case in the proton
flares in which high-energy protons are essentially always accompanied by relati-

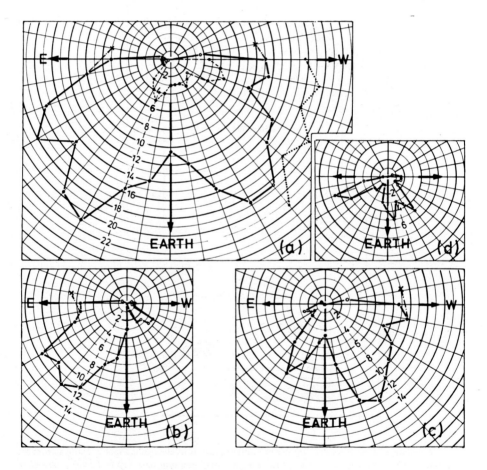

Fig. 92.   Longitudinal distribution of type II bursts, observed from January 1966 to June 1968:
(a) all bursts; (b) bursts not associated with any proton flux near the Earth; (c) bursts associated
with proton events; (d) ambiguous events. (After Švestka and Fritzová, 1974.)

vistic electrons (cf. Table XIV and Section 5.3.2). Besides, the moving type IV (type IVmA, Section 3.2.4.A) indicates the presence of expanding or fast-moving coronal transients above the flare that probably have their origin in the same instability which instigated the acceleration process.

While the association of type IV bursts with proton flares has been known for more than two decades, the correlation between type II bursts and proton acceleration has been proved only in recent years (Švestka and Fritzová, 1974). The very complete *Catalog of Solar Particle Events* (Švestka and Simon, 1975) and improved data of dynamic spectrographs (extended both in time and frequency) made such a proof possible.

Figure 92 shows the distribution of type II bursts in longitude. Graph (a) presents the distribution of all the type II bursts observed, graph (b) shows the distribution of type II bursts that did not produce any particles near the Earth, while graph (c) shows the distribution of those that did. Obviously the proton-producing type II bursts prevail on the western hemisphere, while almost all type II bursts without proton events are on the eastern hemisphere. For particle propagation along the Archimedes spirals this is exactly what one expects if essentially all type II bursts are associated with acceleration of protons on the Sun.

We have shown in Figure 78 that the type II burst originates close to the time of the impulsive microwave and hard X-ray bursts. However, it occurs much less frequently than the impulsive bursts, and its occurrence is clearly related to the importance of the proton event. The correlation between type II and protons near the Earth is at least 73% (and possibly 100%) for the PCA events, less than 38% (and possibly only 15%) for events with >10 MeV proton flux below 10^{-2} protons cm^{-2} s^{-1} sr^{-1}, and essentially zero (i.e., equal to chance coincidence) for pure electron events. This seems to indicate that the type II burst is the manifestation of a second step acceleration process, accelerating to higher energies some of the particles that were pre-accelerated during the impulsive phase.

According to McLean *et al.* (1971) this second step acceleration is accomplished as the type II bursts interact with an arch-shaped magnetic flux tube (cf. Figure 73 and Section 3.2.4.C). McLean *et al.* have based this supposition on the fact that the type IV burst pulsations (Figure 72) are closely related to proton flares. Since, however, type IV bursts generally are associated with proton acceleration, their conclusion might be misleading.

Proton flares are accompanied by intense hard X-ray bursts. Strong events often have multiple maxima and usually a fairly long duration of several minutes. This complexity and long duration of hard X-ray bursts may be considered a characteristic of proton flares. Generally, however, a hard X-ray burst is not a unique indicator of the acceleration of protons; according to Arnoldy *et al.* (1968a) there is a high correlation between all optical flares of importance $\geqq 2$ and hard X-ray bursts in the energy range 10 to 50 keV. Also electron events are associated with hard X-ray bursts, as we shall see in Section 5.3.1.A. On the other hand, a statistical relation exists even between proton flares and the soft (2 to 12 Å) X-ray bursts (Švestka, 1970a; Sarris and Shawhan, 1973): As the ratio of the peak of the X-ray burst to the quiet Sun level increases, the percentage association with protons in space continuously grows from 13% for ratio < 20 to 80% for ratio > 80.

Newkirk (1973) believes to have found differences in the ambient magnetic field configuration between proton and non-proton flares; according to his results proton flares have significantly more open field lines emerging from near their site than do flares of the same importance which do not produce protons in space. Thus the ambient field configuration may significantly influence whether or not protons escape from the solar corona into interplanetary space. One hardly can expect to see protons from a flare that does not have at least some of the basic characteristics of the proton flares; on the other hand there are events (cf., e.g., Fritzová and Švestka, 1966; Švestka and Fritzová, 1974) when the proton flare characteristics are well developed, the flare is favorably located on the Sun and still no protons, or only very few, arrive at the Earth. One can suspect that in such a case the Newkirk criterion has played the most significant role.

B. *Characteristics of the Active Region*

The active regions in which proton flares occur are characterized by several specific properties:

(a) The region is always of a magnetically complex type, i.e., some mixing of magnetic polarities occurs. This is not surprising since we have seen in Section 2.1.2 that essentially all acceleration (i.e., non-thermal) processes occur in active regions which are magnetically complex. Thus probably the formation of a powerful proton flare is just a matter of the scale of complexity.

As Martres (1968) has shown, active regions of magnetically complex type occur only when a new region forms within the remnants of an old one. The complexity of the newly formed active region is greater if the old active region is younger. Thus, there are always some areas on the Sun where proton active regions can, but of course need not, be born, and other areas where the occurrence of such regions is essentially impossible.

(b) Of course, only some of the complex sunspot groups produce proton flares. A necessary condition is a substantial increase of the gradient of the magnetic field in the active region (Gopasyuk *et al.*, 1962; Severny, 1964b, 1969b; see Table IV and Figure 16) manifested by a close approach of large spots of opposite magnetic polarity (Avignon *et al.*, 1963). Such a configuration is particularly favorable for the production of energetic particles (C.S. Warwick, 1966) if a common penumbra is formed embracing spots of different polarities (the so-called δ-configuration, as in Figure 34). The most important particle events appear when high gradients of the magnetic field exist within a large portion of the active region, which is manifested optically by the close approach of two rows of spots of opposite polarity (Figure 93). Such a situation has been called an A-configuration by Avignon *et al.* (1964, 1965). According to Sakurai (1972) high gradients are most likely to occur in a sunspot group with rotating motion. Therefore, the sunspot groups which show a reversed polarity distribution

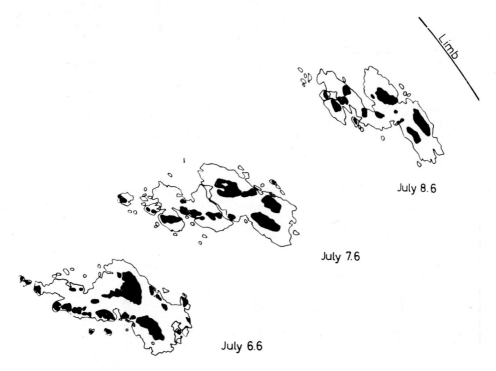

Fig. 93.   Example of the A-configuration in a sunspot group. The upper row of spots had a southern, and the lower row predominantly a northern magnetic polarity. The group produced the cosmic-ray flare of 1966, July 7. (After C. Sawyer, 1968.)

appear to be very productive in proton flares (see examples in McIntosh, 1969; Sawyer and Smith, 1970; Zirin and Tanaka, 1973).

(c) As a rule, the magnetic field gradient increase in the active region is associated with an intensification and a hardening both of the microwave and X-ray spectrum. In the microwave range the radio emission increases strongly (Pick, 1961) and the usual maximum close to 10 cm is submerged in the greatly increased flux at shorter wavelengths so that the microwave spectrum flattens and eventually the maximum shifts toward centimeter wavelengths (Tanaka and Kakinuma, 1964; Tanaka *et al.*, 1969; Tanaka and Énomé, 1975). At the same time, the X-ray flux also increases. This increase often exceeds one order of magnitude in the spectral range of a few angströms, decreases with increasing wavelength, and becomes insignificant for wavelengths above 40 Å (Křivský and Nestorov, 1968; Friedman and Kreplin, 1969; Švestka and Simon, 1969; Horan and Kreplin, 1969). Krüger (1972b) finds the most significant variations for 0 to 8 Å in X-rays and ~3 cm in microwaves.

According to Figure 45 this behavior indicates an increase in temperature in the coronal condensation above the active region, while the microwave flux variations also indicate an increased electron density (Krüger, 1969a). Skylab observations, which unfortunately did not include many big flares, have revealed that the variations

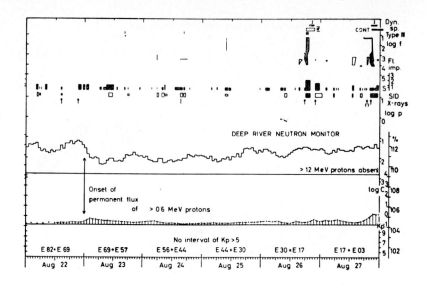

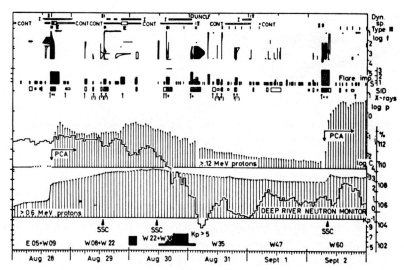

Fig. 94. Schematic summary of activity in the region which produced the proton flares of 1966,
August 28 and September 2. (After Švestka and Simon, 1969.)

From above: radio records by dynamic spectrographs (types I, II, III, IV and continuum); spectral
diagram of the radio activity composed from single-frequency observations at different frequencies f;
Hα flares: the height of the black square marks the importance; SID's and X-ray bursts; flux of
>12 MeV protons; flux of >0.6 MeV protons; neutron monitor record (showing Forbush decreases);
and geomagnetic activity. The longitude of the group is marked at the bottom, above the date.

Note the inactivity of the flares in the radio range before August 26. After that date, when the
region became magnetically complex, more than one half of all flares and subflares were accompanied
by a radio burst. This began to happen about two days before the first big proton flare occurred.

The onset of permanent flux on August 23 was due to particles stored from the preceding rotations
when another region in approximately the same position was a powerful source of protons. Only on
August 27 did the present region begin to contribute to the particle flux.

Note that any small increases in the proton flux, which might have been produced by many flares
in this region, could not be detected after August 29, due to the high remnant-flux after the proton
flare, in particular in the >0.6 MeV energy range.

are due to the appearance of new very bright and hot cores in the coronal condensation.

However, similarly to de Feiter and de Jager's (1973) 'superthermal plasma nodules' (cf. Section 3.1.1), it is difficult to judge whether these hot cores represent a pre-flare emerging flux or whether they are formed during earlier flares and simply survive them.

(d) During the same time, the association of flares with X-ray and microwave bursts is greatly increased (Eliseev and Moiseev, 1965; Eliseeva, 1967; Křivský and Nestorov, 1968; Švestka and Simon, 1969). Figure 94 shows an example of this.

(e) The region becomes a source of 'permanent' emission of protons with peak energy of the order of 1 MeV. As Bruzek (1969a) has observed, the proton flare may be preceded by smaller flares which already possess some of the characteristic features which characterize the proton flare (cf. Figure 94).

Some other characteristics have been noted by Banin et al. (1969), McIntosh (1972), Severny and Steshenko (1969), Simon et al. (1969), Simon and Švestka (1969), and Švestka and Simon (1969), probably of less general significance. It is necessary to emphasize, however, that the effects (a)–(e) are not observed in all cases. Usually only some of them are well developed. Nevertheless their validity is general to the extent that short-term forecast of proton flares can be based on them (Severny and Steshenko, 1969; Simon et al., 1969).

Since the characteristics described above start to develop about two or three days before the proton flare occurrence (cf. Figure 94), a 'proton flare alert' can be issued several tens of hours or at least a few hours before the event. As it appears, the success of these forecasts is certainly not worse than that of weather forecasting on the Earth. Severny and Steshenko (1972) claim promising success in the forecasts made at Crimea; during the 1966 Proton Flare Project a Sparmo balloon was launched 4 h prior to the flare of July 7 following an urgent second alert issued by P. Simon in Meudon. One has to admit that many alerts are in vain; on the other hand, however, with the exception of flares close to the limb, very few proton flares occur without an alert being issued prior to them.

C. Complexes of Activity

It has been proved by several authors that the probability of formation of new active regions is not distributed uniformly with solar longitude. At any time there exist one or more sectors in longitude where the probability is increased, and the increase is most striking for the most important solar phenomena (C.S. Warwick, 1965; Sakurai, 1966; Haurwitz, 1968; Švestka, 1968c; Vitinskij, 1969a, b; Dodson and Hedeman, 1964b, 1969a). Since the most important activity, proton flares included, occurs in complex active regions, and these regions only form within remnants of pre-existing old regions (cf. item (a) in the preceding section), this fact can easily be understood.

Let us suppose that the probability of a new region formation per unit area and time is p_c within the longitudes where the probability is increased (these areas are called 'active longitudes' or 'complexes of activity'), while p_0 is the probability for the rest of the solar disc. Then p_c/p_0 more regions will be formed per unit area in a complex of activity than anywhere else on the disc. If a new region is formed, the probability that it coincides in position with remnants of an old existing region is then proportional to p_c^2 or p_0^2, respectively; thus the relative probability of occurrence (p_c/p_0) increases as $(p_c/p_0)^2$ for magnetically complex regions, i.e., also for the proton flare active regions.

These considerations are confirmed by observations. Complexes of activity producing proton flares at a significantly increased rate were demonstrated by Švestka (1968a, c) and Gros *et al.* (1971). Figure 95a shows the dependence of the intensity

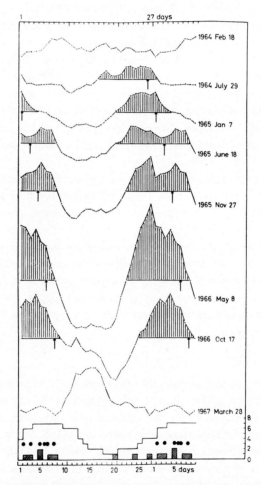

Fig. 95a. Intensity of the 2800 MHz flux (Ottawa) plotted as a function of the 27-day period from 1964 to 1967. Each curve represents a sum of twelve successive rotations and the date gives the first day of this period. Flux above the average is hatched. At the bottom is plotted the number of days on which Pioneers 6 and 7 recorded increased flux of >0.6 MeV protons. Dots and squares show the CMP days of the regions which produced proton flares and permanent proton flux, respectively (After Švestka and Simon, 1969.)

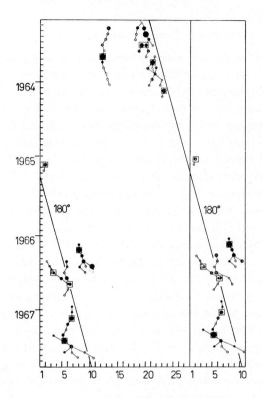

Fig. 95b. Days of CMP of regions that gave rise to proton flares from 1963 to 1967. Proton-active regions are marked with squares, sunspot groups with dots, plages with small open circles. Two lines (180°) denote (After Švestka, 1970b).

of the 2800 MHz radio flux on the 27-day solar rotation. The flux is closely correlated with the degree of activity (cf. the correlation with microwave bursts in Figure 79b) and thus with the occurrence of complex active regions. One can see how the 'active longitude', represented by the shadowed maxima, persisted on the Sun for more than two years. Dots and squares at the bottom of the figure show the CMP days of regions which produced proton flares and permanent proton flux, respectively. The right-hand part of the figure (95b) shows that during the whole period all the proton flares that appeared were close to this active longitude on the solar disc.

Possibly this fact could be used for long-term forecasts of proton flares, or at least for forecasts of the 'quiet' periods. However, as Švestka (1968c) has shown, due to the rather wide cone into which the protons propagate in space (not to mention the many propagation anomalies, cf. Section 5.1.3.A), the practical application of this cumulation of the proton-active regions is fairly small for forecasting purposes. As a matter of fact, as soon as two such active longitudes exist on the disc (which is the most common case, the only exception being the few years close to the solar minimum), particles from the Sun can arrive at the Earth at any time.

D. *Time Variations in the Proton Flare Occurrence*

There are many active regions which produce only one proton flare during their existence (but they may produce many more small particle events, which we discuss in Section 5.1.2). On the other hand, specific regions produce quite a number of very powerful proton flares during their transit over the visible solar disc. Thus, for example, the well-known region of November 1960 produced 6 proton flares, three of them being cosmic-ray flares. Yet another event occurred on the east limb when the region rotated back onto the visible disc (cf., e.g., Švestka and Olmr, 1966).

The time intervals separating the three cosmic-ray flares in November 1960 were 61 and 138 h. An analysis of 27 pairs of successive proton flares which originated in identical regions shows that the shortest time interval between them was 13 h (flares in 1960, March 30, at 02 16 and 14 55 UT), and the median interval was 55 h. Thus, obviously, some tens of hours and about two days on the average are necessary to restore a magnetic situation capable of the production of a new proton flare, and to build up the energy necessary for this process. This agrees with the fact mentioned in Section B that the characteristic changes in an active region start about two or three days prior to a major proton flare occurrence.

Generally, the number of particle events and their importance increase with increasing sunspot number in the solar cycle. During 1 or 2 yr near the minimum of solar activity, flare-associated particle events are extremely rare (e.g., during 1964 (cf. Figure 96) only one very weak PCA was recorded). As the sunspot number rises, particle events occur more frequently, but, somewhat surprisingly, the increase stops before the solar maximum is reached. During solar cycles 18 and 19 the cosmic-ray flares appeared to avoid the cycle maxima completely (Carmichael, 1962; Obayashi, 1964) and the same tendency has been found for strong PCA events (Švestka, 1966b; Sakurai, 1967; Hakura, 1974). The highest number of particle events and the strongest of them occurred about 1 yr before and 2 or 3 yr after the maximum of the sunspot cycle.

This is not a particle propagation effect, since the same avoidance of the maximum has been found for the most intense microwave radio bursts (Boorman *et al.*, 1961; Takakura and Ono, 1962; Fokker, 1963a), type IV bursts (Křivský and Krüger, 1966; Papagiannis *et al.*, 1972), and in a very pronounced form for the intensity of the green coronal line (Gnevyshev, 1963, 1967; Antalová and Gnevyshev, 1965). In the present cycle No. 20 the rule of 'maximum-avoidance' appears to be broken, since three cosmic-ray flares were observed very close to the cycle maximum (Dodson and Hedeman, 1969b; McCracken and Rao, 1970; Švestka, 1970b) and the number of proton flares did not decrease during the maximum of the cycle (cf. Figure 96). Thus, since this maximum was much lower than the two preceding ones, one may conclude that the most important activity phenomena on the Sun do not avoid the phase of the solar maximum, but simply the periods when the sunspot number is extremely high. But even in this rather low cycle a series of outstanding events occurred again in August 1972, i.e., more than 3 yr after the cycle maximum (see

Table III, Dodson and Hedeman (1973) and Fritzová and Švestka (1973) for comparison to the previous cycles).

A plausible explanation might be that the active region does not find a 'quiet' time long enough to build up the conditions necessary for the production of what is actually a big event, if the general activity of the Sun is too high. Due to many sources of disturbances the flare mechanism is triggered before the conditions for a large event can develop. An alternative explanation is that by Gnevyshev (1963), who has suggested that the solar cycle consists of two waves which peak individually before and after the apparent (composite) maximum of the cycle. Gnevyshev and Křivský (1966) and Papagiannis et al. (1972) supported Gnevyshev's idea. However, the fact

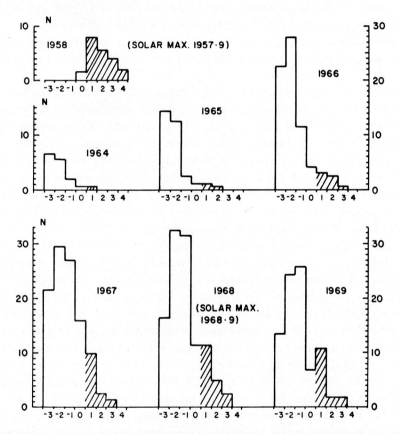

Fig. 96.   Yearly numbers of solar proton events from the solar minimum in 1964 through the maximum of activity in 1968/9. The year 1958, close to the maximum of the preceding cycle, is shown for comparison. (Data taken from Švestka and Simon, 1975.)

The x-axis gives powers of ten in the number of >10 MeV protons $cm^{-2}s^{-1}sr^{-1}$ at the event maximum. Thus, e.g., N in column 2 means 'yearly number of proton events with >10 MeV proton flux between 10^2-10^3 particles $cm^{-2}s^{-1}sr^{-1}$.

Shadowed columns denote events detectable at the Earth. Unshaded columns refer to satellite measurements. The PCA's were transformed into the >10 MeV flux by means of the Shea and Smart classification (cf. Section 5.1.3.C): 110 is taken here as class 1; 120 as 1.5, etc. Uncertain PCA's were classified as 0.5 and that leads to the small unshaded part in the 1958 histogram.

that Papagiannis *et al.* found the second maximum of cycle 20 as early as 1969 while still higher bursts of activity occurred in 1972 (similarly to late bursts in 1942 and 1963) casts serious doubt on this explanation (Fritzová and Švestka, 1973). If the solar cycle is composed of separable waves, there must be more than two impulses of activity. Some of the late impulses may be particularly efficient, since the build-up process is not perturbed by other active processes in nearby parts of the solar atmosphere, as is the case when the total level of the activity is high.

5.1.2. SATELLITE-SENSED EVENTS

The enormous variety of particle emission from the Sun was only fully recognized when direct particle measurements began with satellites and space probes. These measurements now extend our knowledge down to 0.3 MeV for protons, they give us direct information on the proton energy spectrum, and they distinguish different particle species (protons, α-particles, heavier nuclei, and electrons) in the recorded flux.

In order to get some idea of the size distribution of proton events, see Figure 96, which shows the observed number of proton events of different size in the years 1958 (1 yr after the maximum of solar cycle 19, without any satellite measurements available) and 1964–1968, showing the proton activity during the rising part of cycle 20, with satellite observations continuously improving. The PCA events (i.e., the 'proton flares' discussed before) are shaded. The data have been taken from the *Catalog of Solar Particle Events*, 1955–1969 (Švestka and Simon, 1975). This figure shows clearly that the number of low-energy 'satellite-sensed events' exceeds by many times the number of 'proton flares' detectable by ground-based equipment.

The characteristics of large proton flares specified in Section 5.1.1.A are not valid for the majority of the small particle events. Some of the source-flares still possess at least some of the characteristic features, i.e., a two-ribbon shape is indicated in them or a type IV (or type II) burst is observed; but others, in particular those which produce very small effects, cannot be distinguished easily from any other flare. The characteristic that is most persistent even in small events seems to be a strong microwave burst. But this is not a unique characteristic of proton flares; pure electron events are also associated with strong microwave bursts, as we shall see in Section 5.3.1.

Van Hollebeke *et al.* (1975) have studied 185 solar particle events recorded on IMPs 4 and 5. They were able to determine the parent flare for 68% of the events and believe that most of the unidentified increases occurred in the invisible hemisphere of the Sun. This explanation is acceptable, since according to Fritzová and Švestka (1971) the percentage of proton sources behind the solar limb should be $\sim 38\%$ for weak proton events (cf. Section 5.1.3.A). However, Van Hollebeke *et al.* based their statistics on observations of relativistic electrons (with energies from 0.5 to 1.1 MeV) which obviously show a very high correlation with flares. When the statistic is based on low-energy protons only (also counting the events when relativistic electrons are missing), one finds many cases that do not seem to be flare-associated at all. (Compare the *Catalog of Solar Particle Events*, which comprises more than 700 cases (Švestka and Simon, 1975).) The conclusion that there is no association with flares is based on three criteria: (a) There is no known active region behind the solar limb that would be a likely candidate for proton acceleration. (b) The time-development of the proton flux is gradual, atypical for flare events (cf. flare-associated curves with fast rise time

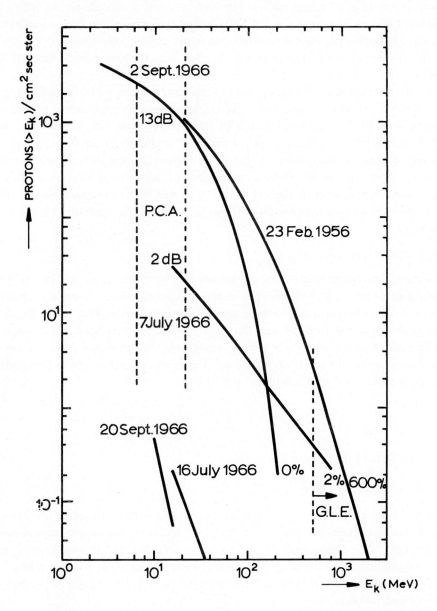

Fig. 97. Energy spectra of three large and two small proton events corresponding to the maximum flux at different energies. The dashed lines show the range of energies most important for the production of a PCA effect, and the lowest energy limit for the production of a GLE. (After Švestka, 1971c.)

in Figure 98). (c) The events recur in 27-day periods. So far, we do not know how these, probably non-flare, proton increases originate.

The small particle events that are flare-associated still originate in the same active regions that are typical for proton event production (Section 5.1.1.B). For example, about one third of all events recorded in 1966 occurred in a single region, which also produced one strong proton (i.e., PCA) flare. This fact implies that some small particle events may be missed, since it is impossible to detect their very small contribution to the greatly increased flux after a proton flare (cf., e.g., the high particle flux in Figure 94).

The concentration of all kinds of proton emission to a few particular active regions indicates that proton acceleration is accomplished only in some active regions of a specific type. McCracken and Rao (1970) are of the opinion that essentially all flares accelerate particles and that it is only a matter of sensitivity of our detectors whether we record them or not. However, the much higher frequency of particle events in specific active regions, the sudden appearance of acceleration processes when the sunspot configuration is changed (Figure 94), as well as the difference between flares with and without the impulsive phase (Section 3.2.1.A) rather indicate that this is not the case.

5.1.3. THE ENERGY SPECTRUM

Spectra of several proton events are shown in Figure 97. These are the 'average' spectra corresponding at all energies to the maximum flux. The 'instant' spectrum varies greatly with time. As the most energetic protons arrive first, while the lower-energy protons are still on the way, the initial spectrum is very hard. Later, when the main bulk of protons of lower energy arrives, the flux at high energies is already decreasing so that the spectrum softens (cf. Figure 98). At high energies the spectrum can be approximated by a differential power law,

$$\frac{\mathrm{d}N}{\mathrm{d}E} = CE^{-\gamma}, \tag{98}$$

or in the form of the integral power-law,

$$N(>E) = KE^{-\gamma'}, \tag{99}$$

where $\gamma' = \gamma - 1$. At lower energies, however, the spectrum becomes flat and thus the power-law approximation fails to fit the whole spectrum. Of course, partial energy ranges can still be approximated by power laws with different values of the exponent γ as many authors often do when only a small energy range is observed by them.

The fit with the spectrum improves if an exponential law in rigidity, P, is assumed (Freier and Webber, 1963),

$$\frac{\mathrm{d}N}{\mathrm{d}P} = \frac{\mathrm{d}N_0}{\mathrm{d}P_0} \exp\left[-P/P_0(t)\right]. \tag{100}$$

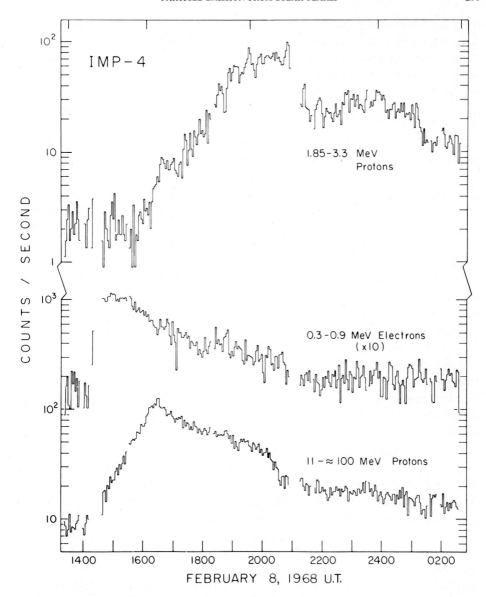

Fig. 98. Records of low-energy protons, mildly relativistic electrons, and >11 MeV protons in the flare of 1968, February 8 (after Simnett, 1972a). Note the differences in the time of the flux maxima for particles with different velocities of propagation.

Here P_0 is the characteristic rigidity, variable in time, which characterizes the hardness of the spectrum, similar to γ in Equation (98). (For energies $\gtrsim 30$ Mev, $P_0 \approx 200$ MV approximately corresponds to $\gamma \approx 2$.) However, even this approximation no longer fits for proton energies below ~ 20 MeV.

As a matter of fact, the spectra do not give us much information on the actual spectrum produced by the acceleration process. Due to the energy-dependent time-shift in propagation, at no time does the spectrum represent the spectrum of the source. Perhaps the best approximation is a spectrum constructed from the maximum flux values (Figure 97), but even this is probably greatly deformed by effects described in the following Sections A and B.

A. *The Propagation Effects*

The fraction of the particle flux which is recorded near the Earth depends strongly on the position of the source on the Sun. Since the transverse particle diffusion across the magnetic field is small in interplanetary space, the particles essentially propagate along the magnetic field lines and significant deviations from this mode of propagation occur only if strong magnetic inhomogeneities in space deflect the particles from their path. Thus theoretically the strongest particle event should be observed when the flare source is close to the root of the magnetic field line which directly connects the Sun to the Earth; from such a position on the Sun the particles have direct access along the magnetic lines of force to the Earth. Under solar wind conditions this root point should be located about 50° W from the central meridian of the Sun; it can come nearer to the center of the disc if strong activity precedes the particle emission, but generally the flares situated at medium longitudes on the western hemisphere are in the most favorable location for efficient propagation to the Earth.

If the particle source is far from the root point, the particles must first get to the field lines which lead to the Earth. Under normal conditions this is a very slow process in interplanetary space, but it probably can be accomplished much faster through diffusion close to the Sun (Reid, 1964; Axford, 1965), where the magnetic field structure is still fairly disordered, with many magnetic field inhomogeneities which permit diffusion in the lateral direction (McCracken and Rao, 1970).

Another way for particles emitted on the eastern hemisphere to get to the Earth is by the corotation effect; however, this causes a delayed event which we have excluded from our consideration. Obviously in any case when the source is far from the root point, only a small fraction of the accelerated particles get access to the Earth, and this fraction may be different for particles of different species and of different energy. The rise time of the proton flux near the Earth is fastest for events that originate between 20° and 80° W (Figure 99) and increases with increasing distance from the Earth-Sun field-line foot point (Barouch *et al.*, 1971; Simnett, 1971; Datlowe, 1971; Reinhard and Wibberenz, 1974; van Hollebeke *et al.*, 1975). The anomalously long rise time for protons originating far on the eastern hemisphere can help when looking in ambiguous cases for their source on the Sun.

It is important to realize that, if corotation is not taken into account, a flare on the central solar meridian is equivalent to a flare 10° behind the western solar limb, as far as the propagation is concerned (cf., e.g., Burlaga, 1967). Therefore, one must expect that a significant fraction of particle events is produced by flares on the invisible solar hemisphere. Statistically, these expectations are confirmed and a scheme of the observed distribution of the sources of strong and weak proton events is shown in

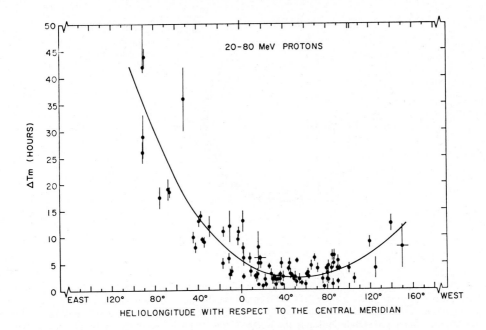

Fig. 99.   The time difference between onset and maximum flux of 20 to 80 MeV protons as a function of the longitude. The solid line is a least square fit through the data, showing a minimum at ∼50° W.
(After Van Hollebeke *et al.*, 1975.)

Figure 100. According to Švestka (1966b) 84% of all strong PCA's before 1963 had their origin between 10° E from the central meridian and the western limb; and about 25% of PCA's are caused by flares behind the limb, the percentage being ≳38% for very weak PCA events and essentially zero for PCA's above 8 dB (Fritzová and Švestka, 1971). For a later period (1955–1967) Figure 101 a shows the dependence of the prompt event occurrence and intensity upon solar longitude. The data are from the Pioneer and IMP spacecraft (McCracken *et al.*, 1967; Lin and Anderson, 1967). Also the time delay between flare occurrence on the Sun and the onset of the particle event at the Earth increases toward the eastern hemisphere (McCracken and Palmeira, 1960; Obayashi, 1964; Haurwitz *et al.*, 1965; Fritzová and Švestka, 1966; Švestka, 1966; Barouch *et al.*, 1971; Figure 101 b). As we shall see in Section 5.3.1.A, the dependence on flare position is still more pronounced for electron events.

However, deviations from this rule may be quite significant in some individual events. Three of these are also indicated in Figure 100. GLE's were observed both from the flare of 1960, September 3, on the eastern limb, and from a supposed flare of 1967 January 28, about 60° behind the western limb (Lockwood, 1968; Dodson and Hedeman, 1969a). On 1966, July 16, protons were recorded even from a flare situated just opposite to the Earth on the invisible solar hemisphere (some ejections from it were observed by Dodson *et al.*, 1969), and there are events when the energetic

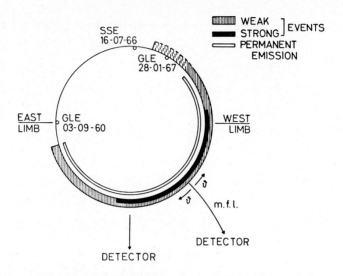

Fig. 100. Schematic drawing of the distribution of the sources of strong (black) and weak (hatched) proton events on the Sun (after Švestka, 1971 c). ϑ measures the angular distance from the root point of the Sun-Earth magnetic field line (m.f.l.). The arrow at the bottom shows the direction to the Earth. The inner semi-circle indicates the range of longitudes from which the permanent particle emission of specific active regions is recorded. See the text for the three anomalous events plotted in the graph.

protons spread essentially over 360° longitude in interplanetary space (Bukata *et al.*, 1972). The causes of such an anomalous particle propagation may be sought both in the corona as well as in interplanetary space.

Palmer and Smerd (1972), Roelof (1973), Duggal and Pomerantz (1973), Simnett (1974), Cherki *et al.* (1974), Reinhard and Wibberenz (1974) and Kirsch and Münch (1974) have presented evidence that accelerated particles should propagate rather fast over large distances in the solar corona or close to the Sun (e.g., over 70° in ~15 min after Duggal and Pomerantz) in order to explain the propagation effects observed. In particular, the following anomalies have been used as support for this supposition: (1) There are cases when low-energy protons reach maximum earlier than those of higher energies; a possible interpretation is that high-energy protons escape fast, but without any direct connection to the observing site, which they can reach only through diffusion; on the other hand, low energy particles are stored for a while and transported in the corona to a position directly connected to the observing site (Palmer and Smerd, (1972; Cherki *et al.* 1974). (2) There are cases when the times of proton flux maximum at different Pioneer space probes surrounding the Sun are not in agreement with the position of the proton flare on the Sun. However, one can improve the agreement if the particles are transported in the solar corona before they are released into space (Simnett). (3) Anisotropy in the particle flux observed at the

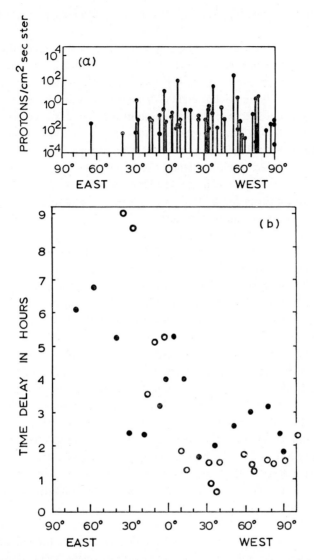

Fig. 101.   (a) The longitude distribution of the flares that were sources of proton events in 1965-1967.
(After McCracken *et al.*, 1967, and Lin and Anderson, 1967.)

(b) The longitude dependence of the delay time between the onset of the flare and the associated
particle flux. Open circles denote delay times as determined in 1969 from the flux of 6–25 MeV
protons by Barouch *et al.* (1971). The filled circles are smoothed means of the delay time for groups
of 10 flares, subsequent in longitude, which produced PCA's between 1956 and 1963. (After Fritzová
and Švestka, 1966.)

Earth from events from behind the western limb can be explained much easier if the
protons are transported to a more favorable position before being released from the
Sun (Duggal and Pomerantz). (4) The flat part of the curves in Figures 99 and 101b on
the western hemisphere can be interpreted as a region in which the particles can propa-

gate across ∼60° in solar longitude in less than 1 h after flare onset (Reinhard and Wibberenz, 1974).

We have seen how much variety one can find in the wave propagation and particle cloud motions in the solar atmosphere; this indicates that the accelerated particles too may possibly propagate to quite some distances from their source before being released into space. Fisk and Schatten (1972) suggest that the propagation takes place along very thin current sheets separating discontinuous field structures in the corona. Palmer and Smerd (1972) have suggested shock waves as the transport mechanism, while according to Simnett (1974) particles are transported through magnetic loops connecting distant active regions. Since we know that shock waves (i.e., type II bursts) are closely connected with the acceleration process (cf. Section 5.1.1.A), the transport mechanism proposed by Palmer and Smerd seems to be the most likely one.

In interplanetary space the real structure of the magnetic field can also deviate, often quite significantly, from the ideal simple configuration of the Archimedes spiral (see, e.g., Keath et al., 1971). As Jokipii and Parker (1969) have pointed out, there should be a stochastic wandering of the magnetic field lines in space because they are anchored in the waving convective cells of the solar photosphere. This process can lead to a significant 'meandering' of the magnetic lines of force (McCracken et al., 1967; Fan et al., 1968) so that the particles guided by the field lines can disperse to a wider cone than a 'combed' Archimedes spiral structure predicts. On a long-time scale this field-line wandering also effectively increases the transverse diffusion, and this can explain the wide cone of the 'permanent' particle emission. It may be that this process does not play any significant role in the prompt events, since their time scale is short. However, one can also expect that the magnetic field configuration is temporarily modified by the shock waves that travel from the Sun (Haurwitz et al., 1965). As Gopasyuk and Křivský (1967) have pointed out, the shock wave (and even more so, a system of two shock waves originating in different active regions) can forge anomalous transient paths for the traveling particles which can then propagate to directions quite inexplicable from the simple Archimedes spiral model.

B. *The Storage Effects*

There are indications that particles may also be stored close to the Sun and/or in space for many hours and even days. Storage in the vicinity of the Sun is implied by several lines of evidence: (a) from the persistence of field-aligned proton flux anisotropies for times long compared to the theoretical expectations for a prompt particle release (Bartley et al., 1966); (b) from delayed flux maxima observed in some events in spite of the fact that the flare source was quite close to the root point of the Sun-Earth magnetic field line (Bryant et al., 1965; McDonald, 1970); (c) from the existence of delayed and even recurrent events (Anderson, 1969; McDonald and Desai, 1971); (d) from the fact that only a small fraction (<1%) of the electrons producing the observed radio and X-ray emission escape into space (Holt and Ramaty, 1969; Lin and Hudson, 1971); and finally (e) from the observed evidence of relativistic electron storage in type IV bursts. McCracken and Rao (1970) have made a rough estimate that while the trapping time is less than 1 h for protons in excess of 10 MeV, it may take many hours for proton energies close to and below 1 MeV. Simnett and Holt (1971) when discussing the period 1968, July 6–16, arrived at the conclusion that protons below 30 MeV and electrons up to highly relativistic energies were stored in the solar corona for at least five days. Elliot (1972), in a detailed discussion of the storage problem, arrived at a similar result.

Some other authors, on the other hand, doubt that storage near the Sun would be so important and suggest that the observed effects, ascribed to storage, must be explained through some continuous acceleration process.

Krimigis (1973) has deduced from the shape of the energy spectra that the low-energy protons do not suffer significant collision losses and thus they cannot be stored nearer to the Sun than ~ 2 to 3 solar radii. However, magnetic loop structure rarely appear at heights $> 2 R_\odot$. Since it is generally assumed that such structures are necessary for long-term particle confinement, Krimigis finds effective trapping and storage of low-energy protons very unlikely.

Newkirk (1973) has calculated ambient magnetic fields near proton flares (assuming a potential field and using Mt. Wilson photospheric magnetic data) and traced particles injected into these fields. A statistical examination of these calculations shows that $\sim 60\%$ of particles find themselves initially trapped and mirroring in closed magnetic loops, 16 to 40% escape immediately into interplanetary space and 1 to 15% impact onto the photosphere. The mean electron density encountered by particles mirroring in the closed magnetic fields permits us to estimate how long the particles can stay in these loops. Newkirk has found mean electron densities smaller than $\sim 3 \times 10^7$ cm^{-3} only for less than 10% of the mirroring particles. Thus, similarly to Krimigis, he concludes that storage for more than a few hours is very unlikely, e.g., only less than 10% of 3 MeV protons can be stored for more than 8 h, and $\ll 1\%$ for more than three days.

There are some ways to prolong the storage time: The loops in which the particles are stored might be abnormally devoid of coronal gas (Newkirk); the storage may occur only at the tops of the arches (Newkirk); or the loop may be predominantly filled with energetic particles (approaching Maxwellian distribution for $kT \simeq 30$ keV) thus reducing the collisional losses (Anderson, 1972b). Still, it seems rather dubious that < 10 MeV protons might be stored near the Sun for periods longer than a few hours.

In space the particles are reflected from the disordered field on the boundary of the solar magnetosphere (at distances of 2 to 10 AU, the estimates vary considerably for different events and authors) so that a significant fraction of them come back to the interior part of the solar system.

During storage the particles lose their energy. Thus in the late phase of a particle event both the particles propagating from the Sun and those stored in interplanetary space contribute to the spectrum, many of them with energies different from the energy they acquired during the acceleration process a few hours or days before.

C. *The Size of an Event*

In spite of all these disturbing factors, some interesting and important information can be deduced from the spectra obtained. Let us consider as an example the three large flares in Figure 97. They all occurred on the western hemisphere, between 48° and 80° W. Thus they were favorably located on the disc and their spectra may be compared.

When the same propagation conditions are assumed, the size of a particle event is determined by two parameters: the total number of accelerated particles in the flare, i.e. the extent of the accelerating region on the Sun, and the slope of the energy spectrum, i.e. the efficiency of the acceleration process. Thus one might conclude from Figure 97 that the acceleration process in the 1966, July 7 flare was extremely efficient (the spectrum is very flat), but limited to a relatively small volume (the total number of particles is small); the acceleration in the February 1956 flare was somewhat less efficient (the spectrum is steeper), but a very large number of particles took part in the acceleration process; whilst in the September 1966 flare the acceleration was relatively weak, although the volume was large.

In an attempt to define the size of a proton event, Smart and Shea (1971) have recently proposed a classification system based on three kinds of observations: GLE, PCA, and the hourly average maximum flux of > 10 MeV protons. Each event is then described by a three-figure number XYZ. Z is 0 for no GLE detected and 4 for GLE $> 100\%$. $Y = 0$ for no PCA detected and 4 for PCA > 15 dB. X equals the characteristic of $\log_{10} N(> 10 \text{ MeV}) \text{ cm}^{-2} \text{ s}^{-1} \text{ sr}^{-1}$. Thus, for example, the flares in Figure 72 would be classified as follows: 1956, Feb. 23: 334; 1966, Sept. 2: 330; 1966, July 7: 121; 1966, July 16, and 1966, Sept. 20: -100. It is evident that the important role of γ (or P_0) is masked in this classification system, but it is difficult to find a better one.

The exponent γ in Equation (98) (or value of P_0 in (100)) varies within wide limits in individual flares, according to van Hollebeke *et al.* (1975) from $\gamma \simeq 1.3$ to ~ 5.0 in the energy range of 20 to 80 MeV, with an average $\bar{\gamma} = 2.9$. Smaller values of γ are found for $E < 20$ MeV and greater ones for $E > 80$ MeV. The hardest spectra are observed in cosmic-ray flares and, as Najita and Orrall (1970) have pointed out, in white light flares. These two categories greatly overlap (McCracken, 1959; Švestka, 1966c). In favourably located white-light flares P_0 exceeds 200 MV. This gives further support to the hypothesis that the white-light emission is produced by particles accelerated to very high energies (cf. Section 2.2.11.B).

In order to estimate the actual efficiency of the acceleration process we should know the peak energy of the accelerated particles. In cosmic-ray flares the proton energy must exceed 500 MeV and a considerable number of particles must be accelerated well above 1 GeV when a strong GLE is recorded. The flare of 1956, February 23 required the presence of protons up to 15 GeV (Carmichael, 1962) and some authors have deduced still higher energies in the same event. This however was the strongest GLE ever recorded. In the GLE of 1967, January 28 (produced by a flare

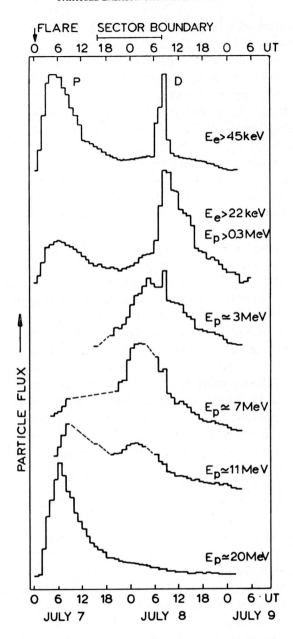

Fig. 102. Prompt (P) and delayed (D) enhancement of the particle flux associated with the flare o 1
1966, July 7. (After Kahler and Lin, 1969). E_e and E_p denote the energies of the electrons and protons,
respectively. Note that the delayed event was completely absent for $E_p \simeq 20$ MeV. This delayed event
was interpreted as storage near the Sun by Lin *et al.* (1968), and by storage near a sector boundary by
Švestka (1968 d).

far behind the western solar limb, cf. Figure 100) Heristchi and Trottet (1971) used the global distribution of neutron monitors as an energy spectrometer to determine a possible upper cutoff in the energy spectrum. They find this cutoff at 4.3 ± 0.5 GeV and argue its reality by the fact that this cutoff value was constant during the whole GLE duration, without any dependence on the GLE time profile.

Another quantity some readers might be interested in is the dosage of the total radiation from a big flare. In the flare of 1966, September 2, which, as we have seen, still does not represent the most energetic of events, Kane and Winckler (1969) measured the highest dosage of about 60 R h^{-1} behind a 0.22 g cm^{-2} thick aluminum wall. The total dosage of a very large flare of 1960, November 12 was estimated by Winckler (1963) to be 700 R. Severny and Steshenko (1972) give for three other major flares doses of 700 to 1000 R behind a shielding of 1 g cm^{-2}. The lethal dose is 300 to 500 R.

Quite often a prompt event is followed by a delayed one from the same flare (Figure 102). In such a case the delayed particles must have been hindered in their propagation by a large-scale magnetic inhomogeneity; the particles appear only when the inhomogeneity passes the Earth. Most frequently this magnetic hindrance is a shock wave originating in the flare (cf. Section 5.5), hence the delayed flux increase coincides with the sudden commencement of a magnetic storm at the Earth. But other cases may also occur in connection with corotation effects (Lin *et al.*, 1968; Švestka, 1968 d). In all cases the energy spectrum of the delayed event is considerably steeper than that of the preceding prompt event: The delayed event is rarely seen at energies >20 MeV, whilst its flux may exceed the flux of the prompt event at energies close to 1 MeV (Figure 102). This shows that protons with energy above ~ 10 MeV do not experience significant storage in interplanetary magnetic inhomogeneities, whilst a significant fraction of particles can be stored by them for tens of hours if one goes down to energies of ~ 1 MeV.

5.2. Heavier Nuclei

Even though observations of nuclei of $Z \geq 2$ have been made only in a few large events, there seems to be good evidence that these heavier particles are always present in the flux from proton flares.

Let us first suppose that atoms of all elements in the flare are completely stripped during acceleration and propagation. In such a case, since $A/Z \simeq 2$ for all nuclei with $A \geq 4$, α-particles and all heavier nuclei moving with the same velocity have essentially the same rigidity; thus they also have the same gyroradii and behave identically during propagation. Consequently, the charge spectrum for $Z \geq 2$ can be considered as a faithful reproduction of the chemical composition of the particles at the point of acceleration and abundance ratios of different nuclei in the particle flux should be the same in all flares.

This expectation has been confirmed in several flares by the comparison of the helium and medium nuclei spectra. These two spectra actually can be brought to a very good fit by multiplying the flux of the medium nuclei ($6 \leq Z \leq 9$) by a factor which appears constant ($\simeq 58 \pm 5$) within the limitations of the experimental errors (Biswas and Fichtel, 1965; Biswas *et al.*, 1966; Durgaprasad *et al.*, 1968; Bertsch *et al.*, 1969, 1972). However, this is true only for particle energies in excess of ~ 15 MeV per nucleon. At lower energies the particle flux is enriched in heavy elements by an amount that increases with charge number Z, decreases with energy E, and varies from flare to flare (Price *et al.*, 1971; Mogro-Campero and Simpson, 1972a, b; Lanzerotti *et al.*, 1972; Crawford *et al.*, 1972; Teegarden *et al.*, 1973; Bertsch *et al.*, 1973, 1974; reviews in Price, 1973, and Crawford *et al.*, 1975). According to Shirk (1974), who detected transiron nuclei in one of the Apollo 16 command-modul windows, the enhancement continues to increase with charge up to at least $Z \simeq 44$ and its increase can be suspected up to $Z \simeq 54$. Figure 103 shows the energy-dependent composition of He, O,

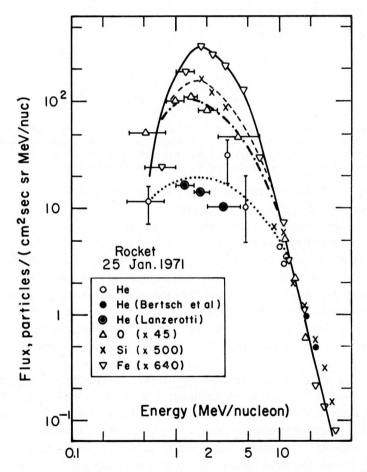

Fig. 103.   Abundances of He, O, Si, and Fe in the particle flux coming from the 1971, January 25 flare as function of energy. Heavy nuclei are more abundant at lower energies, but the relative abundances tend to a constant value above about 15 MeV per nucleon. (After Crawford *et al.*, 1975; *Astrophysical Journal*, University of Chicago Press, Copyright by the American Astronomical Society.)

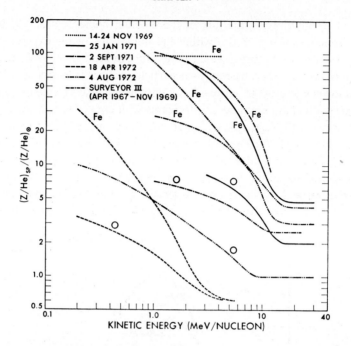

Fig. 104.   Energy dependence of the abundance ratios Fe/He and O/He in various flares, normalized to the solar abundance ratios (i.e., $(Z/He)_\odot = 1$). Note that the energy where the abundance of Fe and O begin to be enriched varies within wide limits. (After Crawford *et al.*, 1975; *Astrophysical Journal*, University of Chicago Press, Copyright by the American Astronomical Society.)

Si and Fe, approaching a constant value above about 15 MeV per nucleon, in the 1971, January 25, flare (Crawford *et al.*, 1975).

Under the assumption that the energy spectra are the same for all nuclei above ~ 15 MeV, so that they become identical when multiplied by constant factors (as in Figure 103), these factors give us the relative abundances of the different species. As Table XV shows, the abundances thus obtained are very similar to those in the solar photosphere and corona. This indicates that large deviations of the high-energy particle population from average solar abundances do not occur. On the other hand, as Figure 104 shows, at 1 MeV nucleon^{-1} the O/He abundance ratio can exceed that in Table XV by a factor up to 10, and the Fe/He abundance ratio by a factor up to 25.

Figures 103 and 104 indicate that the energy, below which the enhancement becomes apparent, is variable. Its mean value is about 15 MeV nucleon^{-1}, but it may vary from ~ 5 to ~ 20 MeV nucleon^{-1} (Crawford *et al.*, 1975). In the major flare of 1972, August 4, Bertsch *et al.* (1974) found an abundance enhancement up to ~ 50 MeV nucleon^{-1}; however, this seems to have been an anomaly, because the 1972, August 4 event was anomalous also from some other points of view (emission of γ-rays and extremely long hard X-ray burst, cf. Figure 118).

Three different mechanisms have been considered to explain the enhancement of low-energy heavy nuclei in the particle flux:

(a) Many years ago Korchak and Syrovatsky (1958) proposed a mechanism for galactic cosmic rays that might perhaps lead to enhancements of abundances of heavy ions also on the Sun. Let us suppose that the pre-accelerated ions are only slightly ionized; that they suffer ionization and excitation losses during the acceleration, comparable to the energy gain by the acceleration process; and that the acceleration process is the Fermi mechanism. Only ions for which the rate of energy loss is smaller than the energy gain due to Fermi acceleration can be accelerated. This threshold depends on Z, while the energy losses need not depend on Z if all ions are only slightly ionized. Thus heavier ions may be preferentially accelerated.

TABLE XV

Relative abundances in the flare-associated particle flux (after Crawford *et al.*, 1975)

| Element | Z | Abundance element/silicon | | |
		Particle streams ($>$15 MeV nucleon^{-1})	Photosphere [a]	Corona [b]
He	2	920[c]	?	(1300)
C	6	5.50 ± 0.20	10.5	9.2
N	7	1.50 ± 0.10	3.2	3.0
O	8	11.0 ± 0.8	19.0	15.6
Ne	10	1.35 ± 0.12	?	1.6
Mg	12	1.72 ± 0.16	0.98	1.1
Si	14	1.00	1.00	1.00
S	16	0.17 ± 0.03	0.46	0.24
Ar	18	0.04	?	0.15
Ca	20	0.07 ± 0.02	0.06	0.04
Ti	22	0.02	0.002	?
Fe	26	0.87[c]	0.71	0.85
Ni	28	0.05 ± 0.02	0.05	0.06

[a] Taken from Withbroe (1971).
[b] Averages of data given by Withbroe (1971), Malinovsky and Heroux (1973), and Walker *et al.* (1974).
[c] Statistically significant variations from flare to flare.

Though this explanation cannot be fully excluded from consideration, it is unlikely on the Sun for two reasons: The ions in the acceleration region on the Sun are probably much more heavily ionized and the Fermi mechanism is probably of a stochastic nature (cf. Section 6.6.2); in a stochastic mechanism the acceleration process may be sufficiently fast that energy losses may be neglected during the acceleration time.

(b) Price *et al.* (1971) suggested preferential leakage of incompletely ionized heavy nuclei from the acceleration region. The mass-to-charge ratio can become a function of Z if the ions are not completely stripped, the heavier ions having greater rigidity at the same speed. Thus they can more readily escape from storage regions on the Sun.

The effect is most pronounced at low speeds, hence the leakage of heavy ions at low energies might be preferred through this mechanism. However, it has not been quantitatively proven that this effect can be large enough to produce the great enhancements observed.

(c) Cartwright and Mogro-Campero (1973) have proposed a two-step acceleration process. First, fully stripped ions are accelerated to about 1 MeV nucleon^{-1} through a process the authors do not specify. These preaccelerated nuclei are then injected into a Fermi acceleration region where particles with rigidity greater than a minimum cutoff rigidity are accelerated to higher energies.

Originally, after the first acceleration step, all the nuclei of the same speed have the same rigidity since they are completely stripped. However, some of them travel to the Fermi acceleration region through relatively dense layers of the solar atmosphere where they pick up electrons. Hence the ionization distribution is changed in the direction of ionization equilibrium so that the relative rigidity of heavier ions increases. Due to the fact that the injection into Fermi acceleration is rigidity-dependent, the acceleration of heavier nuclei is thus preferred.

Irrespective of which explanation one prefers, the enhancement of the abundances

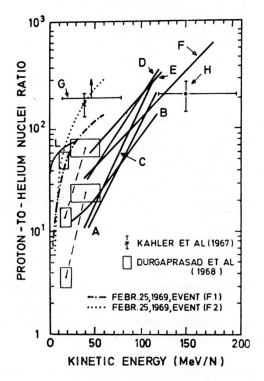

Fig. 105. The proton-to-α ratio as a function of kinetic energy per nucleon in various particle events. The measurements A-H have been summarized by Fichtel and McDonald (1967). (After Wibberenz and Witte (1970) who also added the results for the 1969, February 25 event to the graph.)

The curve labelled *L* is the dependence given by Lanzerotti (1973) in Equation (101).

of heavy nuclei at low energies strongly indicates that the ions do not enter the acceleration process completely stripped, as we assumed at the beginning of this section. On the other hand, according to Mogro-Campero and Simpson (1972) even iron nuclei are fully stripped when recorded near the Earth on OGO 5 in the energy range from 25 to 61 MeV nucleon^{-1}.

The situation is different when we compare α-particles (or other heavier nuclei) with protons. Since the rigidities of protons and α-particles of the same velocity differ by a factor of two, the protons propagate differently. This is actually observed, and the helium/proton abundance ratio α/P varies greatly from flare to flare as well as during the development of a single event. The wide range of P/α ratios and its variation with kinetic energy is shown in Figure 105, which includes data by Fichtel and McDonald (1967), Kahler *et al.* (1967), Durgaprasad *et al.* (1968), and Wibberentz and Witte (1970). The increase of the P/α ratio with increasing energy nucleon^{-1} has as the consequence that the helium spectrum is always softer than the spectrum of protons (Lanzerotti, 1973). Figure 106 shows the time variation of the α/P ratio in the flare of 1969, November 2 (Lanzerotti and MacLennan, 1971).

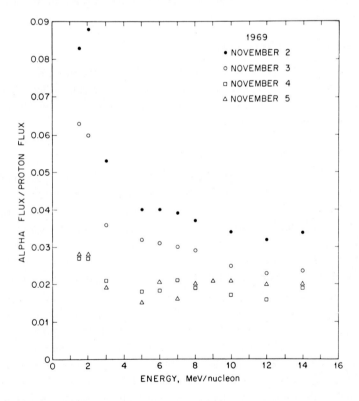

Fig. 106.   Daily average α-to-proton flux ratios for the 1969, November 2 event in dependence on the kinetic energy per nucleon. (After Lanzerotti and Maclennan, 1971.) Note the progressive decrease of this ratio during the first two days after the event occurrence.

Lanzerotti (1973) finds for low-MeV energies

$$\frac{P}{\alpha} \simeq 38 \ (E \ \text{nucleon}^{-1})^{0.2},$$ (101)

with E in MeV (cf. Figure 105). Thus at 1 MeV $P/\alpha = 38$, which greatly exceeds the average in solar wind ($P/\alpha = 22$). However, the abundance of hydrogen relative to helium can hardly be obtained from these highly variable data. If instead one accepts the value of 58 ± 5 for the helium-to-medium-nuclei ratio, and takes the proton-to-medium-nuclei ratio from spectroscopic data (Lambert, 1967), one gets a proton-to-helium ratio of 16 ± 2 (Bertsch *et al.*, 1972).

5.3. Solar Electrons

All observations of electrons in the flare-associated particle flux are of a more recent date than proton measurements, since electrons were first detected by Mariner 4 (Van Allen and Krimigis, 1965). These were non-relativistic electrons with energies > 45 keV. The first relativistic electrons were detected by Cline and McDonald (1968) in the flare of 1966, July 7.

5.3.1. NON-RELATIVISTIC ELECTRONS

Non-relativistic electrons, with $E > 20$ keV or $E > 40$ keV, respectively, have been discussed in detail by Lin and Anderson (1967), Anderson (1969, 1970, 1972a), Anderson *et al.* (1970) and Lin (1970a, b, 1973, 1974). The main results of these studies can be summarized as follows.

A. *The Event Occurrence and the Associated Flares*

During the six and one-half years, 1964 to mid-1970, 230 electron events have been recorded in the vicinity of the Earth, with a threshold of 10 electrons $\text{cm}^{-2} \ \text{s}^{-1} \ \text{sr}^{-1}$ above 40 keV. Many more events have been recorded above 20 keV, but there the survey is not complete, since long-lived streams of solar particles obscure the weak events. The occurrence of the electron events, however, is very unevenly distributed in time. This is due to the fact that some particular active regions tend to generate many flares that accelerate electrons (up to 15 or 20), whilst other regions do not produce any electron events at all. This behavior is similar to that of proton-active regions. These two types of regions also are usually identical, but this need not be always the case (cf. the *Catalog of Solar Particle Events*, Švestka and Simon, 1975).

About 90% of the solar electron events are associated with flares. The ambiguity of this association is usually small (smaller than for the proton events) since the electrons propagate fast and arrive at the Earth within 20 to 60 min. Most of the events not associated with an observed flare are likely to be the product of a flare beyond the limb. The associated flares, however, are usually quite small and many of them are classified as subflares. So far, no optical characteristic of these flares could be found

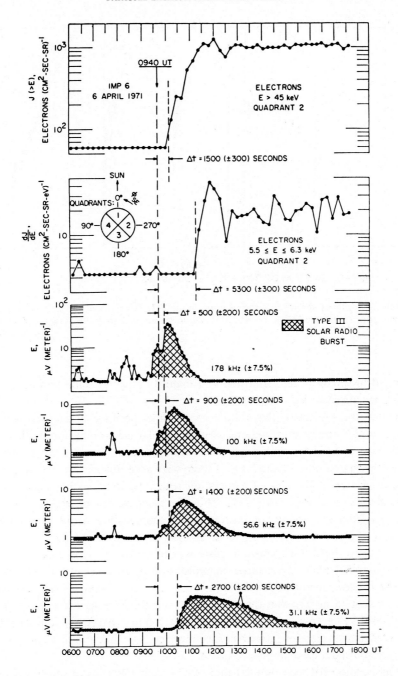

Fig. 107a.   Simultaneous electron and type III burst observations at 1 AU in 1971, April 6. (After Frank and Gurnett, 1972.) If the radio emission is at the fundamental of the plasma frequency then the 5.5 to 6.3 keV electrons would be the burst exciter. However, if the radio emission is at the second harmonic of the plasma frequency (which is the more likely case), higher energy electrons would coincide with the emission (Lin, 1974.).

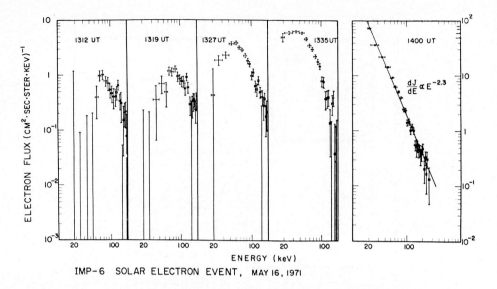

IMP-6 SOLAR ELECTRON EVENT, MAY 16, 1971

Fig. 107b. The energy spectrum of the electrons in the 1971, May 16 event at different times during the onset. Prior to it a type III burst was observed at frequencies from 2.6 MHz to 44 kHz. (After Lin, 1974.)

that might distinguish them from the vast majority of other small flares from which no electrons are recorded in space.

However, these 'electron flares' do possess some specific characteristic features in the radio and X-ray spectral regions. Out of 77 solar electron events studied by Lin, 53 (82%) were associated with a type III radio burst and 54 with a microwave burst. On the other hand, only 9% and 4% of them were accompanied by type II or type IV bursts, respectively. The lack of association with type II bursts has been confirmed also by Švestka and Fritzová (1974). Thus the association with radio bursts differs completely from that found for proton events. Obviously a strong type III and an associated microwave burst characterize an electron event on the Sun.

An impulsive hard X-ray burst is also usually present; this is what one expects, since the impulsive burst is a direct manifestation of an acceleration process which gives rise to electrons of about the same energies as we record in space. Therefore, it is somewhat surprising that there are exceptions to this rule: Kane and Lin (1971) have observed 19 events (out of 60 studied) in which no hard X-ray component could be detected and yet electrons were recorded in space. As a matter of fact, only 37 events of this set showed a clear impulsive phase.

The association between type III radio bursts and electrons in space was recently confirmed by direct observations in space (Alvarez *et al.*, 1972; Frank and Gurnett, 1972; Lin *et al.*, 1973). Figure 107a demonstrates how the type III radio emission from an electron stream propagating from the Sun to the Earth coincides with the observation of an increased flux of electrons on a satellite. Figure 107b shows a

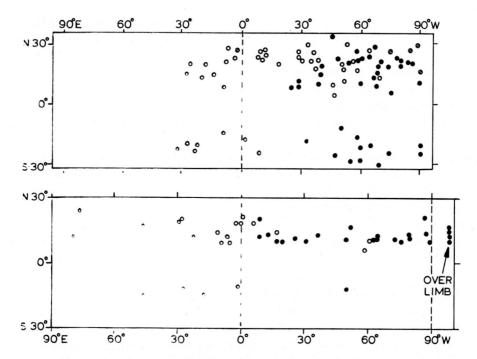

Fig. 108. The longitude distribution of flares producing electron events. The upper graph (after Lin, 1970a) shows the sources of all non-relativistic electron events recorded before the end of 1967. Open circles denote flares which occurred in a region associated with a radio type I noise storm. The lower graph (after Alvarez *et al.*, 1972) shows the sources of the type III bursts which were observed below 350 kHz, i.e., far from the Sun. Filled circles denote the bursts which could be associated with a flux of electrons directly observed near the Earth (like those shown in Figures 107).

sequence of energy spectra during the rising portion of another solar electron event. A type III burst was observed prior to it shifting to progressively lower frequencies. At 13^h07^m UT, when the particle event began, type III burst onset was recorded at 55 kHz (Fainberg, Stone and Evans in Lin, 1973). If the 55 kHz radiation is second harmonic (cf. Section 3.2.6), then the type III emission was produced near the spacecraft close to 1 AU, i.e. by the electrons recorded in Figure 107b.

However, very few type III bursts produce observable electron streams in space. As we mentioned in Section 3.2.6.A, the majority of type III bursts are not flare associated. The bursts that are associated with flares are usually very strong, mostly classified as a series of bursts, and they are often accompanied by a microwave burst, which is not a general characteristic of type III bursts. One could say that while a normal, single type III burst is just one stream of electrons, these flare-associated bursts are often 'showers' of electrons. Even these bursts, however, do not all produce an observable electron flux. Some of these electron streams are probably stopped or decay before reaching the Earth, or perhaps they are deflected to another direction (the streams are fairly narrow); in any case no increase in the electron flux is recorded in the Earth's vicinity.

Similar to the proton flares, the flares producing observable electron events are concentrated in the western hemisphere of the Sun (Figure 108). The concentration of the electron events around the root point of the Sun-Earth magnetic field line seems to be still more pronounced than for the protons. According to Lin (1974) the cones of emission are $\lesssim 60°$ wide; wider cones, up to 90°, are also observed, but only above regions that are associated with a type I noise storm.

With this rather small cone of propagation the total number of electron events on the whole Sun must be quite high; Anderson estimated that at least 500 electron events occurred on the Sun in 1967 and Lin (1974) gives ~ 400 for an active year. Nevertheless, this is still only a small fraction of the number of type III bursts which is more than one order of magnitude higher. For example, during the period plotted in Figure 79a, seven electron events were recorded. Making an allowance for a 60° cone of propagation, the total number of electron events produced on the whole visible hemisphere can be estimated at ~ 25. The number of type III bursts during the same time, however, was ~ 800. Hence only one in about 30 type III bursts produced an observable electron event near the Earth.

The intensity of the electron flux at the Earth ranges from the threshold limit up to $\sim 10^4$ electrons $cm^{-2} s^{-1} sr^{-1}$. A typical moderate electron event produces a few hundred electrons $cm^{-2} s^{-1} sr^{-1}$. Most rise times are in the range of 5 to 50 min, whilst the decrease takes many hours. The total number of electrons above 40 keV can be estimated to 10^{32} to 10^{33} (Lin and Hudson, 1971; Alvarez et al., 1972), which does not exceed the number needed for a type III burst. Holt and Ramaty (1969), Lin and Hudson (1971) and Kane and Lin (1972) estimate that less than 1% of the electrons accelerated in the flare gain access to interplanetary space.

B. *Proton-Electron and Pure Electron Events*

Essentially all proton events also produce non-relativistic electrons. When the narrower cone of propagation for the electrons is taken into account, exceptions from this rule are extremely rare. One of them is demonstrated in Figure 109 (Arnoldy et al., 1968a). It shows records of the proton event of 1965, October 4, which was produced by a flare 30° west of the central meridian. Hence electrons should have come to the Earth, but none were observed. This flare was anomalous also from other points of view. The metric type IV burst was strong (cf. Figure 67D), but its microwave component was extremely weak and, as we can see in Figure 109, no X-ray burst was observed (the flare onset was at 09 37 UT in Hα light).

On the other hand, electron events without detectable protons are quite common and Lin calls them 'pure electron events'. Figure 110 shows the dependence of the electron flux on the intensity of the associated microwave burst, for both pure electron and proton-electron events. One can see from it that when these two kinds of events are considered separately, the electron flux does not depend strongly on the intensity of the radio burst. However, when the two families of events are mutually compared, their distributions in the graph are completely different. This can be considered as an indication that pure electron events represent an essentially different pheno-

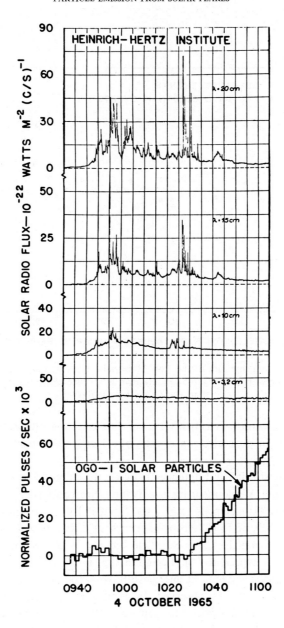

Fig. 109.   Radio records of the type IV burst of 1965, October 4, which was anomalously weak below 20 cm (see also Figure 67D). The onset of the proton flare in Hα light was at 09ʰ37ᵐUT. At the bottom the OGO-satellite ionization chamber recorded an increase in the particle flux about one hour after the flare onset, but no >10 keV X-ray burst at the time of the flare. No electrons were detected in the particle flux either (Lin, 1970a.) Hence, the whole event was greatly deficient in electrons. (After Arnoldy *et al.*, 1968a; *Astrophysical Journal*, University of Chicago Press, Copyright by the American Astronomical Society.)

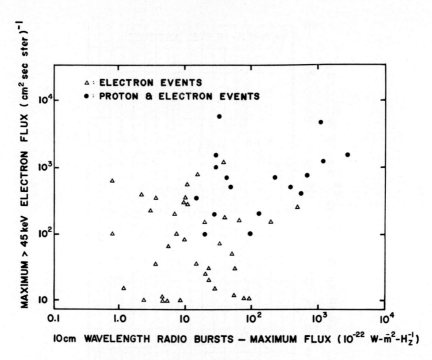

Fig. 110. The maximum intensity of the 2800 MHz radio burst accompanying electron events is plotted here against the maximum >45 keV electron flux. Pure electron events are marked with triangles. (After Lin, 1970a.)

menon than the proton (+electron) events. For example, the acceleration may be rigidity-dependent in pure electron events, but energy-dependent when protons are also observed.

Lin (1974) finds still more differences between the pure-electron (PE) and proton-electron (PR) events. While most PE events have electron fluxes between 10 and 100 and never above 1000 electrons $(cm^2\,s\,sr)^{-1}$, most PR events have electron fluxes between 100 and 1000, up to 10^4 electrons $(cm^2\,s\,sr)^{-1}$. In the longitudinal distribution the rms deviations from the Earth-Sun field line are $\sim 35°$ for the PE and $\sim 55°$ for the PR events. The exponent δ in Equation (85) is 2.5 to 4.6 in PE events below 100 keV, with rapid steepening at energies above ~ 100 keV; nevertheless, electrons up to ~ 500 keV may still be observed. In contrast typical electron spectra for PR events extend smoothly in a power law out to ~ 10 MeV and δ is almost always less than 3.5. (See example in Figure 111. Note that, for particle flux,

$$\frac{dJ}{dE} = v\,\frac{dn}{dE},\qquad(102)$$

hence

$$\frac{dJ}{dE} \sim E^{-\gamma} \sim vE^{-\delta} \sim E^{-(\delta-\frac{1}{2})}.\,)\qquad(103)$$

In both types of electron events the power law fits the spectrum down to below ~ 5 keV, and sometimes even to ~ 2 keV. Whereas, however, the number of > 20 keV electrons is estimated at $\sim 10^{33}$ in the PE events, it can be as large as $\sim 10^{36}$ in large proton events. Sarris and Shawhan (1973) find that in the microwave radio range the PE events are associated mostly with peak fluxes < 50 merg cm^{-2} s^{-1} and short decay time of the radio burst, whereas the PR's are preceded mostly by peak fluxes > 50 merg cm^{-2} s^{-1} and the microwave bursts have considerably longer decay times.

However, all these differences between the PE and PR events can also be explained as being simply due to differences in the intensity of the acceleration process. For example, one can interpret the graph in Figure 110 also in the following way: no protons are observed when the electron flux is lower than 100 cm^{-2} s^{-1} sr^{-1}, in some cases protons are observed when the flux exceeds 100, and the protons are always

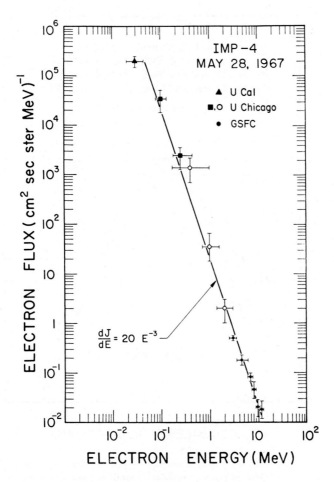

Fig. 111.   The differential electron energy spectrum for the proton-electron event of 1967, May 28. The observed flux fits to a single power over three decades in energy. (After Lin, 1974.)

present when the electron flux is high. In that case protons should always be present in the flux, but we simply do not record them since the energy threshold of proton detectors is higher than 300 keV. Actually, it might be quite difficult to recognize them even if we had measurements at lower energies: While 40 keV electrons travel to the Earth for ~ 20 min, 40 keV protons need about 14 h to cover the same distance. Therefore, the time-association of the two species at very low energies is very uncertain.

The fact that the power law fits the electron spectrum down to below 5 keV without any obvious flattening can be used for an estimate of the altitude at which the electrons are accelerated. For the event of 27 April, 1972, Lin (1973) has estimated that the number of ambient ions the electrons met along their path to the Earth must have been $\lesssim 3.5 \times 10^{19}$ cm^{-2}. For hydrogen this is $\lesssim 60$ μg cm^{-2} of material traversed. This leads to an estimate that the acceleration could not take place below the coronal layer where $n_i \simeq 2 \times 10^9$ cm^{-3}. This is the density in the low corona and Lin places

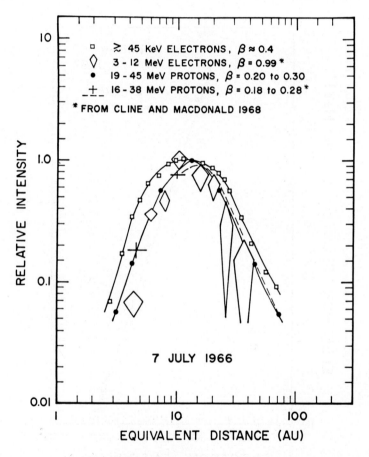

Fig. 112. The normalized flux (i.e., flux at maximum = 1.0) is plotted here against the 'distance travelled' (= velocity × time after injection) for different kinds of particles coming from the flare of 1966, July 7. Note that although the proton and relativistic-electron curves agree within the error limits, the >45 keV electron curve is broader and precedes the others. (After Lin, 1970a.)

it in an active region near 30 000 km above the photosphere. Since no bend has been detected so far in the electron spectra at low energies, this is in fact the minimum altitude of the electron acceleration in the flare region.

C. *Time of Acceleration*

The short travel time of electrons that get from the Sun to the Earth within ~ 20 min make the identification of the particle source on the Sun rather easy. The injection time in the solar atmosphere inferred from the expected travel time then falls in most cases between the start and maximum times of the optical flare, close to the impulsive phase when the hard X-rays, impulsive microwaves and type III burst are observed. Thus there is little doubt that this is the moment when the non-relativistic electrons are accelerated. What is not yet clear is the relation of this (first-step?) acceleration process, producing non-relativistic electrons, to the acceleration of protons and relativistic electrons.

We have mentioned that the white-light flare emission and the γ-rays, which both should be produced by high-energy protons, practically coincide in time with the impulsive phase (Sections 2.2.11 and 5.4.3). Therefore, the acceleration of protons should be accomplished either through the same process that accelerates non-relativistic electrons or in a second-step acceleration that immediately follows the primary one (Section 3.2.5). However, two kinds of electron observations appear to contradict this conclusion. One of them is demonstrated in Figure 112. If one multiplies the particle velocity by the time elapsed after the acceleration, one obtains the distance travelled. Then plotting the particle flux against this distance one gets curves of the type shown in Figure 112, which in some events are identical for different species and energies (Bryant *et al.*, 1962, 1965). This was also the case for the flare of 7 July 1966 for protons and relativistic electrons (Cline and McDonald, 1968). However, when the curve of non-relativistic electrons is added to the graph (Lin, 1970a), it is broader and precedes the others. This indicates that non-relativistic electrons arrive earlier than one would expect according to their speed relative to the other species. The implication of it is that these particles are either accelerated earlier than the protons and relativistic electrons, or they are accelerated higher up in the solar atmosphere where their release is faster. In fact, Lin (1970a, 1974) has presented only one example of this discrepancy, for the cosmic-ray flare of 1966, July 7, so that it might have been an anomaly. However, as we shall see in the next section, relativistic electrons quite systematically show a time delay with regard to non-relativistic electrons.

5.3.2. RELATIVISTIC ELECTRONS

Records of relativistic electrons from flares have been reported and discussed by Cline and McDonald (1968; $E > 3$ MeV), Datlowe *et al.*, (1969; $E > 12$ MeV), Koechlin *et al.* (1969; $E > 3$ MeV), Simnett *et al.* (1970; $E > 3$ MeV), Sullivan (1970; $E > 0.7$ MeV), Datlowe (1971; $E > 12$ MeV), Simnett (1972a; $E > 0.3$ MeV), Dilworth *et al.* (1972; $E > 7.5$ MeV), van Hollebeke *et al.* (1975; 0.5–1.1 MeV), and reviewed by

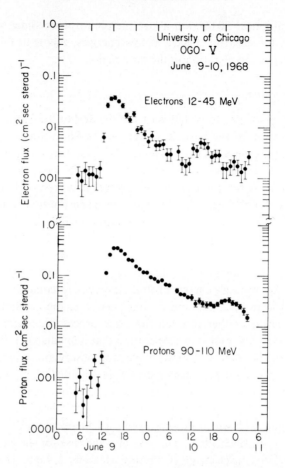

Fig. 113.   Records of highly relativistic electrons and >90 MeV protons in the flare of 1968, June 9.
Note the similarity of both curves and the difference in flux. Although the electron energy is signifi-
cantly lower, the electron flux is by one order of magnitude smaller than that of >90 MeV protons.
(After Datlowe, 1971.)

Simnett (1972b, 1974a). A still earlier observation was reported by Meyer and Vogt
(1962), but it is doubtful whether this was a flare-associated event.

Relativistic electrons generally accompany proton events, in particular the strongest
of them (Cline and McDonald, 1968; Datlowe, 1971; Dilworth et al., 1972). Typical
examples of simultaneous records of protons and relativistic electrons are shown in
Figures 98 and 113. The flux of relativistic electrons, as one can see, is much lower
than that of protons of the same (and even much higher) energy, probably due to
rigidity-dependent injection into the acceleration process (cf. Section 6.6.3) and to the
heavy sychrotron losses that the electrons suffer before leaving the Sun. This is probably also the reason why the number of flare-associated relativistic-electron events is
smaller than that of proton events. According to Datlowe (1971) the ratio of electrons

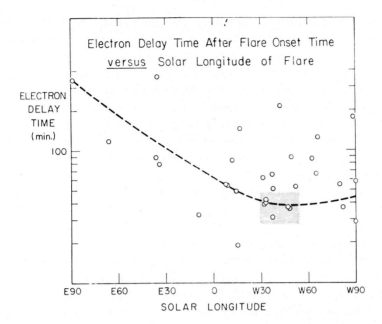

Fig. 114.   The delay time of the onset of the 0.3 to 0.9 MeV electron events from the flare onset on
the Sun, plotted as a function of the flare longitude. (After Simnett, 1971.)

to protons in the 12 to 45 MeV range varies between 10^{-2} to 5×10^{-6}, with a median
value $\approx 3 \times 10^{-4}$. However, Lanzerotti and Maclennan (1971) have found the flux of
>1.1 MeV electrons still comparable to proton flux at the same energy, hence a
decrease in electron number should occur only at higher energies.

Simnett (1974) compared the 0.3 to 0.9 MeV electron flux with $\gtrsim 11$ MeV proton
flux in more than 30 events. The ratio of these two fluxes varied within a very wide
range from ~ 0.2 to >10. Simnett considers this as a natural consequence of the fact
that the electron spectra are very much the same (as we show below, $\gamma = 3.0 \pm 0.4$),
whereas the proton spectrum is widely varying (cf., e.g., Figure 97 or van Hollebeke
et al., 1975).

From a study of 70 events above 3 MeV Simnett (1972a) finds that only $\sim 70\%$
of the relativistic-electron events have a plausible origin in a visible flare. Later studies
confirm this ratio (Simnett, 1974a). Similarly to the proton events most of the events
without flare-association come from flares behind the western limb. However, there
are also relativistic electron events in which no protons are recorded at all. It seems
that in such cases any flare-association is indeed missing. Many of them are observed
during quiet time periods (McDonald, 1970; McDonald et al., 1972; L'Heureux et al.,
1972), and can be explained as being of galactic origin.

The energy spectra (Figure 111) of prompt relativistic electron events below 10 MeV
fit well to the power law (103), with $\gamma = 3.0 \pm 0.4$ (Simnett, 1972a, b); beyond 10 MeV
the spectrum seems to be somewhat steeper (Datlowe, 1971; Simnett, 1972a, b),
thus indicating the possibility of a finite cutoff energy on the Sun at higher energies.

The γ value may be as low as ~ 2 at the very onset of the event, but it increases fast and then stays fairly constant during the whole event (Datlowe, 1971). A combination of non-relativistic and relativistic measurements for a few events indicates only a slight decrease in γ toward the lower energies (Lin, 1970a), and often the whole spectrum may be fit with one γ, as in Figure 111. This rather constant γ value within a wide energy range and the small dispersion of the γ values in different electron events are properties significantly different from those found for the spectra of protons.

A serious problem is raised when the time-delays of the relativistic electrons are considered (Simnett, 1971, 1974a). Figure 114 shows the time-delay between the onset of the flare and the arrival of 0.3 to 0.9 MeV electrons to the Earth. A 660 keV electron moves with velocity $0.9c$; thus, if the length of a typical spiral field line is assumed to be ~ 1.3 AU, electrons of this energy should arrive at the Earth within 12 min. After making allowance for the 8 min which light needs for propagation to the Earth, the theoretical delay should be 4 min. But the delay times in Figure 114 are tens of minutes. This means that the relativistic electrons either are accelerated tens of minutes after the flash phase (which usually takes place a few minutes after the flare onset), or their release from the Sun must be delayed.

We have mentioned in the preceding section that the first explanation is improbable, because white-light flare and γ-ray emission is observed practically simultaneously with the impulsive phase. Another argument against this explanation is that the type IV burst in strong events starts very close to the impulsive phase (Švestka, 1970a), and this burst is certainly produced by relativistic electrons. Therefore, the explanation through storage seems more acceptable.

Simnett (1971) has proposed a model in which disordered magnetic field lines out to 3 to 5 solar radii trap low-rigidity ($\lesssim 50$ MV) flare particles. Flare-produced fast-moving waves disturb the field and thus open a free way to space to the trapped particles. However, the particles must wait for this opening until the flare disturbance reaches a height of 3 to 5 solar radii. At velocities of about 2000 km s^{-1} the disturbance reaches 4 solar radii in ~ 25 min, which is in agreement with Figure 114.

If this were true, then the fact that ~ 40 keV electrons are probably released earlier (cf. Section 5.3.1.C) would imply that the non-relativistic electrons must be accelerated at an altitude exceeding 3 solar radii, and this seems very unlikely. Thus, we have actually no explanation for this time delay.

5.4. Nuclear Reactions in Flares

So far, perhaps with the exception of the white-light flares (Section 2.2.11.B), we have discussed only effects caused by electrons in the solar atmosphere: X-ray bursts produced through bremsstrahlung (Sections 3.1.2 and 3.2.1), continuous radio emission produced through gyro-synchrotron radiation (Sections 3.2.3 and 3.2.4), Hα emission through Coulomb collisions (Section 2.3), and long-wave radio bursts produced through plasma waves (Sections 3.2.5 and 3.2.6). Probably jointly with the nuclei, the electrons produce white-light and EUV emission through collisional heating

(Sections 2.2.11.B and 3.2.2) and contribute to the heating of the high-temperature coronal gas. There is only one effect in solar flares that is due solely to atomic nuclei, and that is the production of nuclear reactions. The nuclear reactions one can expect (Lingenfelter and Ramaty, 1967; Cheng, 1972b; de Feiter, 1974; Ramaty et al., 1974) can be summarized as follows:

(1) The production of neutrons in reactions initiated either by high energy protons or α-particles. Neutrons can be produced through interaction of these nuclei with the most abundant elements, C, N, O, Ne, Fe, Mg, Si, but far most important appear to be the reactions between the hydrogen and helium nuclei themselves. Nuclei must be accelerated to more than 30 MeV to accomplish these reactions. Neutrons might be produced also through a proton-proton reaction, but this requires proton energies in excess of 292 MeV; there are hardly enough protons above this energy present in the flare region.

A great part of the neutrons produced in these reactions escape from the Sun, and they should be observable near the Earth shortly after the acceleration process (cf. Figure 117). The remaining neutrons are captured by ambient ^{1}H and ^{3}He nuclei in the photosphere. While captures on ^{3}He do not lead to photon emission, captures on protons produce deuterons and 2.23 MeV gamma-ray line.

(2) The production of positrons emitted from $\pi+$ mesons and radioactive nuclei like ^{10}C, ^{11}C, ^{12}N, ^{13}N, ^{14}O, ^{15}O. These positron emitters originate again through interactions of accelerated protons and α-particles with the ambient medium. The thresholds for these reactions range from ~ 3 MeV nucleon^{-1} for producing ^{11}C through protons on ^{14}N, or ^{15}O through α-particles on ^{12}C, to 41.4 MeV for producing ^{10}C through protons on ^{16}O. The threshold for the proton-induced $\pi+$ mesons is 292 MeV on ^{1}H and 185 MeV on ^{4}He. The emitted positrons partially escape from the source region while those remaining on the Sun slow down through collisions and finally combine with an electron to form a pair of γ-ray photons of 0.511 MeV photon energy.

(3) The production of nuclei in excited states; these are isotopes which either are abundant in the solar atmosphere or are easily produced in nuclear reactions of energetic protons and α-particles with the abundant constituents. De-excitation of these nuclei to the ground state takes place by the emission of a γ-ray photon of energy corresponding to the energy of the excited level. Expected lines in particular are at 0.431 MeV (from ^{7}Be), 0.478 MeV (^{7}Li), 2.31 MeV (^{14}N), 4.43 MeV (^{12}C), and 4.14 MeV (^{16}O). Unlike (1) and (2) above this emission is prompt, whereas in both (1) and (2) the emission is delayed due to relatively long half-life times of the emitters.

(4) The production of π^0 mesons through proton-proton and proton-alpha reactions. The mesons decay instantaneously into two γ-photons. The γ-ray spectrum resulting from the π^0 meson decay is rather flat between 70 and 140 MeV.

(5) Production of ^{2}H (deuterium), ^{3}H (tritium) and ^{3}He as secondary products of interactions between accelerated protons and α-particles (threshold ~ 25–30 MeV per nucleon) and of some other nuclear reactions.

Thus, summarizing, production of deuterium, tritium, helium-3, neutrons, positrons and γ-rays is to be expected on the Sun when protons and α-particles are accelerated in flares to energies exceeding ~ 30 MeV per nucleon.

5.4.1. Accelerated Isotopes of Hydrogen and Helium

Severny (1957) was the first who tried to detect deuterium produced in solar flares and found the abundance ratio $\Gamma(^2H/^1H) \simeq 1 \times 10^{-4}$ in an active region. In later years several authors used satellite data and more sophisticated methods for determining this ratio, but astonishingly enough after 20 years they have arrived at essentially the same value Severny gave in the year when the first Sputnik orbited the Earth: $\Gamma(^2H/^1H) = (8 \pm 2) \times 10^{-5}$. This value has been deduced by Anglin et al. (1973b) from IMP 5 and IMP 6 measurements by summing data from 25 flares in the energy range ~ 7 to 13.5 MeV per nucleon. The same authors get $\Gamma(^3H/^1H) = (2 \pm 1) \times 10^{-5}$ for the abundance of tritium. The statistical significance of these ratio values is evident from Figure 115.

The $\Gamma(^2H/^1H)$ value found in flares is four times higher than the upper limit found by Epstein and Taylor (1970) in the solar wind, 16 times higher than the upper limit found by Grevesse (1969) in quiet regions of the Sun, and it is even above the current estimate for the 'cosmic' abundance of ^2H (Reeves et al., 1973). Anglin et al. (1973a, b) conclude that these increased values are consistent only with their production in

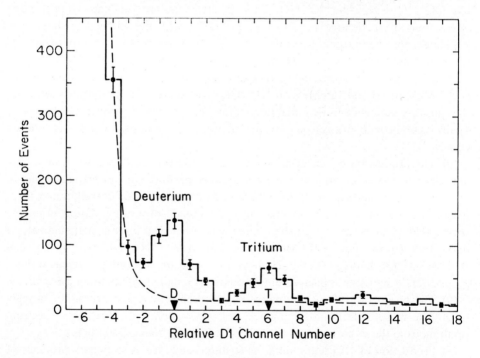

Fig. 115.   Track-sum histogram (i.e., the number of events per channel) for deuterium and tritium collected from a sum of solar flare data on IMP 6 in energy intervals from 6.2 to 13.2 MeV. The dashed curve indicates the background level. (After Anglin et al., 1973b; *Astrophysical Journal*, University of Chicago Press, Copyright by the American Astronomical Society.)

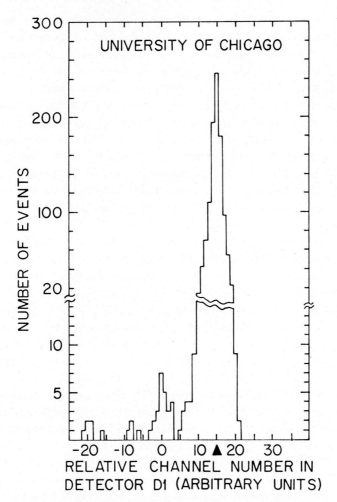

Fig. 116. Track-sum histogram for the ³He (at zero ordinate) and ⁴He collected from five '³He-rich' flares: 1969, November 2; 1969, May 29 and 30; 1970, August 14; and 1971, January 25. (After Dietrich, 1973; *Astrophysical Journal*, University of Chicago Press, Copyright by the American Astronomical Society.)

nuclear interactions associated with flares, even when the yield certainly varies greatly from flare to flare, being very sensitive to the energy spectra of the bombarding nuclei.

While the increase in ²H and ³H abundances becomes statistically significant only as an average in a great set of flare data, the high yield of ³He can be shown even for individual flares. However, the flares with a high yield of ³He are rare; from June 1969 to January 1971, IMP 5 recorded only two flares very rich in ³He, on 1969, November 2, and 1971, January 25, and three other flares with an obvious ³He increase, 1970, March 29, 1970, May 30, and 1970, August 14 (Dietrich, 1973; Anglin *et al.*, 1973a). Of course, the statistics are greatly improved for a sum of flare data; an example of such a sum for the five flares mentioned above is shown in Figure 116.

Hsieh and Simpson (1970) were the first who were able to separate the ^{3}He isotope in particle streams emitted from flares. For the sum of seven flares they found $\Gamma(^3\text{He}/^4\text{He}) = (2.1 \pm 0.4) \times 10^{-2}$ in the 10 to 100 MeV nucleon^{-1} energy range and the exponent in the power law about one power lower for ^{3}He than for ^{4}He energy spectrum. That means that the $\Gamma(^3\text{He}/^4\text{He})$ ratio increases with increasing rigidity. This behavior was confirmed by Dietrich (1973) in the 10 to 50 MeV nucleon^{-1} energy range for the ^{3}He-rich flares of 1969, November 2 ($\Gamma = (7.7 \simeq 2.0) \times 10^{-2}$) and 1971, January 25 ($\Gamma = (2.0 \pm 1.0) \times 10^{-2}$). In the same energy range, the sum of 17 ɪMP 4 flares gives $\Gamma = 2.2 \times 10^{-2}$ and the sum of the five flares yielding the increase in Figure 116 gives $\Gamma = 3.3 \times 10^{-2}$ (Anglin et al., 1973a). All these values are significantly higher than the values $\Gamma = 4 \times 10^{-4}$ found from the solar wind measurements (Geiss et al., 1970) and $\Gamma = 3 \times 10^{-4}$ reported by Cameron (1968) as the universal abundance.

The nucleon production of these isotopes can be directly related to the mean path length of solar material traversed by the accelerated nuclei (Lingenfelter and Ramaty, 1967). According to Anglin et al. (1973b) the observed flare-associated increase in ^{2}H and ^{3}H is consistent with a path length of ~ 0.1 to 1 g cm^{-2} of solar matter, while the ^{3}He-rich flares need a path length up to ~ 8 g cm^{-2}. The ^{2}H, ^{3}H and ^{3}He nuclei must also be accelerated sufficiently after their production to compensate for energy loss by ionization. If the whole process is accomplished during the impulsive phase (~ 100 s), then these path lengths require the acceleration to take place in the chromosphere.

Garrard et al. (1973) obtained still higher ^{3}He abundances. Averaging for 7 flares in 1969, they found ^{3}He/^{4}He $= 0.10 \pm 0.02$ in the 4 to 5 MeV nucleon^{-1} range. This result would indicate that the ratio ^{3}He/^{4}He increases toward lower energies, in contradistinction to the opposite trend found by Hsieh and Simpson (1970) and Dietrich (1973). However, as we shall see in Section 5.4.3, such a high ^{3}He abundance contradicts the fact that the 2.3 MeV γ-ray line has been observed in some flares; since ^{3}He captures neutrons that are capable of producing this γ-ray line, extremely high ^{3}He abundance in the flare region would greatly suppress the γ-ray emission (Ramaty et al., 1974). Of course, with the small number of observed events, one cannot exclude the possibility that only flares with low ^{3}He/^{4}He ratio produce an observable 2.3 MeV γ-ray line.

5.4.2. NEUTRONS

Biermann et al. (1951) were the first to point out that protons accelerated on the Sun could interact with the solar atmosphere and produce nuclear reactions with the emission of neutrons. Detailed calculations of this process were carried out by Lingenfelter et al. (1965) and by Lingenfelter and Ramaty (1967).

Lingenfelter et al. estimated the peak flux of neutrons at the Earth after the major cosmic-ray flare of 1960, November 12 at 30–70 neutrons cm^{-2} s^{-1}. This estimate was based on the measured total flux of protons at the Earth. The computation of the corresponding number of protons accelerated on the Sun and the assumption that the

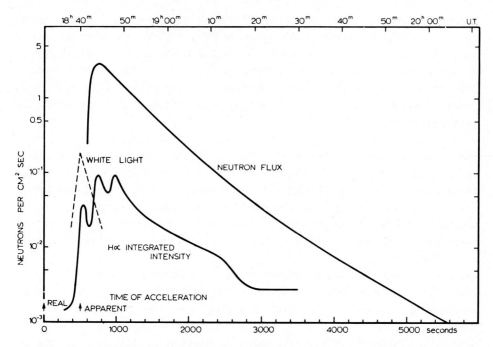

Fig. 117. The minimum *expected* neutron flux at 1 AU that should have been produced by the white-light flare of 1967, May 23. The bottom scale shows the time in seconds after the acceleration in the flare region and the upper scale is the UT time, both under the assumption that the acceleration occurred at the same time as the white-light flare was observed. A comparison with the Hα development curve demonstrates the early arrival of the neutrons. (After Švestka, 1971b.)

proton flux into space and toward the photosphere are equal may easily involve errors of one or two orders of magnitude.

Later on, Lingenfelter (1969) estimated the neutron flux from solar flares under the assumption that the energy of optical emission in flares is completely provided by ionization losses of accelerated particles in the flare region. With this energy estimated to be 10^{29} erg s^{-1} and assuming an exponential rigidity spectrum of the accelerated protons with $P_0 = 100$ MV, he has found that the neutron flux at the Earth should be slightly less than 10 neutrons cm^2 s^{-1}.

A better estimate of the number of protons penetrating to the photosphere can be made if the protons actually produce the white-light emission as has been suggested by Švestka (1970a) and Najita and Orrall (1970) (cf. Section 2.2.11.B). Under this assumption, applying Lingenfelter *et al.*'s theory, Švestka (1971b) has estimated the neutron flux from the white-light (and proton) flare of 1967, May 23 (illustrated in Figures 36 and 38). The resulting neutron flux is shown in Figure 117. Because the threshold sensitivity of the neutron detectors is about 10^{-2} neutrons cm^2 s^{-1}, the neutron flux should have been detectable in this flare. The figure shows that neutrons should be looked for as soon as the flare brightens, since in fact the maximum neutron flux is coincident with the maximum Hα intensity of the flare.

The analysis has also shown another important result: A detectable neutron flux should accompany only flares with extremely hard spectra (with $\gamma \lesssim 2$ in Equation (98) or $P_0 \gtrsim 200$ MV in Equation (100)). Already for $\gamma = 3$, with the same number of particles accelerated to energies above 100 keV, the number of >30 MeV protons in the flare would be too small and the neutron flux at the Earth would decrease to only 2×10^{-3} neutrons cm^{-2} s^{-1}. Therefore, the occurrence of flare-associated neutrons in space should be a very rare phenomenon, like that of white-light (or cosmic-ray) flares.

Attempts to detect neutrons from the Sun were made by many authors in the past 10 yr. (See the review by Chupp (1971), the more recent observations by Cortellessa *et al.* (1971) and Eyles *et al.* (1972), reviewed by Chupp *et al.* (1973), and the OSO-6 measurements reported by Ifedili (1974).) So far the results of all these experiments have been negative. Many of these experiments were made during periods when no neutrons could have been expected from the Sun: in the absence of flares or after flares of minor importance. However, a few experiments, summarized below, covered bigger flare events.

Bame and Asbridge (in Lingenfelter, 1969) did not detect any neutrons after the major flare of 1966, September 2, that produced a PCA effect of ~ 13 dB. However, the proton spectrum of this flare (shown in Figure 97) was very steep at higher energies (with $\gamma > 4$ and $P_0 \simeq 60$ MV after Švestka and Simon (1969) and Lingenfelter (1969)) so that the number of protons with energy in excess of 30 MeV, necessary for the neutron production, was too small. Similarly, Cortellessa *et al.* (1971) made their measurements after a flare with $P_0 \simeq 60$ MV. Ifedili (1974) reports OSO-6 measurements of neutron flux $< 5 \times 10^{-2}$ neutrons cm^{-2} s^{-1} for five flares in 1969, but the characteristic rigidity was small again in all these cases. Only for the December 19 flare was the rigidity higher ($P_0 = 148$ MV), but the flare was small (classified as of Hα importance 1 n).

Thus unfortunately no measurements have been carried out in space after a flare from which an observable neutron flux should be expected. Therefore, Kirsch (1973) has tried to detect solar neutron emission by analyzing neutron monitor data of several mountain stations on the surface of the Earth. While the negative results of the space measurements mentioned above are compatible with Ramaty and Lingenfelter's computations (since P_0 was too small in all the events observed), Kirsch has deduced from the ground-based measurements upper limits of the neutron flux that are in striking contradiction to the theory. He finds 0.04 neutrons cm^{-2} s^{-1} as the highest upper limit to the neutron flux at the Earth even in the case of the flare of 1967, May 23, to which Figure 117 refers.

Kirsch's results may easily be in error since diurnal variation makes the evaluation of the ground-based neutron records difficult. Nevertheless, they indicate that the neutron flux might indeed be lower than expected. Kirsch assumes that the main cause of it is anisotropy of the accelerated protons and α-particles in the solar atmosphere. If the fraction of protons impinging upon the photosphere is significantly lower than in the isotropic model of Lingenfelter and Ramaty, the neutron production will be greatly suppressed. Indications that this might happen have been presented by

Newkirk (1973; cf. Section 5.1.1.B): computed maps of potential fields in proton-producing active regions indicate that only 1 to 15% of the accelerated particles impact upon the photosphere. On the other hand, Newkirk's computations may be irrelevant during the acceleration phase when the scale of the region where the γ-rays are produced is probably much less than the space resolution of his maps. In any case, anisotropy cannot explain the discrepancy with Figure 117 since the neutron flux there was deduced for a known number of protons bombarding the photosphere.

5.4.3. γ-RAYS

Direct observation of neutrons in space is one possibility of detecting their production in solar flares. Another one, as we mentioned in the introduction to this section, is observation of the γ-line at 2.23 MeV produced when neutrons are captured by protons in the solar atmosphere. This line was actually observed by Chupp *et al.* (1973) in the γ-ray spectrum of the flare of 1972, August 4 (Figure 55), together with the 0.511 MeV line produced by positrons and the 4.43 MeV and 6.14 MeV lines due to ^{12}C and

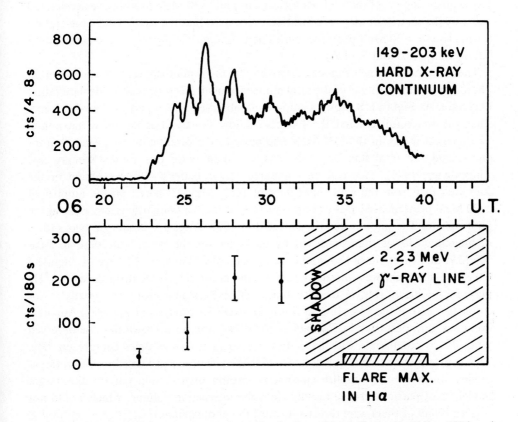

Fig. 118.   A comparison of the time variation in the flux of hard X-rays (de Feiter, 1974) and 2.23 MeV γ-ray line (Chupp *et al.*, 1973) in the flare of 1972, August 4. The γ-ray counting rates are 3 min accumulations. Within this range of uncertainty the two emissions started simultaneously in spite of the fact that the γ-ray line needs >30 MeV protons for its production.

^{16}O nuclei de-excitation. The γ-ray flux at 1 AU (photons cm^{-2} s^{-1}) was given by Chupp *et al.* as

$(7\pm1.5)\times10^{-2}$ in the 0.511 MeV line,

$(2.2\pm0.2)\times10^{-1}$ in the 2.23 MeV line,

$(3\pm1)\times10^{-2}$ in either of the two de-excitation lines.

The γ-ray flux at 2.23 MeV (Figure 118) set in practically simultaneously with the hard X-rays (within the uncertainty of 3 min-accumulation counts). This is important evidence that protons with the energy of $\gtrsim30$ MeV, necessary for the production of neutrons, were accelerated in the flare at about the same time as the electrons producing the hard X-rays through bremsstrahlung. As a matter of fact, some delay in the production of the 2.23 MeV γ-line is to be expected in any case due to the finite neutron capture time in the photosphere. According to Wang and Ramaty (1974) this delay is between 30 s and 3 min, depending on the neutron energy, the density which the neutrons traverse, and the ^{3}He population, since ^{3}He 'kills' the neutrons without emitting any photon. Thus the comparison of the γ-rays and hard X-rays in Figure 118 gives convincing proof that $\lesssim1$ MeV electrons and >30 MeV protons are accelerated either by the same (one-step) acceleration process, or that the second-step acceleration (giving rise to >30 MeV protons) immediately follows the process of pre-acceleration (cf. Sections 3.2.5 and 5.1.1.A).

The fact that many neutrons can decay on ^{3}He in the photosphere without emitting a photon makes the estimate of the total number of neutrons produced in the flare rather uncertain. As Figure 118 shows, Chupp *et al.* observed only a part of the γ-ray burst, since the spacecraft entered the Earth's shadow shortly after the burst maximum. The integrated flux in the 2.23 MeV line prior to the occultation was ~80 photons cm^{-2}, and the total flux probably was less than twice this value (Ramaty and Lingenfelter, 1973). This can be compared to de Feiter's (1971) estimate of the 2.23 MeV flux from the white-light flare of 1967, May 23, based on the number of >20 MeV protons needed for the production of the white-light flare emission in the photosphere. The photon number detected in 1972, August 4 is almost two orders of magnitude lower than predicted by de Feiter for the white-light flare if $P_0=200$ MV, and three times lower still for $P_0=60$ MV. Thus, in the flare of August 4 either the spectrum was unusually soft, or the non-radiative neutron decay on ^{3}He played a significant role, or the distribution of accelerated protons was greatly anisotropic (see end of the preceding section). An extensive analysis of the 1972, August 4 event has been presented by Ramaty *et al.* (1974), who have shown that the observed γ-ray flux is compatible with hard proton spectra if one assumes capture on ^{3}He. This is particularly true if one assumes that the γ-rays are produced in a thin-target model, which actually implies anisotropic proton distribution. (In the thin-target model, most particles escape rapidly from the interaction volume, which would not happen if half of them were directed toward the photosphere.)

Ramaty *et al.* deduced the spectrum of the accelerated protons from the observed ratio of the flux in the 4.4 MeV and 2.3 MeV lines. Figure 119 shows the theoretical variation of this ratio for the thin and thick target models, assuming ^{3}He/H $= 5\times10^{-5}$

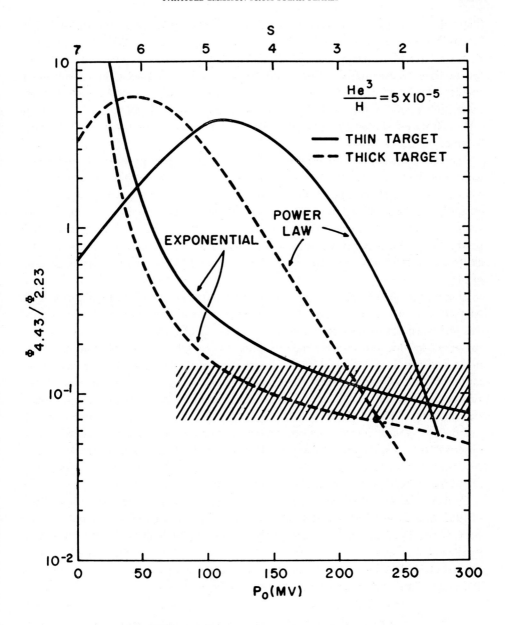

Fig. 119. Ratios of the 4.4 MeV line intensity to the 2.2 MeV line intensity in the thin and thick target models, with power-law and exponential spectra, and with ^{3}He/H $= 5 \times 10^{-5}$ in the photosphere. The shaded area represents the ratio actually observed in the 1972, August 4 flare by Chupp *et al.* (1972). (After Ramaty *et al.*, 1974.)

(the value deduced from solar wind). The value of 0.11 ± 0.04 for the 4.4 MeV 2.3 MeV^{-1} flux ratio observed by Chupp *et al.* (1973) is shaded in the figure. Evidently P_0 must have been higher than ~ 200 MV for the thin-target model and $\gtrsim 120$ MV for the thick target interpretation. With $^3He/H = 0$ the P_0 values would result much smaller. On the other hand, $^3He/H$ might be still higher, and if so, P_0 might become quite high even for the thick target model (according to Section 5.4.1, $^3He/H$ in '^3He-rich flares' should be $\sim 2 \times 10^{-3}$, i.e., 40 times higher than Ramaty *et al.* have assumed).

Ramaty *et al.* estimate the total number of protons above 30 MeV released downward into the Sun in the thick target model at $\sim 10^{33}$. For the thin target model the estimate is more uncertain: about 2×10^{33} protons of energies greater than 30 MeV escape from the flare region if relativistic particles traverse about 1.5 g cm^{-2} in the solar atmosphere. When a comparison is made with the number of electrons that produced the hard X-ray continuum through bremsstrahlung (cf. Figure 55), the electron number is lower by about a factor of 10 to 100 than the number of protons of the same energy.

Chupp *et al.* (1973) also observed strong indications of γ-rays in the flare of 1972, August 7; in this case, however, the most important initial phase of the γ-ray burst was lost when the spacecraft was in the Earth's shadow. The remnants seen after the spacecraft left the shadow give the flux equal to $(3.4 \pm 1.1) \times 10^{-2}$ photons $cm^{-2}\,s^{-1}$ in the positron line, $(4.8 \pm 1) \times 10^{-2}$ in the neutron line, and $< 2 \times 10^{-2}$ for the de-excitation lines at 4.4 and 6.1 MeV. Independent observations made on Prognoz 2 by a French-Soviet group (Talon *et al.*, 1975) seem to confirm that these observations were real and that the white-light flare of August 7 was indeed another source of γ-rays (maybe even a more intense one, since the γ-ray remnants Chubb *et al.* recorded were observed 40 min after the flare onset). Unfortunately, the Prognoz-2 measurements cover only the energy ranges < 340 keV and > 2.90 MeV, thus yielding no information about the most important γ-lines at 0.5 and 2.3 MeV.

Gruber *et al.* (1973) suspect that γ-rays were emitted also by the white-light flares of 1967, May 21 and 23 (the latter one being the subject of Figures 36, 38 and 117). On May 21 two MeV channels containing the 0.5 and 2.3 MeV lines showed statistically significant increased rates between 19^h25^m and 19^h28^m, while the white-light flare peaked at 19^h26^m. On May 23, however, the rate was increased about 1 h after the white-light flare, which contradicts the other observations.

5.5. Particle Clouds and Interplanetary Shock Waves

The association of geomagnetic storms at the Earth with flares on the Sun has been known for a long time, and it was actually suspected since 1859 when the first observed solar flare was followed shortly afterward by a severe geomagnetic storm (Carrington, 1859). The sudden commencement (s.c.) which initiates the storm follows tens of hours after the flare occurrence on the Sun, hence the exciting agent must travel through space with a speed of the order of 1000 km s^{-1}.

5.5.1. CHARACTERISTIC PROPERTIES OF INTERPLANETARY SHOCKS

We now know that the exciting agent is a shock wave which originates in the flare. The mean delay time between the flare and the associated s.c. is 55 h, which corresponds to a mean speed of 730 km s^{-1} (Hundhausen, 1970). The sonic Mach number is generally near 3, indicating that the typical interplanetary shock is of intermediate strength. The mean propagation speed near the Earth is 500 km s^{-1} so that most shockwaves are obviously decelerated during their transit from the Sun; but this is not always the case (Hirshberg et al., 1970). If two shocks subsequently propagate from the same active region within a few days, the second shock travels much faster (Obayashi, 1964; Caroubalos, 1964), probably because it is not decelerated as much as the first one. Some exceptional shocks can propagate at significantly higher speeds: According to Dryer (1974), the shock ejected from the 1972, August 4 flare moved with a speed in excess of 3000 km s^{-1} near the Sun, and its average speed might have been as high as 2700 km s^{-1}. For reviews see Hundhausen (1972a, b) and Dryer (1974).

Since the flare-associated travelling interplanetary shocks show high correlation with types II + IV radio bursts (Hundhausen, 1972a, b), there is little doubt that the shocks in space are simply continuations of shocks giving rise to type II and moving type IV bursts in the solar corona (cf. Sections 3.2.4.A and 3.2.5). However, this fact, though suspected by many authors for more than two decades, could be proved only in recent years.

Malitson et al. (1973) were the first who succeeded in tracking a type II radio burst (in 1972, August 7) from the Sun through interplanetary plasma right to the Earth (Section 3.2.5). Its source was a white-light flare, also emitting particles and γ-rays (Section 5.4.3). Maxwell and Rinehart (1974) have described the type II + IV burst on the Sun, with fast pulsations in the 100 to 200 MHz type IV continuum, which McLean et al. (1971) associate with the acceleration process (cf. Section 3.2.4.C). The mean propagation velocity of the shock was 1270 km s^{-1}. Palmer (1974) believes to have linked another type II observed in the corona in 1971, June 30, with radio emission in space recorded down to 0.3 MHz. The correlation, however, is not fully convincing.

Stewart et al. (1974a) observed a flare in 1973, January 11 that produced a spray, white-light disturbance in the corona and type II + IV radio burst. They do not relate these solar phenomena directly to an interplanetary shock (it was a limb event, and shocks from limb flares rarely reach the Earth), but give evidence that the energy contained in this coronal disturbance was of the same order as in the shock waves recorded in interplanetary space. With the total number of particles in the expelled gas estimated to $\sim 10^{40}$, total mass is $\sim 1.7 \times 10^{16}$ g, and $\sim 10^{31}$ erg are involved in the kinetic energy of the ejection. In comparison to this, Hundhausen (1972b) derived 5×10^{15} to 1.5×10^{17} g and 5×10^{30} to 5×10^{32} erg for 22 shocks near the Earth.

Finally, Gosling et al. (1975) using the Skylab data could confirm Stewart et al.'s result for another big flare (of 1973, September 7) and present, in addition, a direct correlation with a shock wave recorded in space. The flare occurred at 11^h41^m and was associated with a white-light disturbance in the corona and with type II+IV radio burst. Speed of the disturbance was estimated at 960 to 1300 km s^{-1}, mass of the ejected material at 2.4×10^{16} g and kinetic energy (for 960 km s^{-1}) at 1.1×10^{32} erg. At 07^h30^m on September 9 (44 h after the flare onset) Pioneer 9 detected a strong shock in the solar wind, moving with a speed of 720 km s^{-1}. Since the travel time of 44 h corresponds to a mean speed of 950 km s^{-1}, the shock was obviously decelerated during its travel from the Sun to Pioneer 9 so that the solar velocities of 960 to 1300 km s^{-1} are in good agreement with the Pioneer 9 data. Pioneer 9 was at that time at 1AU from the Sun and in a position, from which the flare would appear at the center of the solar disc. The disturbance in space carried mass of 4×10^{16} g and energy of 1.2×10^{32} erg, comparable to the values deduced from the coronal phenomena.

Essentially all clearly defined s.c.'s are caused by hydromagnetic shocks (Burlaga and Ogilvie, 1969). Many other s.c.'s, however, are simply the result of tangential discontinuities in the solar wind and do not imply the existence of a shock wave or a flare (Gosling *et al.*, 1967). This explains why many s.c.'s (and magnetic storms) are observed without any obvious association with a flare on the Sun and why such storms mostly recur with a 27-day period.

A discontinuity in the solar wind appears when there is a high gradient of velocities, a high-speed stream outrunning a slow one. For many years the source of the high-velocity streams was looked for in the active regions as a consequence of the high temperature in the coronal condensations. However, more recent data strongly indicate that it is not the coronal temperature, but the structure of the magnetic field, that determines the solar wind velocity. Krieger *et al.* (1973, 1974) have found a correlation between high-velocity solar wind streams and coronal holes, i.e., regions of open magnetic fields. Also Gulbrandsen (1974) confirms from a correlation between the intensity of the green (λ 5303 Å) coronal line and geomagnetic activity that the high velocity streams originate from regions of low density and open magnetic field structures. Bell and Noci (1975) arrive at similar conclusions from a study of Fe xv (284 Å) isophotograms on OSO 7.

5.5.2. Shock-wave-associated variations in the particle flux

Behind the shock a 'cloud' of slightly accelerated particles moves to the Earth, and these particles are recorded in the form of a solar wind enhancement for many hours after the shock has passed the Earth. For instance, in an example presented by Hundhausen (1972a), the solar wind flow speed increased from 400 to more than 500 km s^{-1}, proton density increased from 20 to about 50 protons cm^{-3}, and proton temperature rose from some 5×10^4 to 5×10^5 K.

Apart from this, low energy particles up to about 20 MeV for protons are stored behind the shock, and the shock also sweeps diffused particles of these energies ahead as it propagates. Some authors believe that the shock itself can also accelerate particles to mild energies, for example, due to Fermi acceleration between the shock and ap-

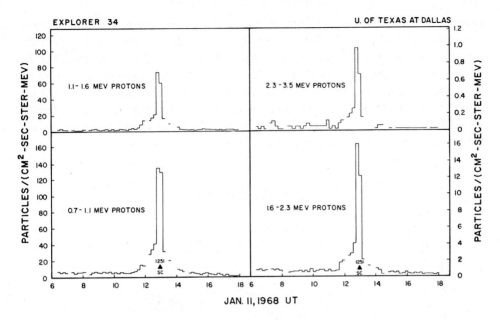

Fig. 120a. Example of a short-lived increase in the flux of low-energy protons associated with a sudden commencement. Note the peak-flux decrease toward higher energies. (After Palmeira *et al.*, 1971.)

proaching magnetic field inhomogeneities (Parker, 1965; Van Allen and Ness, 1967; Rao *et al.*, 1967; Vernov *et al.*, 1971; Vernov and Lyubimov, 1972, and references herein). Thus, when the shock arrives near the detector we generally observe a sharp increase in the flux of low-energy particles. The rise may begin up to one hour before the shock arrival and last for a few hours afterward, but in most cases the increase is very sharp and short-lived (Figures 120a, b; Palmeira *et al.*, 1971; Palmeira and Allum, 1973). In cases like that in Figure 120b the solar energetic particles appear to be confined in the region just ahead ($\sim 10^6$ km) of the advancing shock front.

For flares situated more than about 50° west from the central meridian a decrease in the flux of low-energy particles is often recorded after the shock wave passage (Reid, 1964; Kahler, 1969). Reid explains it in the following way: Behind the shock wave the solar wind speed is increased and the root point of the Sun-Earth magnetic field line shifts nearer to the center of the solar disc. Thus, for all flares situated to the east of the root-point the access of particles to the Earth is improved behind the shock wave. For flares more than $\sim 50°$ to the west, however, the opposite is true: the distance between the flare and the new root-point increases.

A similar model has been presented by Kahler (1973b). Figure 121a shows the quiet Archimedes spiral structure of the magnetic field lines in interplanetary space. Accelerated particles propagate along the field lines (shadowed cone); they get to an observer (cross) who has the flare on the western hemisphere, but miss another observer (circle) who has the flare more to the east. A propagating shock sweeps the magnetic field lines and removes, thus, the particles from the site of the observer (cross) who saw them before (Figure 121 b). Therefore, the particle flux drops (or decreases) for flares far on the western hemisphere when the shock arrives. On the other hand, the shock brings the particles to the site of the other observer (circle) who did not see the particles before (Figure 121 c). Thus, when the shock arrives, the particle flux sets in (or increases) for flares to the east of $\sim 50°$ W.

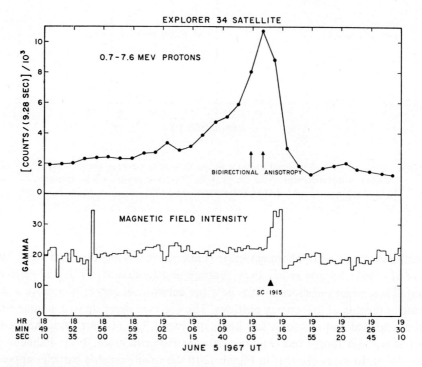

Fig. 120 b.   High time-resolution record of another short-lived increase in the flux of >0.7 MeV protons associated with a sudden commencement. Time variations in the interplanetary magnetic field close to the Earth are shown in the lower graph. (After Palmeira and Allum, 1973).

An important phenomenon associated with shock waves in space is the Forbush decrease, i.e., a decrease in the neutron monitor counts at the Earth caused by deflections of galactic cosmic ray particles in the shock (an example may be seen in the neutron monitor record in Figure 94).

The Forbush decrease is deeper when the shock speed is higher (Haurwitz et al., 1965). The greatest Forbush decrease so far was recorded in 1972, August 4: 27% decrease at McMurdo and 35% at the South Pole (Pomeranz and Duggal, 1973); before that the greatest decrease was 24% (in July 1959). The parent flare occurred

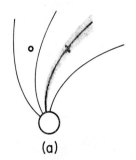 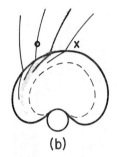 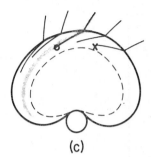

(a)   (b)   (c)

Fig. 121. Schematic drawing of an expanding plasma cloud with a shock front propagating through space, after Kahler (1973 b). The shaded region indicates the low energy protons, the dashed line is the front of the expanding plasma cloud, and the heavy line is the shock front. The observer at ○ (for whom the flare is on the eastern hemisphere) sees the protons only after the s.c. For the observer at × (for whom the flare is on the western hemisphere) the proton flux drops when the shock arrives.

on August 2 at $20^h 33^m$ in the same active region that produced the γ-ray flares on August 4 and 7 (Section 5.4.3). The Forbush increase set in 29 h later, which corresponds to mean shock speed of 1437 km s^{-1}.

However, the depth of the Forbush decrease also depends on the position of the parent flare on the Sun and strong Forbush decreases do not occur if the source flare is situated more than 40° or 50° to the west of the central meridian (Haurwitz *et al.*, 1965; Švestka, 1966 b). This is demonstrated in Figure 122 which shows the distribution of strong PCA's (> 3 dB) and deep Forbush decreases (> 5%) during 1956–1963 with respect to position of the flare on the solar disc.

One can suggest for this the following explanation: As the shock wave propagates through the normal Archimedes-spiral structure of the interplanetary magnetic field, it compresses the magnetic field lines toward the west (as indicated in Figure 121), thus building a strong magnetic boundary at the western border of its propagation; on the other hand, the field is fairly open toward the east (cf. Haurwitz *et al.*, 1965). This strong magnetic boundary plays the main role in the particle deflection; if it is too far from the Earth, as is the case with flares far on the western hemisphere, the resulting Forbush effect is small. On the other hand the PCA's are produced by high-energy particles that move much faster than the shock and follow the pre-shock spiral magnetic field lines; thus they have much easier access to the Earth from the western hemisphere.

The completely different distribution of the flares favorably located for the production of a Forbush decrease or a proton event near the Earth, demonstrated in Figure 122, may be helpful if there is an ambiguity in the determination of the source of the event on the Sun.

5.5.3. INCREASED HELIUM CONTENT BEHIND INTERPLANETARY SHOCK WAVES

In the last few years many authors have reported an increased helium content in the solar wind plasma behind interplanetary shock waves. Such an increase is observed

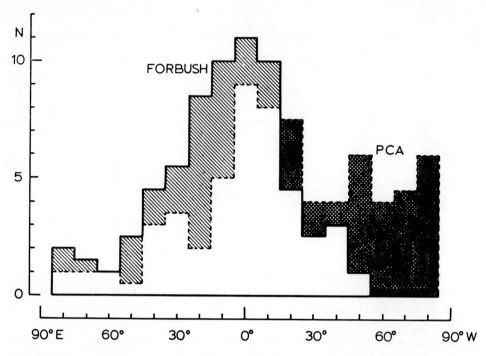

Fig. 122.   The distribution of sources of strong PCA effects (>3 dB) and deep Forbush decreases (>5%) on the Sun. While most strong PCA's come from flares between 10° E and the western limb, the strong Forbush effects are only produced by flares between the eastern limb and 50° W. (After Švestka, 1967.)

5 to 12 h after the shock and in some cases the helium/hydrogen density ratio exceeds 0.20, while its average value in the quiet solar wind is about 0.04 (Gosling *et al.*, 1967; Bame *et al.*, 1968; Ogilvie *et al.*, 1968; Ogilvie and Wilkerson, 1969; Lazarus and Binsack, 1969; Hirshberg *et al.*, 1970, 1971, 1972).

Hirshberg *et al.* (1972) reported 12 well-identified cases of the association between helium enhancements (He/H > 15%) and major solar flares. These cases indicate very strongly that a helium enhancement characterizes the plasma ejected into the interplanetary medium by the flares. If this is true, then high helium abundance should be a characteristic property of the solar corona in the flare region; or alternately, helium should be preferred by the process which accelerates the flare plasma.

However, one has to be aware of the fact that no helium enrichment has been detected in fluxes of particles accelerated to much higher energies in proton flares (J. W. Warwick, 1973). According to Lanzerotti (1973), the He/H ratio in proton events decreases with increasing energy and even at 1 MeV it is still smaller than the average in solar wind. Equation (101) might indicate that the helium abundance increases toward very low energies, which are observed behind shock waves, but this is a crude extrapolation because this equation was deduced only from the energy interval between 1 and 12 MeV nucleon^{-1}.

The problem is confusing and yet unsolved. One has to bear in mind that the He/H ratio variations are generally very large in the solar wind plasma, the ratio varying between the values 0.01 and 0.25, and the reasons for this are not known. However, the occurrence of very high values is fairly rare, and the fact that Hirshberg *et al.* (1972) were able to associate 75% of them with flare shock waves is convincing evidence that the flare plasma behind a shock wave is helium-enriched.

CHAPTER VI

FLARE MODELS

First of all, let us summarize the basic properties of flares which we learned in Chapters II through V, and which an ideal flare theory should explain.

6.1. Basic Properties of Flares

(1) As far as the present resolving power of solar observations reveals, a necessary condition for the occurrence of a flare is the existence of a bipolar magnetic structure in the solar atmosphere.

(1a) Wherever a bipolar field emerges, the coronal response is a heating above the newly emerged flux. The area thus becomes visible as an X-ray bright point. Flare-like X-ray brightenings can occur in such bipolar structures a few hours after the structure is born.

(1b) In newly born bipolar structures that develop into active regions, small flares (visible not only in X-ray but also in Hα) can occur about one day after the birth. Greater flares, however, are usually observed only after the first appearance of sunspots.

(1c) Flares occur most frequently during the phase when the active region approaches the maximum of its development.

(1d) Rare, but often quite extensive flares also occur very late in the active region development, when only a weak plage and a quiescent filament are present (the so-called 'spotless' flares).

(2) The most important flares (in particular from the point of view of the intensity of non-thermal impulsive bursts and energetic proton production) occur in regions where the originally bipolar structure becomes magnetically complex; i.e., where opposite magnetic polarities are mixed (γ-configuration), and in particular where both polarities occur within a common penumbra (δ-configuration).

(3) The first flare brightenings in the chromosphere (for the most part brightenings of already existing structures of the chromospheric plage) occur close to the zero line of the longitudinal magnetic field; when two or more bright areas form, they are distributed on both sides of the zero line.

(3a) The expansion of individual flare parts always proceeds within the region of one magnetic polarity.

(4) In larger flares individual bright patches subsequently merge and form more or less regular rows along the zero line. Thus a characteristic two-ribbon chromospheric

flare is formed. It is particularly conspicuous in the cases when the zero line is straight. This is typical for old 'spotless' active regions, but it often can also be seen in well-developed sunspot groups.

(4a) In the case that a quiescent filament (which always extends along the zero line) is visible in the active region, the filament often disappears and the chromospheric flare brightenings form along both sides of it.

(5) Two-ribbon flares in old, spotless active regions develop along both sides of a disappeared dark filament and the two ribbons travel with a fairly constant velocity of 1 to 10 km s^{-1} outward from the zero line; the expansion proceeds until the chromospheric flare fades.

(6) Two-ribbon flares in young active regions (embedded within sunspots) usually begin as one unresolved bright ribbon which subsequently separates into two, moving outward from the zero line with decreasing velocity. The speed may be up to 100 km s^{-1} during very short earliest phase of the ribbon separation; it is usually about 10 km s^{-1} later on, but the velocity dramatically decreases as soon as the travelling ribbon encounters regions with a strong longitudinal field. In such a case the ribbon motion is often completely stopped.

(6a) Two-ribbon flares of this type are the typical sources of greatly accelerated protons. This is not true for the two-ribbon flares mentioned in (5), in which the protons are accelerated only to low energies or they are not accelerated at all.

(6b) In cases (5) and (6) the separating ribbons in the chromosphere mark the travelling foot-points of growing coronal flare loops.

(7) As a rule, flares when expanding or forming new bright patches, tend to avoid sunspot umbrae, i.e., regions with extremely strong magnetic field. However, on some occasions, flare parts do penetrate above sunspots. This seems to happen in particular when the longitudinal field gradient near the spot is very steep.

(7a) There are indications that flares which extend over sunspots are very productive in X-rays and radio microwave radiation.

(8) In some parts of particular active regions, flares show a tendency to recur in nearly identical form. This proves that the general structure of the magnetic field must be preserved or regenerated between flare occurrences.

(9) On the other hand, magnetographic measurements with high resolution in space and time indicate that changing magnetic fields are characteristic of active regions producing flares.

(9a) Flares tend to occur when new magnetic flux emerges inside the active region and this effect amplifies if the new flux has magnetic polarity opposite to the ambient field.

(9b) After great flares, the magnetic field structure seems to be somewhat simplified and the magnetic field gradient at the flare position may decrease, at least temporarily.

(10) When a quiescent filament is present close to the position of the flare, it becomes activated minutes or tens of minutes before the flare occurrence. This gives evidence that in many (and possibly all) cases changes in the magnetic field start tens of minutes before the flare onset.

(10a) These pre-flare changes are of a non-eruptive nature. Disruption of the filament only occurs when the flare proper begins.

(10b) EUV lines in the transition layer reveal the presence of large non-thermal motions in active regions prolific in flares.

(10c) Soft X-ray measurements sometimes show a gradual preheating in the active region for about 10 min prior to the flare.

(11) Besides these pre-flare effects, the earliest manifestation of the flare is an increase in the soft X-ray and EUV flux reflecting a heating in the corona or transition layer. Skylab X-ray measurements reveal that this starts as a point-like brightening probably located in a coronal loop.

(12) Shortly after the onset of the soft X-ray burst (but the timing is uncertain) the first flare brightening in the Hα line is observed, thus manifesting increased excitation of hydrogen in a limited part of the chromosphere.

(12a) The enhancements (11) and (12) are common characteristics of all flares, and in many flares only these two kinds of emission are observed, sometimes being associated with a weak gradual microwave burst. All these effects can be explained as thermal or quasi-thermal phenomena, in arcades of loops rooted in the chromosphere, where Hα is emitted, and extending high into the corona, where the X-ray flare is observed.

(13) Generally, the sequence of maxima keeps to the following scheme: first, maximum in temperature in the coronal region; then, maximum intensity in the Hα light; after that, maximum in the soft X-ray flux; and finally, a maximum in emission measure.

(13a) The maximum temperature deduced from soft X-rays is within the limits 1 to 4×10^7 K, the emission measure is of the order of 10^{48} or 10^{49} cm^{-3}, and mean electron density can be estimated at 3×10^{10} cm^{-3}. However, the quasi-thermal flare appears to be greatly heterogeneous both in temperature and in density.

(14) In some flares a short-lived non-thermal phase, giving rise to hard X-rays, EUV, and impulsive microwave bursts, is superimposed on the thermal development described in items (11)–(13). Flares of this type occur predominantly (and may be exclusively) in magnetically complex active regions.

(14a) This impulsive phase (if present) always starts at or after the onset of the soft X-ray increase and peaks (and mostly ends) before the flare maximum in the Hα light, i.e., it takes place during the flash-phase of the flare.

(14b) The onset of the impulsive phase is sometimes delayed by several minutes behind the beginning of the quasi-thermal flare. Therefore, it cannot be the primary process giving rise to the flare.

(15) The non-thermal phase is the manifestation of an impulsive acceleration of electrons to a power-law energy spectrum with exponent between 2 and 5. Its duration extends from a few seconds to several minutes in the most outstanding events. The number of electrons accelerated to energies above 100 keV is as high as 10^{36} in strong bursts.

(15a) Major impulsive bursts are composed of many 'elementary bursts' with lifetimes of ~1 to 20 s. These elementary bursts may be produced subsequently in different parts of the active region.

(16) Positions of the impulsive Hα kernels and the white-light flare patches, which most probably are produced by energetic particles penetrating downward during the impulsive phase, indicate that the acceleration process (giving rise to hard X-rays, EUV, and impulsive microwave bursts) is restricted to a small volume and is located close to the zero line between large spots of opposite polarities, at the place where the apparent gradient in the longitudinal magnetic field is the highest.

(16a) Also the sources of the elementary bursts while changing position track the zero line.

(17) Another effect, sometimes associated with the impulsive phase is a type III radio burst. When a hard X-ray, impulsive microwave, and type III burst coincide in time, the type III can often be traced up to the Earth's distance and fluxes of non-relativistic electrons are then recorded in space near the Earth.

(17a) Type III is obviously caused by the same electrons that we later observe *in situ* in space; the type III is observed as these electrons pass through the coronal and interplanetary plasma.

(17b) However, most type III bursts are not associated with flares. They occur in high numbers, often in impulsive series, in a few specific active regions, and these regions need not be identical with those that produce the non-thermal flares.

(18) This implies that the type III burst represents another acceleration process, different from that which produces the impulsive phase. However, in regions that are prolific in both type III bursts and non-thermal flares, the two acceleration processes show a strong tendency to be linked.

(19) There is no definitive evidence that protons and relativistic electrons are accelerated during the same time and by the same process as the non-relativistic electrons which produce hard X-ray and microwave bursts. However, the γ-ray burst recorded during the flash phase of the flare of 1972, August 4, needing >30 MeV protons for its production, indicates strongly that the relativistic electrons and protons must be accelerated either simultaneously with the non-relativistic electrons, or in a second-step acceleration which *immediately* follows (cf. Item 26b).

(19a) The same conclusion can be drawn from the coincidence of the white-light flare with the impulsive phase, since the white-light emission is probably caused by >20 MeV protons and/or high-energy electrons.

(20) In some flares the non-thermal burst is followed by a type IV radio burst. This indicates the presence of clouds of mildly relativistic electrons which produce continuous radio emission through gyro-synchrotron radiation.

(20a) The type IV burst in microwaves starts with the impulsive phase, while its onset at lower frequencies is usually slightly delayed. The very short time difference between the impulsive phase and the metric type IV onset is further evidence for acceleration of relativistic electrons in the course of, or immediately after, the impulsive phase.

(20b) At metric wavelengths the early component of the type IV burst often moves from the Sun up to several solar radii. The later component is always stationary.

(21) In association with some flares a type II radio burst is recorded, which indicates the passage of a shock wave through the corona, with velocities of the order of 1000 km s^{-1}. The shock originates close to the maximum of the impulsive phase.

(21a) Type II bursts show very high correlation with the acceleration of protons to high energies in flares. Therefore, type II appears to be a manifestation of the second-step acceleration which accelerates the particles, pre-accelerated during the impulsive phase to ~ 1 MeV, to higher energies.

(21b) In some flares both type II and IV occur. In that case type II usually travels in front of the moving type IV. However, flares in which only type IV or only type II is observed are frequent.

(21c) In flares where all the types II, III, and IV occur, the most common sequence on metric waves is: type III first, then type II, and finally type IV. One can interpret this as follows: type III coincides with the impulsive phase when particles are accelerated to energies ~ 1 MeV. A process associated with type II and moving type IV bursts accelerates these particles to higher energies. Some of the relativistic electrons are trapped in closed magnetic fields in the corona and produce the stationary type IV.

(22) On some occasions a flare-produced wave can be seen directly as it affects the visible structures of the chromosphere. Such a wave travels with rather constant velocity close to 1000 km s^{-1} and it originates, like the type II burst, close to the impulsive phase. It is most probably a fast mode MHD wave-front in the corona which may develop into a shock wave that generates the type II burst.

(22a) There are indications that simultaneously a slower wave may travel from the flare with a velocity close to 100 km s^{-1}. This wave may be responsible for many of the 'disparitions brusques.'

(23) The shock wave seen in the corona as a type II burst and/or in the chromosphere as a flare wave may continue its propagation through interplanetary space up to the distance of the Earth, where it causes the 'sudden commencement' of a geomagnetic storm.

(23a) These interplanetary shock waves show-high correlation with the type II+ IV bursts; as a rule, all flares producing intense flux of energetic protons also give rise to shock waves of this type.

(23b) In many cases the plasma travelling behind the shock is helium-enriched.

(23c) Energy carried in such a shock can be as high as 5×10^{32} erg.

(24) Strong fluxes of energetic protons in space appear to be associated with big flares, most of them showing two expanding ribbons embedded among large spots, and accompanied by type II burst. They are also characterized by type IV bursts with a strong microwave component.

(24a) Protons up to energies of at least 5 GeV and possibly 20 GeV have been recorded coming from a flare.

(24b) Protons in space are accompanied by relativistic electrons, but the electron flux is orders of magnitude smaller than the proton flux at the same energy.

(24c) Protons in space are almost always accompanied by non-relativistic electrons. On the other hand, many non-relativistic electron events are not accompanied by protons.

(25) Heavy nuclei are more abundant in the streams of accelerated particles than in the solar atmosphere. Below ~ 15 MeV nucleon^{-1}, the abundance increases with decreasing energy per nucleon and with increasing Z, showing great variations from flare to flare.

(25a) This indicates that heavy nuclei are not fully stripped when entering into the acceleration process on the Sun. On the other hand, they appear to be fully stripped when they are recorded near the Earth.

(26) When protons and α-particles are accelerated to energies in excess of ~ 30 MeV, nuclear reactions begin to produce neutrons, positrons, deuterons, tritons, ^{3}He, and excited nuclei. Through capture, decay, and de-excitation of these elementary particles γ-rays are produced in the flare region.

(26a) Only flares with extremely hard proton energy spectra ($P_0 \gtrsim 200$ MV) produce enough > 30 MeV nuclei to give rise to detectable γ-ray emission. The same condition applies to the white-light emission. This indicates that the white-light emission, like the γ-rays, may be produced by energetic protons.

(26b) The γ-rays and white-light emission occur practically simultaneously with the flare impulsive phase. (See Item (19)).

(27) It is estimated that only 1 to 15% of the accelerated particles go downward toward the photosphere and $\lesssim 40\%$ of them (for non-relativistic electrons possibly $\lesssim 1\%$) leave the Sun. Most accelerated particles are captured in coronal arches.

(27a) Storage of particles near the Sun for periods in excess of one day seems unlikely.

(28) The basic process of energy transfer from the corona to the chromosphere is heat conduction. In some flares, and only in restricted regions of them, streams of particles also contribute to the chromospheric heating.

(28a) The chromospheric layer where energy is deposited is only a few hundred kms thick.

(29) This is in agreement with the analysis of the chromospheric flare spectrum which leads to very small linear thicknesses for chromospheric flares. Nevertheless, in order to account fully for the observed spectra, one needs to suppose, in addition, that the chromospheric flare also has a heterogeneous fine structure.

(29a) Similar inhomogeneities, possibly on a larger scale, have also been detected in the coronal flare.

(30) The electron density in flare elements in the chromosphere is of the order of 10^{13} cm^{-3} in the maximum phase of flare development. The highest value ever deduced from hydrogen lines is 4×10^{13} cm^{-3}.

(30a) The density of hydrogen atoms in the elements of a major chromospheric flare must exceed 10^{14} cm^{-3}.

(30b) The density decreases with height; it is of the order of 10^{12} cm^{-3} just above the chromosphere, and of the order of 10^{10} cm^{-3} at a height of about 20 000 km above the chromospheric base.

(30c) The density in flares appears to be about two orders of magnitude higher than in the quiet Sun throughout the chromosphere, transition layer, and low corona.

(31) The electron temperature in the flare elements emitting in hydrogen is < 10 000 K (in some cases even as low as 7000 K) in the chromosphere, and slightly higher than 10 000 K in the cold flare parts seen in hydrogen above the limb.

(31a) Flares must be greatly inhomogeneous in temperature in the lower corona, since at the same height one observes flare parts corresponding to temperatures of the order of 10^4 K and 10^7 K, respectively.

(32) There is no evidence for non-thermal motions in the chromospheric part of a flare. However, non-thermal motions seem to contribute (at least in some flares) to the line broadening in higher layers of the flare.

(33) Emission lines in the chromospheric spectrum of a flare show a striking asymmetry that is most pronounced close to the flare maximum in the Hα line, the long-wave wing being broader and more intense than the short-wave one. This asymmetry affects all lines in the spectrum, and it must be due to some kind of internal motion in the flare region.

(33a) In some flares (and maybe in all) an opposite asymmetry appears for a short time at the very onset of the Hα flare.

(34) Some flares are accompanied by spray prominences in which the gaseous material is ejected into the corona with speeds exceeding the gravitational escape velocity.

(34a) In some very rare cases a whole portion of the flare is expelled in a spray-like form into the corona.

(34b) There are indications for an association of sprays with the flare waves and type II bursts, but generally, sprays appear to be less common than the other effects associated with flare-produced waves.

(35) Many flares are accompanied by surge prominences, in which dense material first ascends and subsequently descends with velocities of the order of 100 km s^{-1}.

(35a) However, unlike the sprays, surges do not seem to require that a flare occur and actually most surges occur without flares.

(36) Two-ribbon flares, in particular in sunspot groups, are often accompanied in Hα light by loop prominences which span the two bright ribbons, rise upward and survive for many hours after the flare decay. The very broad spectral lines in the loop prominences appear to be a visible manifestation of non-thermal particles trapped in flare-associated magnetic loops.

(37) The total amount of energy released in the most common flare phenomenon, a subflare, is $\sim 10^{28}$ to $\sim 5 \times 10^{30}$ erg, depending on whether the non-thermal phase is accomplished or not; the upper limit is an estimate for flares producing strong type III and electrons in space. In an average flare the energy is 10^{30} to 10^{31} erg. In the most outstanding events it may reach several times 10^{32} erg. It seems that most energy is then deposited in the shock wave, which is a manifestation of a powerful

process that accelerates atomic nuclei to 10 to 100 MeV energies and ejects large amounts of gaseous material from the Sun.

It is quite clear than no flare model can explain all of these aspects of the extremely complex flare phenomenon. All the models so far proposed present only the basic mechanisms which can give rise to a flare in the solar corona and chromosphere. The items (1) through (37) listed above, however, may serve as a basis for judging which of the various models and mechanisms proposed is closest to the situation that is actually observed.

6.2. Synthesis of a Flare

We will now use these known properties of flares (i.e., as of the beginning of 1975) for suggesting a tentative synthetic model of a flare.

A. *Pre-Flare Phase*

A few minutes or tens of minutes prior to a flare the magnetic field in the flaring active region begins to be rearranged. Maybe this rearrangement is due to an emergence of new magnetic flux from below the photosphere, to build-up of a current sheet, or to gradual relaxation of the non-potential magnetic field in the active region.

The rearrangement manifests itself through three effects: pre-heating of the gas in a magnetic loop (or possibly several magnetic loops) in the corona, visible as a gradual X-ray enhancement; increased non-thermal motions visible in the EUV lines; and activation of a quiescent dark filament extending along the $H_\parallel = 0$ line. As it seems, all these effects can also happen without giving rise to a flare; but many flares (and maybe all flares) are preceded by such effects (the filament activation in particular).

B. *Quasi-Thermal Flare*

The flare itself starts with a fast increase in coronal temperature as manifested by a soft X-ray burst. The reason for this temperature increase may be thermal dissipation of the current sheet built in the phase (A), sudden increase in the relaxation rate of the non-potential magnetic field in the active region, or any other plasma instability with purely thermal (or quasi-thermal) consequences. The temperature reaches its peak value of $\gtrsim 10^7$ K pretty fast, in one or a few minutes; as it declines exponentially the emission measure increases further, indicating that the flare pushes the transition region lower into the chromosphere, thus increasing the number of electrons at high temperature. Through conduction, quite fast due to the high temperature in the corona and the steep temperature gradient in the compressed transition layer, the flare energy is transported to the chromosphere and heats it to $\sim 10^4$ K to a depth of only a few hundred km. In most flares (and as it seems these are predominantly flares in bipolar active regions) nothing else happens. We observe a soft X-ray flare and sometimes a gradual microwave burst, both produced through bremsstrahlung of electrons with a Maxwellian distribution of velocities. There is also an EUV burst and an Hα flare emitted from transition and chromospheric layers heated through conduction.

C. *Impulsive Flare*

When some additional condition is fulfilled (probably magnetic complexity in the flare region), this heating leads in a short time to an impulsive instability that accelerates electrons and nuclei to energies of the order of 10 to 100 keV. This instability sometimes occurs at the onset of the flare when the temperature begins to rise, but more often it is delayed, anywhere up to ~ 10 min. In any case, however, it is accomplished during the flash phase of the flare (i.e., during the time when energy is supplied to the flare volume).

The accelerated electrons produce hard X-rays through bremsstrahlung, impulsive microwave bursts through gyro-synchrotron radiation and impulsive EUV bursts together with Hα impulsive kernels through Coulomb collisions of the downward-deflected particle streams with the ambient gas, by heating it to higher temperature. The region where such an impulsive acceleration process occurs is much smaller than the total volume of the quasi-thermal flare, but the acceleration can successively occur in different locations, all along the $H_{\parallel} = 0$ line. Wherever particles are impulsively accelerated, their influence on the transition layer and chromosphere prevails when compared to the heat conduction. The cause of this impulsive instability is unknown, but current-associated plasma instabilities seem to be the most likely candidates for it.

D. *Electron Flare*

In addition to the impulsive instability mentioned in item (C), we encounter on the Sun another acceleration process that gives rise to streams of non-relativistic electrons producing the type III bursts. This process occurs only in some parts of specific active regions, very often without any associated flare, and it probably takes place at higher levels in the corona, e.g., through the tearing-mode instability at the base of a coronal streamer (as in Figure 123b). However, if there are flares in the active region where the type III bursts occur, the type III tends to be associated with the flare. In particular, if there is the impulsive acceleration mentioned in item (C), the type III instability tends to coincide with it exactly in time. Thus, in such a case, the two instabilities are linked; we may conjecture that while current interruption occurs in the low corona, a tearing-mode instability is accomplished at higher layers (as in Figure 123d). When this happens the type III burst is very intense, it can be tracked at very low radio frequencies down to the distance of the Earth, and the non-relativistic electrons of the type III-producing stream can be recorded directly on spacecraft. As far as we know, there are no protons in these streams, hence the type III acceleration process does not seem to accelerate protons as efficiently as electrons.

E. *Proton Flare*

In some flares of the categories (C) and (C + D) a blast wave originates either simultaneously with the impulsive phase or very close to it in time. The instability that produces it must be much more powerful than the instabilities mentioned in items (B), (C), and (D), since the blast itself contains more energy than any of the other flare phenomena. It must be some sort of powerful explosion.

The blast wave sometimes can be seen as its skirt sweeps across the chromosphere and, more often, as a shock propagating upward through the corona in the form of a type II radio burst. The shock can propagate through interplanetary space up to the Earth and far behind it. The electrons and protons pre-accelerated in the process (C) are accelerated in this explosive instability to much higher energies – to tens and hundreds of MeV. The association of a shock with the acceleration process strongly suggests that this second-step acceleration is of the stochastic Fermi type. Either the shock wave itself, as it crosses the magnetic field lines, or the explosion that gave rise to it, produces turbulent fields, in which the pre-accelerated particles are accelerated again, this time to very much higher energies.

The energy spectrum of the accelerated protons varies within wide limits. When the spectrum is hard enough and many $\gtrsim 30$ MeV protons bombard the photosphere we observe products of nuclear reactions in the flare ejecta and in the associated γ-emission, particularly in the 2.3 MeV and 511 keV lines. The same protons, possibly jointly with energetic electrons, heat the uppermost photospheric layers through Coulomb collisions. These layers than become visible for a short time as the white-light flare.

The relativistic electrons are accelerated less efficiently due to the rigidity-dependent injection into the acceleration process and they also lose energy very fast through synchrotron radiation. Therefore, their flux in space is diminished relative to protons of the same energy. Electrons that are captured behind the shock become visible as a moving type IV burst. Other electrons trapped in magnetic arches produce the stationary type IV emission which may last for many hours, until the electrons decay. Some of the particles are captured in low-lying magnetic arches, thermalize there, and thus make these arches visible as loop prominences.

F. *Problems*

While items (1) through (37) in Section 6.1 are generally valid until new, more sophisticated observations eventually change them, the model suggested in Section 6.2 is purely tentative, and its only purpose is to demonstrate one way in which the complex flare data can be interpreted. Thus, any flare theory should be tested by questioning how many of the items (1) to (37) in Section 6.1 it can explain; but even if this test is good it does not exclude the possibility that the theory will significantly deviate from the model presented in Section 6.2.A–E.

Generally, there are four fundamental questions, to which a reply should be found in any theory of flares:

- What is the source of the flare energy (up to several times 10^{32} erg)?
- Is this energy stored in the solar atmosphere? If so, in what form?
- How is this energy released during the very short time required?
- What are the heating and acceleration mechanisms in a flare?

6.3. The Energy Source

It is not particularly difficult to suggest an ultimate energy source for flare production, since in fact the energy released in a flare, at most of the order of 10^{29} erg s^{-1}, is quite small when compared to the total energy output of the Sun, i.e., 4×10^{33} erg s^{-1}. The more so if we suppose that the flare energy builds up over a relatively long time. In such a case, after de Jager's (1972) estimate, we need a continuous energy supply of the order of 10^{26} erg s^{-1}. This, for example (de Jager, 1969), is only a small fraction of the missing energy flux in a sunspot ($\sim 10^{29}$ erg s^{-1}). We encounter a serious problem only later on, when we want to specify to which form this ultimate energy is transformed in order to become available for the actual creation of flares. This then leads us to the problems of energy storage.

6.4. The Energy Storage

Essentially all hypotheses on the origin of flares presuppose an important role for the strong magnetic field existing in the active region. This supposition follows quite naturally from the occurrence, location, and shape of flares, which clearly show an intimate relation to the magnetic field. Hypotheses which ignore these facts (e.g., Unsöld, 1968) are hardly acceptable. However, two problems arise:

(a) Does the flare energy come into the atmospheric layers from below during or immediately before the flare process itself, or is it brought there earlier and stored then in the solar atmosphere for tens of minutes, hours, or even days prior to the flare?

(b) If the flare energy is stored in the solar atmosphere prior to the flare, is the role of the magnetic field passive or active? In other words, is the flare energy stored in other non-magnetic forms in the active region while the magnetic field only plays the role of a container, guiding path, or catalyst? Or is the magnetic energy itself the primary source of the flare?

Those who suggest the magnetic energy as the flare source (Giovanelli, 1947, 1948, was the first) argue in the following way (e.g., Syrovatsky, 1972): There is no observational evidence for the accumulation of energy in any other form before the flare, and nobody can prove any sizable energy store prior to the flare except the energy of the magnetic field. It is true that so far no striking alterations in the magnetic field have been detected after a flare occurrence; but this may be due to the fact that magnetic field measurements refer to the photosphere, while the changes occur in the chromosphere and corona; but in particular, it is due to the size of the expected field changes which are nearly at the observable limits with our magnetographs (Rust, 1975b).

Nevertheless, the fact that we still do not have any definite evidence that the magnetic field changes *as a consequence* of a flare (cf. Section 2.1.2.C) is used as an argument by all those who oppose the active role of the magnetic field in flares. In addition, they point out (e.g., Sen and White, 1972) that it is difficult to reconcile the existence of homologous flares with any drastic change in the topology of the magnetic field. Piddington (1973) also argues that the fact that many flares occur in a simple dipole configuration (bipolar groups) is incompatible with the field annihilation mechanism.

One can immediately attack Piddington's argument by pointing out that the term 'bipolar group' does not necessarily mean a simple dipole configuration. Even in a bipolar group the field can significantly deviate from the minimum-energy configuration, for instance, through twisting or shear of field lines. A bipolar region can also emerge into relatively strong ambient field and then even a current sheet can be formed in the location where the polarity of the region is opposite to the polarity of the ambient field (cf. Figure 123e).

Another of Piddington's (1973) arguments is that all the models assuming storage of magnetic energy are imperfect: some authors propose for the origin of current sheets very artificial configurations that observations do not confirm (cf. Figure 123); currents should create coronal magnetic fields greatly in excess of any observed (also see Anzer, 1973). The second objection, however, can be opposed: as a matter of fact, we have little information on the magnetic field strength of the corona and none whatever on the field strength in narrow filaments or sheets that might carry high currents.

Some of the opponents of the active role of the magnetic field in flares still admit energy storage in the solar corona, but in a form different from the magnetic energy. Others, like Piddington, emphasize the fact that there is no evidence for any energy storage prior to flares other than accumulation of magnetic energy. Due to the objections raised above, these authors then deny any energy storage in the solar atmosphere at all and suppose that the flare energy is brought from sub-photospheric layers during the flare process.

At the present time we cannot decide whose opinion is correct; however, more and more theoreticians incline to the conclusion that energy is stored in the active region prior to the flare and that the magnetic field itself is the primary source of the flare energy released.

6.4.1. ALFVÉN WAVES AS THE SOURCE OF FLARES

Osterbrock (1961) showed that Alfvén waves are too strongly absorbed in the photosphere and therefore cannot contribute to the coronal heating. However, this is true only for small magnetic field strengths: if the photospheric field is about 2 G, the Alfvén waves decay within a distance of less than 1 km. For 10 G the decay length stretches to ~ 100 km, i.e., close to the scale height in the photosphere, and for 100 G the decay length exceeds 10^5 km (Uchida and Kaburaki, 1974). Thus, any place where an appreciable magnetic field exists provides a 'transparent window' for the Alfvén waves. Through these 'windows' energy can then be supplied by means of Alfvén waves from sub-photospheric layers to the solar corona.

The important role of Alfvén waves is further supported by the fact that the magnetic field we measure in the photosphere is most probably much smaller than the true magnetic field strength. The field appears to be concentrated into small areas that we cannot resolve (see, e.g., Stenflo, 1973) and in these areas the damping of Alfvén waves can be very small.

The Alfvén waves are attractive from two points of view: They are expected to be generated very effectively directly in the convection layer under the photosphere in magnetic regions (Uchida and Kaburaki, 1974) and once they get through the photosphere they are not easily dissipated so that they can travel a significant distance in the corona without attenuation (Wentzel, 1974). The fact that their flux through the photosphere is increased in regions of strong magnetic field makes them very likely candidates for the heating of active regions.

Piddington (1973) argues that the only fact we know for sure about the flare occurrence is that flares are intimately associated with strong magnetic fields protruding through the solar surface. This is where the Alfvén waves are effective. Therefore, one tends to associate the flares and the Alfvén waves supposing that the flare energy emerges from the sub-photospheric layers in the form of Alfvén waves during the progress of the flare. This, according to Piddington, also represents the simplest way in which the missing conductive flux in sunspots (i.e., the most probable ultimate source of the flare energy) can be transformed to energy of flares.

Piddington assumes that Alfvén waves flow at all times into a closed magnetic field structure where, since they propagate along the field lines, they are trapped and dissipate their energy. Thus energy is fed into the structure to provide hot plasma condensations in the corona. This is an acceptable hypothesis that was also proposed earlier by Pneuman (1967; cf. Section 6.4.2.D). Wentzel (1974) has shown how non-linear interactions among Alfvén waves can dissipate them when the fluxes meet 'head on' near the top of the magnetic loop.

However, Piddington extends this picture further: He assumes that the Alfvén waves also produce the flares when their flux suddenly increases. This Alfvén wave model of flares also requires energy storage, but contrary to Sections 6.4.2 and 6.4.3, the energy is not stored in the solar atmosphere, but in the sub-surface layers. Piddington (1973) first suggested that the energy is stored in the form of hypothetic transverse oscillations in and below sunspots. Later on (Piddington, 1974) he has proposed a storage in the form of helical twists in the magnetic fields of sunspots. One needs up to $\sim 3 \times 10^{32}$ erg to be released in $\sim 10^3$ seconds, i.e., $\sim 3 \times 10^{29}$ erg s^{-1}. Taking a flare area of 3×10^{19} cm^2, the flux needed is $\sim 10^{10}$ erg cm^{-2}s^{-1} as compared to sunspot energy deficit estimated to $\sim 5 \times 10^{10}$ erg cm^{-2}s^{-1}.

It should be mentioned that a somewhat similar point of view was expressed also by Mullan (1973). However, the existence of the coherent oscillations observed by K. Tanaka (unpublished) prior to a flare, which forms the basis of Mullan's considerations, has not been confirmed.

6.4.2. MODELS WITH A PASSIVE ROLE FOR THE MAGNETIC FIELD

Before discussing flare models that consider storage and annihilation of the magnetic field energy as the source of flares, let us first briefly mention other models that also assume pre-flare energy storage in the corona, but in which magnetic field plays only a secondary role.

A. *Elliot's Model*

According to Elliot (1964, 1969) the surplus energy is stored in the form of the kinetic energy of elementary particles which are continuously accelerated in an active region. A similar idea was suggested earlier by J. W. Warwick (1962). While electrons easily lose their energy and cannot survive, the accelerated protons may continuously accumulate, being trapped in magnetic (banana-like) loop structures in the low corona above the active region. These stored protons are the energy source of the flare, which occurs when the protons splash over into the chromosphere (cf. Section 6.5.1.B).

From the energetic point of view this hypothesis appears to be reasonable, since the energy density stored can approach within an order of magnitude the energy density of the containing field, $H^2/8\pi$. The problem is whether the time needed for the complete build-up of energy to the order of 10^{32} erg, which can be estimated to be at least one day, is actually available. Magnetic field inhomogeneities in the active region, and in a rapidly changing, complex one in particular, certainly scatter the protons and thus will remove them from the magnetic trap. Unless the traps extend to distances in excess of 2 solar radii, collisional losses will de-energize the protons within less than one day (Newkirk, 1973).

Another objection is that there is no definite evidence that energetic particles are continuously accelerated and stored in active regions prior to flares. Elliot's argument that the permanent acceleration is proved by the existence of the permanent particle streams in space is not fully convincing. We have seen in the introduction to Chapter V that these streams can also be due to series of impulsive acceleration processes, the particle flux at lower energies being smoothed through long-time storage after the events. On the Sun, the only observation which might indicate continuous acceleration is the increase in X-ray and radio flux prior to some major, usually proton, flares. This increase may be due to the continuously accelerated and decaying electrons. However, it can also be explained by increasing temperature in the coronal condensation even when only thermal electrons are considered.

On the other hand, one has to admit that the limited possibility of particle storage demonstrated by Newkirk (1973) gives renewed support to the idea of continuous particle acceleration in some active regions (cf. Section 5.1.3.B). One also has to emphasize that it is essentially impossible to see any effect of the stored protons in the corona, since even the neutron and γ-ray emission produced by them, which Elliot expects to occur, would probably be below the observational limit. Thus one can raise many objections against Elliot's suggestion (also see Section 6.5.1.B), but one cannot prove it definitely wrong.

B. *Schatzman's Model*

The storage of free energy by trapping of particles accelerated in the pre-flare state was also considered by Schatzman (1967). While Elliot did not mention any particular process for the continuous acceleration, Schatzman assumes stochastic acceleration by random Alfvén waves which, with $\sim 10^4 \, \text{erg cm}^{-2} \text{s}^{-1}$, may produce particles in excess of 1 MeV within one day. Of course, there is some incompatibility between the presence of large stochastic motions and magnetic traps stable for one day in the region. Therefore, the region of the acceleration, and the region of storage, must be completely separated. Schatzman has supposed that the particles are stored in a magnetic bottle above the flare, quite high in the corona. A sudden opening of the bottle releases the particles and increased turbulence then accelerates the particles to still higher energies. A decompression wave caused by the change in the coronal magnetic field triggers *pre-existing* filamentary structures in the chromosphere, where the magnetic field is thus decreased and the heat balance of the chromospheric structures is thus changed.

Again, there is no evidence for the existence of such chromospheric condensations in the pre-flare state. The time variation of the electron density in flares (Section 2.2.4.A) rather indicates that the flaring elements in the chromosphere are formed *during* the flare.

C. *Carmichael's Model*

The Schatzman model already supposes that a part of the flare energy is due to magnetic field annihilation. The same is true for the model proposed by Carmichael (1964), who has supposed that the surplus energy is brought into the active region by the solar wind and stored there in the form of excess gas pressure which must be compensated for by a strengthened magnetic field. The quiet-Sun solar wind might store about 10^{30} erg per day over the area of an active region. Therefore, one has to assume that the actual outflow of the mechanical energy in the solar wind in an active region must be two orders of magnitude higher. Whether this is true is a question, in particular because of the increasing evidence that active regions are not the sources of solar wind enhancements (cf. Section 5.5.1). A similar model has been proposed later on by Sturrock (1968), and we shall come to it in Section 6.5.2.B.

D. *Pneuman's Model*

Pneuman (1967), similarly to the considerations we described in Section 6.4.1, assumes that a significant fraction of the mechanical energy flux entering a plage is in the form of Alfvén waves which dissipate at a slower rate than the energy is injected. Since these waves propagate along the magnetic field and cannot re-enter the photosphere, their energy begins to be stored if the magnetic configuration is closed. Thus, in contradistinction to Piddington (1973, 1974; Section 6.4.1) Pneuman assumes that the surplus energy, continuously brought into the corona above active regions by means of Alfvén waves, can be stored there in various forms. Eventually, when the small-scale energy density exceeds the energy in the mean field, an instability occurs, and this leads to a flare event. Such an instability, e.g., might be a thermal runaway as proposed by Kahler and Kreplin (1970).

With a mechanical flux of about $10^8 \, \text{erg cm}^{-2} \text{s}^{-1}$, which does not suffer significant dissipation, energy of 10^{32} erg can be collected over an area of 10^{20} cm^2 in a few hours. Schmidt (1969) argues that the flux of Alfvén waves entering the corona is overestimated since Pneuman neglects the reflection by the strong density gradients between the chromosphere and corona. This probably is not correct, because actually it is more difficult to dissipate the Alfvén waves than to bring them into the corona (Uchida and Kaburaki, 1974; Wentzel, 1974).

Thus one could possibly explain the quasi-thermal flares by Pneuman's mechanism and this idea is particularly attractive since it does not require magnetic complexity in the active region, in agreement with observation. Problems appear only when the impulsive instabilities and acceleration processes are to be explained. One cannot exclude the possibility that these processes are due to completely different effects and are superposed on Pneuman's basic flare in a complex magnetic-field configuration, as also Kahler and Kreplin (1970) suggested.

E. *Model by Sen and White*

Sen and White (1972) have proposed a dynamo mechanism for the production of solar flares, in which the surplus energy available for the flare is stored in the Hall current caused by the crossed electric and magnetic fields; but ultimately it comes from the convective velocity field, the magnetic field acting more or less as a catalyst. The dynamo mechanism can work only in a region where the relative velocity between the conducting medium and the magnetic lines of force is large, i.e., in a thin layer in the immediate vicinity of the photosphere. In this region the solar gas is only weakly ionized. Close to a sunspot, where $H \simeq 1000$ G and $V < 1$ km s^{-1}, Joule dissipation of the current gives rise to plages in active regions. When, however, the velocity of flow exceeds $V \simeq 1$ km s^{-1}, a two-stream instability triggers a flare.

This model certainly leaves many flare phenomena unexplained. It demonstrates, however, that there are still new ways to model a flare while leaving the magnetic field structure in the active region essentially untouched. A somewhat similar model with an active role for the magnetic field has been proposed by Obayashi (Section 6.5.2.C).

6.4.3. MODELS WITH AN ACTIVE ROLE FOR THE MAGNETIC FIELD

If the source of the flare energy is the magnetic field itself, up to about 10^{32} erg must be made available through annihilation of a part of the field. Therefore, two fundamental questions arise. Is there enough energy in the magnetic field, and if so, how much does the observed photospheric field change when such a large amount of magnetic energy disappears?

In principle, we can calculate only the energy of the potential or force-free field corresponding to an observed photospheric flux distribution. According to Howard and Severny (1963) and Severny (1969) the magnetic energy of the potential field of an active region which produces a large flare (e.g., that in Figure 14) is in excess of 10^{33} erg. As the energy available for the flare process must be stored only by departures from a potential field, it appears that relatively small departures can provide the energy needed.

Of course, this energy was deduced from the fields in the photosphere. If one goes higher, where the flares actually occur, the energy decreases substantially. Thus one would expect that in the case of major flares we cannot obtain all the flare energy only from relaxing the deviations from the potential field in the chromosphere and corona, but that necessarily the photospheric magnetic field also must be influenced by a flare occurrence. However, this influence may be fairly small if high enough field strengths are involved (de Jager, 1969). If the initial flare volume is V, the energy of 10^{32} erg is gained when the field strength H decreases by

$$\Delta(H^2) = 8\pi \times 10^{32}/V.\qquad\qquad(104)$$

Thus, for example for $V = 10^{29}$ cm^3 a decrease of only 26 G is required in a field originally of 500 G. All this shows that in principle flares can derive their energy from the magnetic field.

It is not easy, however, to make this energy available to the flare process through magnetic field annihilation. The problem can be summarized in the following way. In order to annihilate a part of the field, its field lines must reconnect so that a part of the field disappears and its energy is transformed into other energy forms. Thus the model must fullfil the following requirements:

(a) The configuration of the magnetic field in which the energy is stored must make the reconnection possible;

(b) the configuration must be metastable, i.e., stable against small amplitude disturbances, but unable to maintain the existing stability condition against certain types of triggering mechanisms; and

(c) when the instability sets in, the annihilation must proceed at such a rate that all the flare energy is released within 10^2 to 10^3 seconds.

In principle, two types of configuration have been proposed: Twisted fields, in the form of force-free twisted tubes of magnetic flux (Gold and Hoyle, 1960; Alfvén and Carlqvist, 1967) or sheared force-free magnetic fields (Tanaka and Nakagawa, 1973); and current sheets, which assume magnetic cumulation near neutral lines in the magnetic field (in particular Sweet, 1958; Petschek, 1964; Syrovatsky, 1966; Sturrock, 1968, Friedman and Hamberger, 1969a; Coppi and Friedland, 1971; Priest and Heyvaerts, 1974).

A. *Model by Gold and Hoyle*

In an active region the opposite magnetic polarities are linked by flux tubes ascending into the chromosphere and corona and anchored in the photosphere. Gold and Hoyle (1960) supposed that each such tube may be twisted at its feet with the twist distributed evenly along the whole tube. According to Anzer (1968) the reason for this twist may be corotation of the field at its feet with the convective rotation of the parts of the sunspots to which it is linked. The degree to which the flux tube is twisted is the measure of the excess energy stored in the tube compared to the energy content in an untwisted potential field. In a static case the strengthened field must be force-free; because if Lorentz forces were present, the small kinetic pressure of the plasma in the upper chromosphere and corona could not balance them and the configuration could not survive. This means that the electric currents $\mathbf{j}$ in the tube are very nearly parallel to the magnetic field $\mathbf{H}$, in order to make $\mathbf{j} \times \mathbf{H} = 0$.

Let us now suppose that there are two such tubes close together, with the initial longitudinal field in opposite directions, and that they are twisted in opposite senses. Then the axial currents are in the same direction and the tubes must attract each other. Thus, if there is a system of such twisted flux tubes, and if they can migrate, it will happen that from time to time oppositely directed fields come into contact. In such a case a neutral point is formed where magnetic field lines reconnect and some part of the magnetic field is annihilated.

An attractive feature of this model is the chance to produce many small flares with a fairly high frequency of occurrence. However, it is fairly difficult to imagine how a major extensive flare is formed by this mechanism. There are also serious problems with the stability of such a configuration. It appears to be immediately unstable, and one has to make additional assumptions in order to stabilize it (Anzer, 1968; Syrovatsky, 1969a).

B. *Model by Alfvén and Carlqvist*

Alfvén and Carlqvist (1967; Carlqvist, 1969) have pointed out that instability may occur in a single twisted tube if the effects of electric currents are considered. Since the field is force-free, the currents go along the tube. Each current system always represents a certain energy W stored in the induced magnetic field,

$$W = LI^2/2, \tag{105}$$

where L is the inductance, proportional to the length of the loop-tube in the atmosphere. For a semi-circular loop connecting two places of opposite polarities $\sim 10^4$ km apart, the energy $W = 10^{32}$ erg, needed for a major flare, corresponds to a current $I \simeq 10^{12}$ A.

One can be sure that currents are produced in the flare regions. The storage of magnetic energy can be accomplished only through departures from a potential field, since the potential field represents the minimum energy state with no extractable energy. The potential field is current-free and any deviations from it must be associated with the existence of currents. Since in an active region the magnetic-field energy density exceeds the plasma energy density, the magnetic field should be largely force-free, as we mentioned in Section A. The problem is whether the currents can be as strong as Alfvén and Carlqvist need for their mechanism.

Severny (1965) who measured both the longitudinal and transverse components of the magnetic field, could compute the distribution of electric currents and found vertical currents of the order of 10^{11} A in the neighborhood of sunspots. Since the resolving power of the magnetograph is rather poor, the magnetic field, and hence the currents, too, may be still larger in individual loop-tubes of small dimensions. In addition, as we already mentioned in Section 2.1.2, Moreton and Severny (1968) and Zvereva and Severny (1970) have found a coincidence between bright knots appearing during the initial phase of flares and areas with increased vertical currents. In a rope of ~ 5000 km diameter which corresponds to the size of the bright flare patches, the current should exceed 2×10^{11} A. Heyvaerts (1974) has shown how photospheric motions can drive strong coronal currents responsible for storing magnetic energy. Thus both the measurements and the theoretical considerations indicate that the flare energy might be stored as magnetic energy in the current system. If this is true, then according to Alfvén and Carlqvist this energy may be released very fast, and almost completely, if the current is interrupted (cf. Section 6.5.1.E).

C. *Sheared Fields*

As we described in Section 2.1.2.C, Zirin and Tanaka (1973) reported the existence of great shears in the magnetic field of the active region that produced the big 1972 August flares. The shear was large at the beginning of a flare and relaxed during the flare development (cf. Figure 15). Tanaka and Nakagawa (1973) have assumed that the magnetic field was force-free and computed the amount of energy stored in such a sheared force-free magnetic configuration. They have shown that up to 6×10^{32} erg can be stored there over the area of 3×10^{18} cm^2 and that such a build-up of magnetic energy can be achieved by the proper motions of sunspots in the time interval between successive large flares.

Tanaka and Nakagawa have pointed out two advantages of twisted force-free fields in the interpretation of flares: First, the energy is supplied by the magnetic field over the whole flare region, whereas in the current sheet models only a very thin sheet is involved in the energy release. Second, with force-free fields, one only needs to untwist the field lines, whereas opposing magnetic flux must be continuously supplied to the current sheet with speed of ~ 0.1 Alfvén speed, and observations do not show such motions. However, also the observations of the shear relaxation are very rare and as we mentioned in Section 2.1.2.C, they also might be interpreted as an altitude-dependent and not time-dependent relaxation. Zvereva and Severny (1970) observed shears in other flare-producing regions (e.g., Figure 14) that did not relax at photospheric level after flares.

D. *Force-Free Flare Build-up*

A different approach to the problem of flares and force-free fields has been taken by Low (1973). He considered the resistive diffusion of force-free magnetic fields in a compressible medium which is free to move to accommodate the changing magnetic configuration. In other words, instead of keeping constant α in the equation (cf. Section 2.1.2.C.5)

$$\nabla \times \mathbf{H} = \alpha \mathbf{H}, \tag{106}$$

Low assumes that α varies with time. In consequence, the lines of force change their shapes with time and the plasma yields to accommodate this change. The force-free configuration is kept due to the fact that the magnetic field energy greatly exceeds the kinetic energy of the plasma.

In such a case the force-free field evolves slowly for an extended period of time, but this evolution speeds up and abruptly develops steep gradients in the magnetic field. Low does not suppose that this process itself produces flares; he suggests, however, that this is the 'flare build-up' process which sets the stage (by suddenly developing steep magnetic field gradients) for the onset of flares. The flare proper, according to his opinion, occurs than in a current sheet.

For the specific case of an open current sheet (i.e., current sheet in a coronal streamer) Barnes and Sturrock (1972) presented a model of how the sheet can be built from a force-free field configuration. They first assume that an initially current-free (potential) configuration is distorted by photospheric motions, leading to the production of a force-free magnetic field. If the photospheric motions are sufficiently large, energy stored in the force-free field may ultimately exceed the energy of an open-field configuration. In such a case, because there is an accessible state of lower energy, the force-free magnetic field configuration becomes unstable or metastable and can change into an open-field configuration that develops a current sheet. Barnes and Sturrock computed the magnetic-field pattern and the energy involved by numerical methods and these computations indeed showed that it is possible to store in a force-free magnetic-field structure an amount of energy exceeding the energy of the corresponding open magnetic field structure.

As a practical example, Barnes and Sturrock have considered a sunspot of one polarity surrounded by a magnetic field region of opposite polarity. Without any initial twisting the magnetic field has a current-free configuration. Then, however, the sunspot begins to rotate in consequence of differential rotation. The magnetic field configuration develops currents, but, because of the low plasma density, the field adopts a force-free structure. The energy stored in the magnetic field will then increase as the rotation angle increases.

The authors further suppose that the force-free configuration is metastable and that the instability that converts the metastable force-free magnetic field into an open field is explosive. In the practical example they associate this explosive instability with the pre-flare activation and disruption of a filament. This, however, does not fit the observations. As we showed in Section 4.1, the filament dissolves gradually and disrupts only at the very onset of the flare. Thus, a gradual build-up of the current sheet from the force-free configuration, with disruption of the filament and abrupt onset of the flare-associated instability in the current sheet would better fit the situation.

The Barnes and Sturrock study is important because it demonstrates that a current sheet might be built from a force-free configuration that probably represents a common configuration in active regions. However, the specific open current sheet model for which this evidence has been presented meets with difficulties when trying to explain common flares. As we shall see in Section 6.5.2.B, it may explain some flare-associated phenomena (like type III bursts), but hardly the whole flare process.

E. *Current Sheets*

A very large group of flare models has been built on the assumption that the interaction of magnetic fields of developing sunspots or the interpenetration of sunspot groups leads to the creation of neutral lines in the magnetic field. Several different possibilities for the location of such neutral lines are schematically drawn in Figure 123. One can see from it that a neutral line, if it is formed, coincides in projection with the $H_{\parallel} = 0$ line close to which the flares originate, and this makes these theories very attractive.

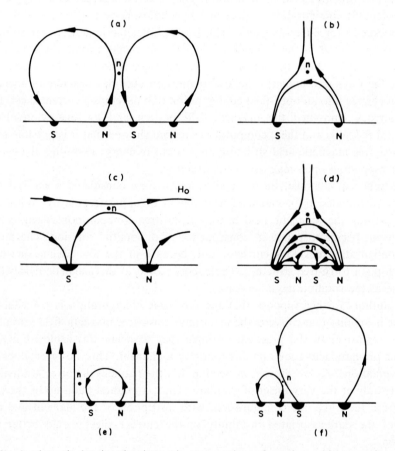

Fig. 123.   A schematic drawing showing various magnetic configurations which may give rise to a neutral line in the magnetic field: (a) Sweet's (1958) model of two approaching sunspot groups (an *X*-type neutral line); (b) Sturrock's (1968) model of an *Y*-type neutral line between the closed and open field in a bipolar active region; (c) Syrovatsky's (1968a) model of an *X*-type neutral line over a bipolar group covered by an overlying external field H_o; (d) Sturrock's and Syrovatsky's models combined; (e) the most simple case when a new bipolar field emerges into a unipolar region; (f) case (a) as modified by Priest and Heyvaerts (1974): new flux of opposite polarity emerges at the edge of an active region.

Of course, it is necessary to note that strictly neutral lines can hardly exist in the topology of the field, since arbitrarily small changes in the field destroy them. However, lines connecting two neutral points in the magnetic configuration can exist as limiting magnetic field lines for two systems of adjacent independent magnetic fluxes, with properties similar to a neutral line (Sweet, 1969). In particular, if a neutral sheet is formed, starting from the neutral points, it extends along such a limiting quasi-neutral line.

The formation of a neutral sheet can be demonstrated in Sweet's (1958) model, shown in Figure 123a. Two bipolar sunspot groups approach each other and this motion brings the oppositely directed field lines close together. They flatten, like tires of a heavily loaded car, and form a neutral sheet going through the neutral line between them. In order that the magnetic field H should reverse, a current j must flow in the boundary layer. Therefore, the term 'current sheet', which is perhaps better, is often used. Physicists also use the term 'current pinch', since in the absence of magnetic pressure the plasma pressure alone must hold the magnetic walls apart.

When the two oppositely directed fields merge they annihilate one another and release magnetic energy which is transformed to kinetic energy of the plasma. The rate of this annihilation, however, is greatly insufficient if no additional mechanisms are considered. Nevertheless, as we shall see in Section 6.5.2, there are ways in which the dissipation rate can be enhanced by several orders of magnitude so that a flare origin in a neutral current sheet appears to be a realistic model.

According to Sweet's original idea the current sheet forms from an X-type neutral line between two approaching sunspot groups (Figure 123a). While this is rather an artificial conception, we know at least one configuration on the Sun where the existence of a neutral sheet has been proved; this is the configuration of a coronal streamer above an aging bipolar active region. This had led Sturrock (1968) to the supposition that flares can originate at the bases of neutral sheets of this type (Figure 123b). As a matter of fact, this model is a modification of Carmichael's idea (Section 6.4.2.C), since solar wind is obviously responsible for the build-up of neutral sheets of this type.

Another modification of Sweet's original idea has been presented by Syrovatsky (1966, 1969a). He also assumes the existence of an X-type neutral line, but supposes that such a configuration can appear in one single bipolar group, if its closed field is overarched by another field of opposite orientation (Figure 123c). This external (more or less horizontal) field may be produced by neighboring spots or be remnants of an old active region. It is of interest to note that, for example, such a configuration also forms if new polarities emerge inside Sturrock's configuration in Figure 123b. Then we may meet with two neutral lines, one of Sturrock's Y-, and the other of Syrovatsky's X-type (Figure 123d). Other modifications of the Syrovatsky's configuration can also be found (e.g., Křivský, 1968) as soon as the group becomes magnetically complex, which is exactly the situation when important flares occur.

Priest has drawn attention to the fact that even a simple bipolar configuration can give rise to a current sheet when it emerges into a unipolar field. Then, on one side or the other, the polarity is opposite to the polarity of the ambient unipolar field and a neutral line might form there (Figure 123e). Perhaps the most realistic situation appears when new bipolar flux of opposite polarity emerges on the periphery of an active region (Figure 123f). A current sheet is then formed along the contact surface (Priest and Heyvaerts, 1974) and these are some of the situations actually observed (Rust, 1973b, 1975a).

6.5. The Energy Release

It is quite clear that at the present time we do not have any comprehensive and generally accepted theory of flares, and as a matter of fact, we can hardly expect it. An observer on the Sun, equipped with our observational techniques, could not understand the processes in the Earth's atmosphere either, and he would have no knowledge at all about the most devastating processes at the Earth, since their dimensions are too small to be resolved. We need much better spatial resolution in order to recognize the intimate relationship between the flare and the magnetic field, and we also need knowledge of the field structure in the chromosphere and corona. So far, the magnetic field configuration at these altitudes can be deduced only from the observed coronal structures and from mathematical extrapolations of the measured photospheric fields under restricting current-free or force-free assumptions. The scales of the fields thus implied greatly exceed the scale involved in flares.

Another difficulty arises from the fact that a flare can represent a set of different instabilities, which need not all be accomplished in all flares (cf., e.g., Piddington, 1974; Švestka, 1975; Coppi, 1975). We have seen that many observations indicate that this is actually true (cf. Sections 6.1 and 6.2). Thus several of the models proposed may approach the actual situation in flares, from one or the other point of view: for example, Pneuman (1970) might have explained the thermal flare, Alvén and Carlqvist (1967) the pre-acceleration during the impulsive phase and Sturrock (1968) the type III burst.

Nevertheless, most theoreticians incline to the opinion that the basic flare configuration is a neutral current sheet. Therefore, we shall take the liberty to give some preference to the current-sheet theories. Before doing this, however, we shall describe briefly the mode of energy release in some other models that point out interesting ideas, of which at least some bear a close resemblance to situations that actually develop in flares or in some phase of the flare evolution.

6.5.1. Models without a current sheet

A. *Piddington's Model*

In the model proposed by Piddington (1973, 1974) the energy is stored below the photosphere in the form of helical twists in the magnetic field of sunspots, and it is brought into the corona by hydromagnetic disturbances during the flare (Section 6.4.1). The helically twisted magnetic field unwinds into the atmosphere thereby releasing kinetic energy by the upward propagation of the disturbances. Piddington assumes that these disturbances are Alfvén waves which are generated by interactions between magnetic flux strands or by kinks in the strands.

The basic idea is certainly interesting, since it would offer a direct transformation of the sunspot magnetic field into the flare energy. However, as Altschuler (private communication) has pointed out, helical twists do not produce waves in any simple way.

Piddington also does not clearly explain what is the cause of the sudden increase in supplied energy causing the flare. His explanations of the associated processes like impulsive phase, radio bursts, and particle acceleration, are as follows: The impulsive phase should occur when the emerging flux strands are pressed together; wave phenomena occur as the twisted field expands upward; and particles are accelerated in shocks that are formed as the waves propagate into weak coronal fields. All these suggestions need a more quantitative approach to make the model convincing.

B. *Elliot's Model*

The attractive point in Elliot's model (Section 6.4.2.A) is the easy way in which the presence of flare-produced high-energy particles in space is explained. All other theories struggle with the acceleration (or pre-acceleration) problem. It is not easy to accelerate protons to GeV energies when one injects into the acceleration process particles with thermal energies; if pre-acceleration is present, the problem is much easier. However, Elliot's idea has never been worked out in detail. According to him, the flare may set in when the magnetic field at the base of some coronal arches weakens and releases protons into the chromosphere. We have no data on the fine structure of the magnetic fields, but it seems rather that just the opposite is true; a flare starts when the field, and its gradient in particular, becomes too strong. One could also suggest another alternative, namely that a disruption of the storage region happens when the density of accelerated protons exceeds some critical limit. In any case, however, the model fails to explain the two different phases in the flare development, the thermal and the impulsive one. Also, one should recall that in the flares of 'purely thermal' type there is no evidence at all for any catastrophic release of energetic particles in the flare region.

C. *Kahler and Kreplin's Model*

Kahler and Kreplin (1970) have suggested that the solar flare 'trigger' mechanism is a thermal runaway which is a consequence of the fact that the radiative power loss in plasma decreases with increasing temperature above $\sim 2 \times 10^5$ K (cf. Figure 40).

Let us denote L the heat loss function, defined as energy losses minus energy gains per gram and second in a plasma of density ρ that loses energy only through radiation. Then, as we can see from Figure 40,

$$\left(\frac{\partial L}{\partial \rho}\right)_T > 0 \qquad (107)$$

and, for $2 \times 10^5 \text{ K} \lesssim T \lesssim 2 \times 10^7$,

$$\left(\frac{\partial L}{\partial T}\right)_\rho < 0, \tag{108}$$

which, according to Field (1965), are conditions leading to thermal instability in a perfect gas. Thus, if additional energy is supplied to upper atmospheric layers within the range of the above-mentioned temperatures and the heat conduction is unable to compensate for the energy excess, temperature begins to rise at an increasing rate: the higher the temperature, the lower are the radiative losses (cf. Figure 40). Kahler and Kreplin do not specify the source of the heating, but the energy might be supplied, e.g., by means of Alfvén waves trapped in closed magnetic fields (cf. Section 6.4.2.D).

Kahler and Kreplin take $T \simeq 4 \times 10^6$ K as the pre-flare temperature and assume that heating above this temperature produces the thermal instability. In order to explain a rise in T by $\sim 5 \times 10^6$ K in ~ 100 s they need $n_e \simeq 5 \times 10^{11}$ cm^{-3}. This is a rather high value, since we have seen that thermal X-rays yield n_e's one order of magnitude lower. Thus one might need still another unknown 'catalyzing' factor to make this process more efficient.

Kahler and Kreplin suppose that the Hα emission is a secondary effect, due to heat conduction from the corona and believe that their mechanism might fully explain small 'thermal' flares. Larger flares and in particular flares with an impulsive phase need further energy sources (possibly conversion of magnetic field energy), but even these flares should start with the mechanism Kahler and Kreplin suggest.

One should realize that according to (107) and (108) the variation of radiative energy losses with temperature, shown in Figure 40, gives rise to a thermal instability any time when $L < 0$ and heat conduction cannot compensate it. This most probably happens during a flare. Thus, the question is whether the effect discussed by Kahler and Kreplin actually can *cause* a flare or whether it only *contributes* to the increase in the flare temperature.

D. *Evaporating Flare Models*

Several authors (Sturrock, 1973; Hudson, 1973; Hirayama, 1974; Hirayama and Endler, 1975) have suggested that the coronal flare is hot gas that 'evaporated' from the chromosphere. The main reason for this supposition is the mass excess in a coronal flare; the emission measure of the low-temperature X-ray component may be as high as 10^{50} cm^{-3} (cf. Section 3.1.4) and the mass contained in loop prominences exceeds that in a corresponding volume of non-flaring corona (Section 4.4).

Sturrock (1973) might be right when explaining the loop prominence systems through an evaporating process. Streams of high-energy particles impinge upon the chromosphere, heat the upper layers to $\sim 10^7$ K and the heated gas expands rapidly into the corona. Some of this gas cools fast and becomes visible in Hα in the form of loop prominences. On the other hand, Hudson's and Sturrock's efforts to explain the whole flare through a similar mechanism are less acceptable. Hudson suggests that 10 to 100 keV electrons are accelerated in the corona, stream downward, heat the chromosphere, and then the hot chromospheric gas rises to the corona and becomes the hot condensation which emits soft X-rays. The Hα is then due to heat conduction from this evaporated chromosphere to the cold chromospheric remnants.

This hypothesis is untenable, since energetic particles are not the primary source of the flare as we demonstrated in many places in this book (Sections 2.3, 3.1.3, 6.1), and a similar objection can be raised against Sturrock's approach.

A much more elaborate 'evaporating flare model' has been proposed by Hirayama (1974). His model starts with a prominence (dark filament) that extends along the $H_{\parallel} = 0$ line. When the gradient of magnetic field increases, also the electric currents running parallel to the $H_{\parallel} = 0$ line increase. Due to kink instability the currents start to bend and the prominence begins to rise. A magnetic cavity will appear below the prominence and gas will tend to fill it up. The field will then collapse in a lateral direction forming a reconnecting point (line) below the rising prominence.

Through Joule dissipation a heat flow is generated near this point; as it flows down, it heats the top of the chromosphere and the upper chromospheric layers evaporate. Many of the evaporated particles escape into the high corona. However, as soon as the magnetic field reconnects, the particles can no longer escape from a magnetic loop which bridges the Hα bright ribbons. They are then observed in the form of soft X-rays.

According to Hirayama, the top height of the chromosphere is by now ~ 1000 km lower than in the pre-flare active region. As a matter of fact, this is a situation similar to that which we described in Sections 2.3.1.C and 3.1.4.A when the transition layer was pushed down into the chromosphere. However, there is one significant difference. In the earlier model we supposed that the heating of the upper chromospheric layers is the consequence of thermal conduction from a primary coronal flare. In Hirayama's model the heating of the upper chromosphere gives rise to the coronal flare which is a secondary phenomenon.

Hirayama also explains some additional effects. The prominence continues to rise; thus, also the reconnecting point rises, and heat flows farther apart from the neutral line. In such a way, the expansion of the Hα bright ribbons can be explained. As the gas cools, the cooling process proceeds earlier at smaller altitudes, thus giving rise to loop prominences, visible first in the lower and later in the higher layers of the corona. A shock wave can be generated in front of the rising prominence, and this is the cause of the type II and moving type IV radio bursts.

There are several points Hirayama's model does not explain. First, similarly to Hudson, there should first be the reconnection (i.e., the impulsive phase) and only then the soft X-ray flare, which contradicts the observations. Another problem is with the red-ward asymmetry observed in chromospheric lines very soon after the flare onset (Section 2.2.9 and Figure 31). Hirayama explains it through mass streaming down the loop prominence system. But these prominences are only very rarely observed (Section 4.4) and they do not seem to be present in all flares. Finally, the model presupposes that limb flares do not exist. Everything we observe on the limb must be prominences, because chromospheric flares evaporate completely above the level that we see on the limb. However, we have shown in Section 2.2.10 that this is not true and some arguments for the existence of limb flares were given by Hirayama (1961) himself.

Hirayama and Endler (1975) made some numerical calculations related to this flare model. They have found that, as long as the reconnecting point is in the corona, the corona is heated first. Only when the conducting energy from this coronal core increases to the point that the chromosphere cannot radiate it away, the top chromosphere is heated to the coronal temperature, and the gas expands upward. It is evident that this numerical model is much nearer to the models we described in Section 2.3.1.C than Hirayama's original morphological description.

According to Hirayama and Endler's computations, the coronal temperature rises to $1-3 \times 10^7$ K within 10^2 to 10^3 s, and the density corresponding to this temperature increases from $n_e = 10^8-10^9$ cm^{-3} to $10^{10}-10^{11}$ cm^{-3}. Only gas with $n_e \lesssim 10^{13}$ cm^{-3} is left at chromospheric temperatures. Upward velocities are 100 to 1000 km s^{-1}. The evaporating process would not be seen in Hα since evaporating gas will be too hot.

E. *Alfvén and Carlqvist's Model*

The rather simple model proposed by Alfvén and Carlqvist (1967) has attracted the greatest attention of all the 'non-current-sheet' models. It is still considered, with some restrictions and modifications, as a likely process to be accomplished in flares.

The model is based on current interruption in a twisted loop-like rope (Section 6.4.3.B). An interruption will occur if the twisted rope, for reasons not specified in detail, decreases its thickness. Then the current density increases, and if it exceeds some limit, the ionized particles become unable to compensate local density inhomogeneities in the rope. Then, if even a small disturbance occurs in the form of a dip of the plasma density and the particle velocity is enhanced in order to keep the current density constant, the plasma in the region becomes progressively rarefied and eventually a high-impedance space charge region is created. As the current has to pass the large potential drop, the stored magnetic energy is released and transformed into kinetic energy of the elementary particles which are thus accelerated in the region.

This model has the advantage that it immediately involves and explains the particle acceleration, and it has been considered by many authors as a likely candidate for the pre-acceleration of particles to relatively low energies during the impulsive phase. Similarly to Elliot's model, it does not give an explanation for the thermal phase of the flares, when no impulsive acceleration is observed. However, in contrast to Elliot's approach, Alfvén and Carlqvist's model can be used as an explanation for the impulsive flare phase in combination whith some other 'thermal' models.

D. F. Smith and Priest (1972) raised a serious objection against the Alfvén-Carlqvist model, by pointing out that the ion-sound current instability may develop faster than the Alfvén-Carlqvist instability in a fully ionized plasma. Thus, in most cases the ion-sound current instability will be the current dissipation mechanism instead of the mechanism that Alfvén and Carlqvist suggested. The ion-sound instability leads to turbulent resistivity which prevents any efficient acceleration of particles to suprathermal energies. Instead, heating rather than acceleration is expected. Thus, contrary to our previous conclusion, current limitation should be a good source of thermal X-rays and heating, but not a suitable mechanism for particle acceleration.

This problem remains open. Spicer (1975) claims that particle simulation experiments confirm the validity of Alfvén and Carlqvist's model and contradict the objections of Smith and Priest. Thus, the current interruption is still a valid candidate for the first-step acceleration process in flares.

6.5.2. CURRENT SHEET MODELS

When a neutral sheet is formed between two oppositely directed magnetic fields (Section 6.4.3.E), the magnetic field begins to reconnect due to the finite electrical conductivity of the plasma, and the magnetic energy released is transformed into kinetic energy of the particles. However, the resistive reconnection of magnetic field lines is extremely slow. In the case that the electric conductivity σ (or the resistivity $1/\sigma$) is uniform in space, the changes in the magnetic field are described by the equation

$$\frac{\partial \mathbf{B}}{\partial t} = \nabla \times (\mathbf{v} \times \mathbf{B}) + \frac{c^2}{4\pi\sigma} \nabla^2 \mathbf{B} \tag{109}$$

which, for plasma at rest, reduces to the diffusion equation

$$\frac{\partial \mathbf{B}}{\partial t} = \frac{c^2}{4\pi\sigma} \nabla^2 \mathbf{B}. \tag{110}$$

The quantity $c^2/4\pi\sigma$ is the magnetic diffusivity, and the characteristic time of decay of the field due to this resistive (or Joule) diffusion is

$$\tau_r \simeq L^2/(c^2/4\pi\sigma), \tag{111}$$

where L is a characteristic length of the current sheet. Since σ/c^2 is approximately 10^{-5}, even for L of the order of only 10^3 km we get $\tau_r \simeq 10^{12}$ s, i.e. more than 10^4 yr. Thus, the resistive diffusion in a fully ionized plasma is insignificant for any process of solar activity, and another mechanism must be sought.

The diffusion can become faster, by many orders of magnitude, if the gas is not fully ionized, since then the conductivity is reduced. We speak then about ambipolar diffusion. According to Gold and Hoyle (1960) this ambipolar diffusion might lead to characteristic times of field annihilation as short as 10^2 s. However, as Parker (1963) has pointed out, their assumptions were not realistic. The characteristic time τ_a for ambipolar diffusion is proportional to the product of the ion and neutron atom density. Therefore, the density must be low in order to significantly reduce the ratio τ_a/τ_r. But when the density is low, the gas is highly ionized and ambipolar diffusion does not work any more. It might only work in the chromosphere, but there, due to high density, the ratio τ_a/τ_r cannot be sufficiently reduced.

A significant reduction of τ_r follows from the Sweet's (1957) mechanism, which takes into account plasma flow in the neutral sheet. Since the field vanishes at the neutral surface between the two approaching magnetic configurations in Figure 123a, the plasma itself must balance the external magnetic pressure, and it responds to it by flowing out along the lines of force. As it squeezes out, the two fields approach the magnetic boundary closer and closer and the field gradient increases progressively. In a region of very high electric conductivity σ the motion is initially given by Equation (109) with the second term neglected. However, as the gradient across the neutral plane increases, the second (diffusion) term in Equation (109) becomes more important and it eventually becomes comparable to the first (dynamic) term. When this happens, the diffusion sets in at a rate much faster than in the static case, essentially being determined by the rapidity with which plasma can escape from the neutral sheet. A detailed discussion of this process can be found, for example, in the review by Parker (1963) and in the books by Tandberg-Hanssen (1967b) and Piddington (1970). The characteristic time τ_s is significantly shorter than in the static case, but it still amounts to days, or at least tens of hours under acceptable physical conditions.

A further, very substantial improvement in the rate of the conversion of magnetic energy to plasma energy in neutral sheets is due to Petschek (1964). He has shown that as soon as the two fields are not strictly antiparallel, the reconnection of nearly antiparallel field lines forms kinks in the field lines which can be considered as hydromagnetic waves that propagate away from the reconnecting region, thus further reducing the magnetic energy. The waves propagate at the Alfvén speed, which is independent of electrical conductivity, and therefore this wave propagation mechanism greatly increases the rate of field annihilation in plasma in which high conductivity brakes the magnetic diffusion.

Thus, summarizing all the effects so far discussed, we can schematically draw the model shown in Figure 124: Two oppositely directed fields (MF^+ and MF^-) approach each other, and plasma P flows towards the neutral sheet (N) which has been built from an X-type neutral line (NL). In the vicinity of NL the magnetic diffusion dominates. However to both sides of the neutral line, within the neutral sheet, plasma flows outwards in a direction roughly perpendicular to the initial flow between two stationary fronts (F), at the Alfvén speed V_A. According to Petschek this mechanism can work effectively provided that the initial flow velocity in the regions NF does not exceed $\sim 0.1\ V_A$. Thus the characteristic time for the magnetic energy to disappear is $\tau_p \gtrsim L/0.1\ V_A$. With the characteristic length $L = 10^4$ km we get $\tau_p \gtrsim 10^2$ s for $V_A \simeq 10^8$ cm s^{-1}. Since

$$V_A = H/\sqrt{4\pi\rho}, \tag{112}$$

where ρ is the gas density, this corresponds to the situation in the upper chromosphere ($\rho \simeq 10^{-11}$ g cm^{-3}) with $H \simeq 100$ G, and the mechanism works better, the higher in the atmosphere the neutral sheet has formed.

Some further modifications of the Petchek mechanism were published later on by Green and Sweet (1967) and Petschek and Thorne (1968). Perhaps the most serious difficulty in this model is the result that the thickness of the sheet near the neutral line becomes less than the mean collision distance and even less than the gyroradius so

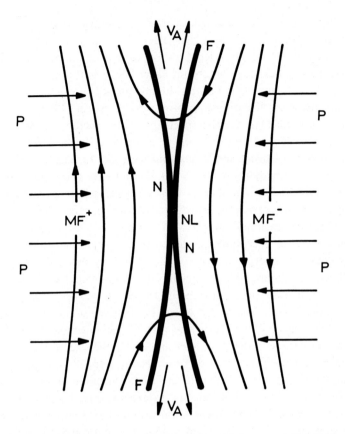

Fig. 124. A schematic model of the neutral sheet, after Friedman and Hamberger (1969a). See text for explanation.

that the region is too thin to be hydrodynamic. In order to remove this difficulty, Friedman and Hamberger (1968, 1969a) have drawn attention to the role of turbulence in the neutral sheet. They have shown that the region near the neutral line, whose dimensions are governed only by diffusion processes, would be subject to current-induced instabilities. Experiments indicate that the passage of a sufficiently dense current through a plasma leads to turbulence through the extremely rapid growth of plasma oscillations; this obstructs the electron drift motion and leads thus to an anomalously small plasma conductivity. Therefore, as soon as turbulence appears in the neutral sheet, the rate of magnetic field diffusion greatly increases, and it would also lead to physically more meaningful dimensions of the sheet near the neutral line.

This leads us to the problem of the instabilities that can occur in a current sheet and that can change the whole conception of the energy release in a very significant way. Of all of them the most important seems to be the tearing mode instability first analyzed by Furth *et al.* (1963) and applied to the flare process by Jaggi (1964) and

Sturrock (1968). The tearing mode is essentially a disintegration of the current sheet into a number of filaments by the pinch effect. According to Jaggi the steady state conditions which have been considered above, cannot be established at all since the tearing mode instabilities set in before the sheet becomes thin enough. According to Sturrock the sheet then would become highly turbulent which, as we mentioned above, greatly increases the diffusion. However, as Sweet (1969) has pointed out, enhanced diffusion thickens up the sheet again and thereby tends to remove the tearing mode instability. Therefore, one is inclined to believe that a quasi-steady state which includes tearing mode instabilities (and turbulence) may eventually develop, and such a situation may be very favourable for a very fast annihilation of the magnetic field. The current sheet is then torn into separate closed loops so that the merging of the fields can occur at reduced intervals between these loops.

Thus the Sweet-Petschek mechanism, with tearing mode instabilities and turbulence included, might be an acceptable model of a quasi-stationary current sheet, fast enough to release the energy needed for flares. A very detailed analysis of such a model, modified for compressible plasma, has been presented by Coppi and Friedland (1971). Other discussions were more recently given by Parker (1973a, b), Sonnerup (1973) and Coppi (1975). Yeh and Axford (1970) tried to eliminate completely the role of the electrical conductivity in the rate of reconnection; according to their approach the electrical conductivity determines only the size of the resulting central reconnecting region while the rate of reconnection depends on the input velocity, which can attain any value up to the Alfvén speed. This point of view was attacked by Sonnerup (1970) who showed that Yeh and Axford's approach has only one physically relevant solution with little practical implication. It was also criticized by Priest (1973) who showed that actually the admissible maximum input velocity is less than in Petschek's model. On the other hand, Bratenahl and Baum (1974) believe that their laboratory experiments confirm Yeh and Axford's results. Thus the whole problem is still open to many modifications and even doubts, in particular as far as the establishment of a quasi-stationary state is concerned.

We shall not go into any further details in the general treatment and instead we will briefly describe some particular models based on the annihilation of magnetic field in current sheets, in order to demonstrate the practical applicability of this process to detailed phenomena observed in flares. One of the models, that of Syrovatsky (1966, 1969a, 1972), we intend to describe in detail; not because we believe it is necessarily the correct one (as we shall see, several objections have been raised against it), but because it demonstrates in the most illustrative way how a current sheet model might explain the various types of phenomena we observe in flares.

A. *Priest and Heyvaerts' Model*

As we mentioned in Section 6.4.3.E the most simple way for a current sheet to form is the case (*f*) in Figure 123: when a new field of opposite polarity emerges at the periphery of an existing active region. This is a phenomenon often observed (Rust, 1973b, 1975a) and Priest and Heyvaerts (1974) propose it as an explanation for the appearance of dark filaments at peripheries of active regions, possibly associated with type III bursts (Axisa *et al.*, 1973 a, b; cf. Section 3.2.6.A).

When a new field emerges, it presses against the field that is already present and thus a neutral current sheet is formed (Figure 123 f). Priest and Heyvaerts argue that a neutral current sheet is the most favourable place for the occurrence of a thermal instability, since the plasma is effectively insulated there by the magnetic field against the thermal conduction that in other configurations tends to suppress the instability. Therefore, Priest and Heyvaerts suggest that a thermal instability occurs in the sheet as soon as the neutral sheet is sufficiently long. The instability produces locally dense, cool plasma that becomes visible in the form of a dark quiescent filament. The associated compression of the neutral sheet stimulates a collisionless tearing mode instability in a central thin slab of the sheet. The resulting electric field grows and eventually accelerates electrons in the central slab to energies > 10 keV. Some of these electrons are channeled along open field lines high into the corona, producing the type III radio bursts, and eventually they escape into interplanetary space.

While Priest and Heyvaerts propose this configuration only as favourable for the production of Type III radio bursts, one might also extend their model to explain a flare. As a matter of fact, Martres *et al.* (1968) and Rust (1968, 1973b) associate emerging new magnetic field with flares. Magnetic energy can be stored between two topologically separate fields pressed together, in a 'quiet' neutral sheet in which the magnetic field dissipates at a very slow rate. This 'quiet' neutral sheet can become an 'active' one, with very high dissipation rate, after the compression exceeds some limit and a Petschek-like mechanism or microinstabilties set in. However, as Priest and Heyvaerts emphasize, the details of the trigger which converts the quiet neutral sheet into an active one are by no means clear.

B. *Sturrock's Model*

Let us now start with quite a different magnetic configuration, schematically shown in Figure 123 b. At low heights the magnetic field lines are closed but at sufficiently great heights the lines are drawn out into an open structure comprising a neutral sheet related to a coronal streamer. Following Carmichael (1964), Sturrock (1968) has proposed that the energy stored in the open-field region is derived from the non-thermal energy flux which heats the corona and drives the solar wind.

At the bottom of the current sheet one meets with a Y-type neutral line. The configuration of the magnetic field in its neighborhood is very sensitive to the equality or inequality of the gas pressures in the closed- and open-field regions. If the pressure in the closed configuration is higher, the magnetic field lines take the form of a cusp which is strongly reminiscent of the helmet shape often seen in coronal structures on the solar limb. Thus configurations of this type obviously exist on the Sun; the problem is whether flares really develop in current sheets of this type.

Sturrock supposes that reconnection of the magnetic field in the current sheet (Figure 125) is effected by the tearing-mode resistive instability which we already mentioned above. According to Furth *et al.* (1963) the growth rate of the instability depends critically on the thickness of the current sheet, y_c. The time scale τ_t of the most rapidly growing wave of the instability is

$$\tau_t = (\tau_d y_c V_A^{-1})^{1/2}, \tag{113}$$

where τ_d is the diffusion time and V_A the Alfvén velocity. However, the value of y_c is not known. In order to estimate it, Sturrock applies Spitzer's (1962) formula

$$\tau_d \simeq 2 \times 10^{-13} \, T^{3/2} y_c^2, \tag{114}$$

and identifies τ_d with the life time of a coronal streamer. Thus τ_d is of the order of days. Taking $\tau_d \simeq 10^5$ s and $T = 3 \times 10^6$ K, one finds $y_c \simeq 10^4$ cm. Thus for $V_A \simeq 10^8$ cm s^{-1} the time scale τ_t of the tearing mode instability is only ~ 3 s.

This very short time immediately draws one's attention to the type III bursts, where the acceleration has a time scale of about 1 s and indeed Sturrock's configuration seems to be very favourable for the production of type III burst electron streams which may propagate along the neutral sheet to very large distances from the Sun where some of them are observed. According to Sturrock, his calculations show that a sufficient number of electrons will be accelerated to high enough energy to explain the principal characteristics of type III bursts.

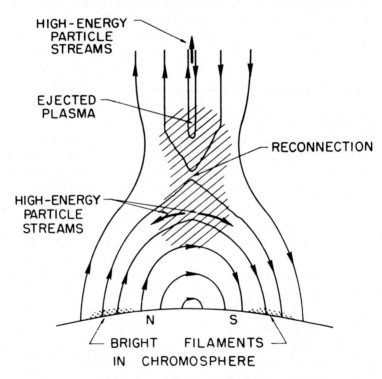

Fig. 125. Situation in Sturrock's model after the reconnection of the magnetic field lines. (After Sturrock, 1968.)

However, Sturrock goes further in order to explain the whole flare phenomenon by this mechanism. In his concept the type III burst represents the initial build-up of the tearing mode instability. After that the instability further develops to the highly non-linear level of amplitude which brings the plasma into turbulence. Then, as we have already seen above, the magnetic diffusion is greatly increased. Sturrock adopts the Bohm diffusion coefficient as characteristic for a highly turbulent plasma and finds the following characteristics for the process: Magnetic field before the flare $H \simeq 100$ G, particle density $n \simeq 10^{10}$ cm^{-3}, plasma temperature $T \simeq 10^8$ K, thickness of the turbulent sheet $y_c \simeq 100$ cm, and the diffusion velocity $v \simeq 4 \times 10^6$ cm s^{-1}.

Let us follow on Figure 125 the further consequences of the model. Sturrock supposes that the stream of accelerated particles as well as heat conduction along the magnetic field lines propagates towards the chromosphere, producing there the two-ribbon brightening typically observed. The acceleration of particles occurs by stochastic processes in the highly turbulent region (Sturrock, 1966), but the accelerated protons and electrons escape only slowly from the 'magnetic bag' formed by the disordered magnetic field. Such a moving cloud of electrons may give rise to the moving type IV burst. A collisionless shock formed ahead of the outward-moving plasma may be the source of the type II burst. The ejected plasma then moves through interplanetary space to the vicinity of the Earth giving rise to a geomagnetic storm. A model of the source of thermal X-rays in Sturrock's flare configuration has been formulated by Strauss and Papagiannis (1971), who suppose that this emission comes from the transition region between the closed and open magnetic field lines as it is heated by a beam of accelerated particles.

A great advantage of Sturrock's model is the easy way in which the flare ejecta of different types can get into the high corona and into space. In a closed configuration like that of Syrovatsky (cf. Section 6.5.3) this is much more difficult. On the other hand, however, observations do not confirm a direct relation between flares and coronal streamers, and one can raise also several other objections against Sturrock's model.

The diffusion velocity of $\sim 4 \times 10^6$ cm s^{-1} is still somewhat low; in order to release the flare energy within less than 10^3 s, one needs a diffusing area of 10^{22} cm^2 in order to produce energy of the order of 10^{31} erg. Another objection is that turbulence broadens the sheet and thus tends to remove the instability (Sweet, 1969). One should also bear in mind that the vast majority of flares appear without any associated type III burst, while according to Sturrock's concept a type III burst should be the most typical characteristic of all flares. And last but not least, the bright flare ribbons in the chromosphere form very close together at the flare onset, whereas in Sturrock's model the ribbons must be formed quite far from the zero line even at the flare's beginning. Still, Sturrock's model remains very attractive as an explanation of type III bursts.

C. *Similarities with Magnetospheric Substorms*

It has been pointed out (e.g. by Ness, 1965 and Akasofu, 1967) that the problems of magnetic field reconnection and annihilation in a neutral sheet on the Sun might have significant similarities to the situation that gives rise to magnetospheric substorms in the Earth magnetospheric tail. If this were true, the similarities might be used for better understanding of both these kinds of instabilities.

An example of such a similarity-study has been given by Obayashi (1975). He suggests that a current system analogous to the auroral electrojet in the Earth's ionosphere is also operative in the solar flare region. A somewhat similar idea was proposed earlier by Sen and White (1972; Section 6.4.2.E). Obayashi reexamines this problem and finds that for the HSRA atmosphere the electrical conductivity across the magnetic field has a maximum near the chromosphere-photosphere boundary, at an altitude of about 500 km above the level where $\tau(5000 \text{ Å}) = 1$. He finds there currents of 10^{12} to 10^{13} A which yield sufficient Joule heat to activate chromospheric Hα flares.

The model is hardly acceptable (like that of Sen and White) since it leaves many flare features unexplained and in some respects it is not consistent with observations; e.g., it requires that electrons penetrate to the bottom of the chromosphere; that Joule heating dominates in the chromosphere over thermal conduction and collisional heating; that the chromospheric emission is produced much lower in the atmosphere than in the models deduced from observations; and, since it starts with Sturrock's (1968) configuration it cannot give rise to the initial bright flare ribbons close to the $H_{\parallel} = 0$ line. Nevertheless, Obayashi's proposal is some sort of challenge to other solar and magnetospheric physicists to look into the problem of whether one might profit from comparisons between the magnetosphere and the Sun and, in particular, from the existence of current sheets both in the magnetosphere and on the Sun.

D. *Pustilnik's Model*

In Figure 84 we have shown the Kippenhahn-Schlüter model of a quiescent filament and we have indicated there a filament instability caused by stretching out the magnetic field lines. The ultimate cause was a rearrangement of the magnetic field that followed some external perturbation (e.g., emergence of new magnetic flux). Contrary to it, Pustilnik (1973) suggests that such a filament, suspended on magnetic field lines above the more rarefied coronal plasma, may become unstable in the solar gravitational field without any external perturbation, if its thickness exceeds a certain critical limit.

Pustilnik also adopts the Kippenhahn-Schlüter (1957) model for the filament structure (Figure 126a) and shows the reason why the filament thickness d is small: if d exceeds a critical limit, the filament is no longer stable and collapses downwards in the form of troughs. However, this instability can take two different forms.

Let us assume that at some place the prominence thickness slightly exceeds the critical limit. Then a trough begins to develop at that place in the form of a tongue pointed toward the photosphere. If there is no oppositely directed magnetic field below the filament, the trough stops at an altitude where the density of the surrounding plasma is equal to the density of the trough and the prominence plasma 'quietly' diffuses into the chromosphere. If, on the other hand, there is oppositely directed

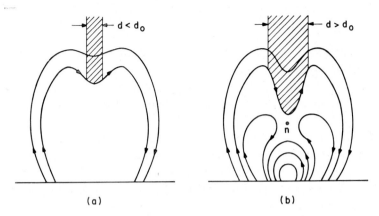

Fig. 126.   Pustilnik's (1973) model of filament destabilization. (a) Stable situation, when thickness d is smaller than a critical value d_o. (b) If $d > d_o$, a trough instability develops, eventually giving rise to neutral line n in Syrovatsky-like configuration (cf. Figure 123c).

magnetic field below the filament (Figure 126b) a current sheet develops as soon as the trough approaches the neutral line n. Pustilnik's model then merges with the model suggested by Syrovatsky (next section), with the only difference being that Pustilnik's current sheet is formed after an instability in the filament suspended above the neutral line, whereas in Syrovatsky's case the sheet is formed after an emergence of new flux from below.

The time between the onset of the trough instability and the onset of a flare in the newly formed current sheet (about 30 min) agrees well with observations. During this period the trough may be several times stabilized and destabilized again, which leads to oscillations in the penetration of the trough downwards. Pustilnik estimates the period of these oscillations at 1 to 3 min and identifies them with the observed 'winking' of filaments.

In spite of these apparent agreements with observations, however, one hardly can accept Pustilnik's model as a general explanation of flares. Quite big flares often occur while no filament is present in the active region (yet), hence they cannot be caused by instabilities in a filament. Pustilnik's conclusion that 'the power of a cosmic ray flare should be determined by the weight of the filament' is definitely unacceptable: The cosmic ray flare of 1966, July 7, e.g., developed without any filament activation at all. Also Pustilnik's other suggestions concerning the cosmic rays and filament disruption are untenable. Nevertheless, one cannot exclude the possibility that Pustilnik's trough instability might have some resemblance to the filament activation prior to flares, if the triggering mechanism is changed; magnetic-field and density variations can give rise to the same kind of instability as the postulated increase in prominence thickness. Then a current sheet may be formed, similar to Syrovatsky's configuration, which we shall discuss in the next section.

6.5.3. SYROVATSKY'S MODEL

As we promised before, we intend to show in detail the build-up of a current sheet and its flare-related disintegration on the example of Syrovatsky's theory. This theory, like the others, can be attacked from several points of view. However, among all the theories so far presented it tries to explain the highest number of the facts observed and summarized in Section 6.1.

Let us suppose that a new bipolar magnetic field emerges into an older existing field of opposite polarity. Then we encounter a situation that can be approximated by the configuration that is schematically drawn in Figure 123c (Syrovatsky, 1966, 1969a, 1972). We assume that n is a neutral line perpendicular to the paper and forming a semi-circle with radius

$$r_0 = (m/H_0)^{1/3} \tag{115}$$

with its ends in the photosphere; here m is the magnetic moment of the bipolar magnetic configuration and H_0 is the strength of the external field, which is parallel to the moment m. For example, for $H_0 = 100$ G, the neutral line is formed at an altitude of 10^4 km if $m = 10^{29}$ G cm^3, or 10^5 km if $m = 10^{32}$ G cm^3.

A. The Pre-Flare Activation

Let us approximate the neutral line by a straight line along the z-axis of local coordinates, perpendicular to the paper. In its vicinity the main part of the vector potential A of the magnetic field is then also directed along the z-axis. If the magnetic moment changes in time by $\Delta m = m(t) - m(0) \ll m$, the neutral line shifts its position over a distance $\Delta r_0 = \delta(t)/r_0$, with $\delta(t) = r_0^2 \Delta m/3m$. Thus the potential in the vicinity of the shifted neutral line becomes

$$A(x, y, t) = \tfrac{1}{2} h_0 [x^2 - y^2 - \delta(t)], \tag{116}$$

where $h_0 = 3H_0/r_0$, provided that we restrict out considerations to the solar chromosphere and low corona, where $H^2/8\pi \gg nkT$. It follows from Equation (116) that in the vicinity of the neutral line there is a uniform electric field along this line which equals

$$E = -\frac{h_0}{2c} \frac{\partial \delta}{\partial t}. \tag{117}$$

This electric field produces a drift motion of plasma outside the neutral line and a strong electric current near this line, as one can see when applying Ohm's law for plasma moving with bulk velocity v,

$$\mathbf{E} + \frac{1}{c} (\mathbf{v} \times \mathbf{H}) = \frac{1}{\sigma} \mathbf{j}. \tag{118}$$

Since the plasma conductivity is very high, Equation (118) shows that an electric field simply produces a drift motion of the plasma with drift velocity

$$\mathbf{v} = c(\mathbf{E} \times \mathbf{H})/H^2 \tag{119}$$

everywhere in the field with the only exception being the places where $H = 0$, i.e. near the neutral line. There the electric field produces a strong current in the plasma.

The change of the vector potential propagates to the neutral line in the form of a converging cylindrical wave, which travels with nearly the local Alfvén velocity and rearranges the field in conformity with the new value of the potential A. As it approaches the neutral line, the amplitude of the disturbances in velocity, field current and density grow without limit. Thus the wave has a specific cumulative property and as soon as it reaches the neutral line, a current sheet begins to develop.

One can estimate that the time difference between the onset of the disturbance and the moment when the neutral line is reached is of the order of minutes. No flare effects are yet observed, but the magnetic field changes and plasma motions occur. Therefore, one can identify this phase with the pre-flare activation of quiescent filaments.

B. *The Thermal Phase*

The next procedure involves numerical integration of the full non-linear system of magnetohydrodynamic equations. An example of one particular solution is shown in Figure 127 (left), assuming $\delta(t) = 0.1\, t$, for $\delta(t) = 0.18$. The neutral line passes through

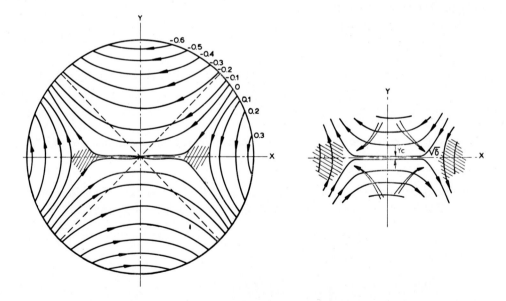

Fig. 127. The development of a current sheet in Syrovatsky's model. Left: Example of one particular solution of the magnetohydrodynamic flow ($\delta = 0.18$). The values of the potential A at a given magnetic line of force are given on the circumference. The regions of plasma compression are hatched. Right: The developed current sheet of width $2\sqrt{\delta}$ and thickness y_c. Hatched areas again show the regions of plasma compression and the arrows indicate the plasma flow. (After Syrovatsky, 1972.)

the origin of the coordinates and the current sheet forms along the x-axis. The curves show magnetic lines of force ($A = \text{const}$) and the A values are given on the circumference in units of $h_0 r_0^2$. The dashed lines show the initial position of the $A = 0$ lines of force through the original X-type neutral line.

The width of the current sheet is $\simeq \sqrt{\delta(t)}$, hence it increases with time. On the edges of this region the plasma is compressed due to the squeezing of plasma out of the sheet (hatched areas in Figure 127) and the two ropes of the compressed plasma separate progressively one from the other as time increases; this may explain the appearance of the two bright ribbons in flares and their separation with time. In addition the plasma in the outer space flows towards the sheet, at a speed of the order of Alfvén velocity close to it, and it is deviated along the current sheet towards the regions of the condensed plasma, as is indicated in Figure 127 (right).

The magnetic field in the current sheet is nearly uniform, with the strength

$$H \simeq h_0 \sqrt{\delta}, \tag{120}$$

but its sign changes abruptly in the neutral plane inside the current sheet. For infinite conductivity σ, the thickness y_c of the sheet (Figure 127) is determined by the particle gyro-radius

$$y_c \simeq \frac{cp_i}{eH} \simeq \frac{c(2m_i kT_i)^{1/2}}{eh_0 \sqrt{\delta}}, \tag{121}$$

where p_i, m_i, and T_i are the momentum, mass, and temperature, respectively, of the ions. The current density in the sheet is

$$j_c = \frac{c}{4\pi} \frac{h_0 \sqrt{\delta}}{y_c} = n_c ev_c, \tag{122}$$

where n_c is the particle density; hence the current velocity v_c, with regard to (121), is

$$v_c = \frac{h_0^2 \delta}{4\pi n_c (2m_i kT_i)^{1/2}}. \tag{123}$$

Let us suppose that the current sheet can be considered quasi-stationary so that

$$n_e kT_e + n_i kT_i = H^2/8\pi. \tag{124}$$

Since the electron energy in the current sheet acquired from the electric field is much greater than the proton energy, we can assume $T_e \gg T_i$, hence with regard to (120)

$$n_e kT_e \simeq h_0^2 \delta/8\pi. \tag{125}$$

Thus Equation (123) takes the form

$$v_c \simeq \left(\frac{2kT_e}{m_i}\right)^{1/2} \left(\frac{T_e}{T_i}\right)^{1/2} \gg \left(\frac{2kT_e}{m_i}\right)^{1/2}, \tag{126}$$

with the result that the current velocity is much greater than the iono-acoustic velocity $(2kT_e/m_i)^{1/2}$. As a consequence iono-acoustic turbulence must appear and in fact, strictly speaking, the quasi-stationary state can never be reached.

We have already mentioned (in Section 6.5.2) a similar argument by Friedman and Hamberger (1969a) who have shown that the turbulence decreases the conductivity to a finite value which then determines the actual thickness of the sheet instead of the gyro-radius in Equation (121). With a finite value of σ,

$$y_c = \frac{c^2}{4\pi\sigma}\bigg/v, \tag{127}$$

where v is the plasma input velocity into the current sheet, after Syrovatsky (1969b),

$$v = \frac{\partial\delta}{\partial t}\bigg/2\sqrt{\delta}. \tag{128}$$

The power delivered in a sheet of length πr_0 is

$$P = \pi r_0 y_c \sqrt{\delta}\,(j_c^2/\sigma), \tag{129}$$

where according to Ohm's law (118) and Equation (117)

$$j_c = \sigma E_c = \frac{\sigma h_0}{2c}\frac{\partial\delta}{\partial t}. \tag{130}$$

Therefore, with regard to (127) and (128)

$$P = \pi r_0 \frac{h_0^2 \delta}{8\pi}\frac{\partial\delta}{\partial t}, \tag{131}$$

and P is independent of conductivity. This means that the sheet thickness y_c automatically takes the value which gives the needed rate of magnetic energy dissipation. The quantity which determines the rate of dissipation is the product $\delta(\partial\delta/\partial t)$, but the ultimate cause is the change of the magnetic source configuration.

This result shows that, depending upon the value of $\delta(\partial\delta/\partial t)$ the dissipation may vary from slow heating of some regions of the solar atmosphere to the very powerful energy release observed in major flares. The development of plasma turbulence in the sheet may be the cause of the rise to high temperatures which is reflected in the different aspects of the thermal flare radiation in the corona and chromosphere. The chromospheric, low-temperature emission can be explained as due to thermal conductivity along the magnetic lines of force from the corona. The fine structure of the chromospheric emission is then caused by the inhomogeneous distribution of the conducting field lines. Therefore, the formation of a current sheet, turbulent, but still close to a quasi-stationary state, explains the thermal phase of a flare, which in many flares is the only phase observed.

As an illustration, we present in Table XVI one particular set of flare characteristics which result from Syrovatsky's model. All these parameters can be varied within wide limits (cf. Syrovatsky, 1972). However, since the model certainly greatly simplifies the actual situation, it is meaningless to study all these variations and their consequences in detail. The set of data presented here clearly demonstrates that conditions similar to those in flares can be built up. The volume and emission measure of the current sheet itself are very small when compared with the values observed ($n_c^2 V_c = 4 \times 10^{44}$ cm^{-3} in the example presented in Table XVI). This, of course, is not astonishing since the observed emission must come from a larger volume than that occupied by the current sheet.

TABLE XVI

Flare characteristics resulting from Syrovatsky's model (for one selected set of parameters – modifications are possible)

Initial plasma density	n_0	10^{10} cm^{-3}
External magnetic field	H_0	300 G
Magnetic moment of the bipolar configuration	m	10^{31} G cm^3
Magnetic gradient	h_0	3×10^{-7} G cm^{-1}
Top height of the neutral line	r_0	3×10^9 cm
Width of the current sheet	$2\sqrt{\delta}$	2×10^9 cm
Time variation of δ	$\partial\delta/\partial t$	3×10^{15} cm^2 s^{-1}
Plasma input velocity	v	1.5×10^6 cm s^{-1}
Speed of bright ribbons expansion	v	1.5×10^6 cm s^{-1}
Thickness of the current sheet	y_c	5×10^2 cm
Current velocity	v_c	2×10^7 cm s^{-1}
Particle density in the current sheet	n_c	2×10^{11} cm^{-3}
Electron temperature	T_e	1.5×10^8 K
Surplus magnetic energy	ΔW	6×10^{31} erg
Power delivered in the sheet	P	10^{29} erg s^{-1}
Characteristic time of dissipation [a]	τ	10^3 s

[a]   Computed from $\tau = m/(\partial m/\partial t) = 1/3\, r_0^2\, (\partial\delta/\partial t)$.

C. *The Impulsive Phase*

Even in the thermal phase there will be some run-away particles, in particular protons and heavier nuclei. However, an impulse acceleration process, giving rise to large numbers of particles with subrelativistic or relativistic energies, as is observed in many flares, needs quite a different mechanism, and it probably occurs only in some flares. Syrovatsky explains this phase in very much the same way as Alfvén and Carlqvist (1967) do. According to him this impulsive phase corresponds to the transition from the quasi-stationary picture considered above to a highly non-stationary one due to the rupture of some part of the current sheet and interruption of the local current. Syrovatsky's conception differs from that of Alfvén and Carlqvist in three aspects: It is an interruption of a current sheet instead of a twisted rope connecting opposite magnetic polarities; the strong electric field that appears is of electromagnetic instead of electrostatic origin as Alfvén and Carlqvist assume; and the whole phenomenon is only one phase of the flare development (often absent), while Alfvén and Carlqvist made it the basis of all flares.

According to Syrovatsky the plasma density in the current sheet increases to approximately

$$n_c \simeq n_0 \ln(\sqrt{\delta}/y_c),$$ (132)

where n_0 is the original uniform plasma density. However, one has to expect many inhomogeneities in the current sheet density. In such a case, as we already have mentioned when discussing the Alfvén-Carlqvist model, a current break-off can occur as the current velocity (inversely proportional to the density according to (123)) greatly increases in that part of the sheet where the plasma density is anomaloulsy low. Another reason for the current sheet rupture may be the tearing mode instability discussed by Jaggi (1964) and Sturrock (1968) (cf. Section 6.5.2.B).

In Syrovatsky's interpretation of the current sheet interruption an electromagnetic wave appears which transforms the magnetic field H into the electric field E, which can be estimated to be

$$E \simeq (V/c) H,$$ (133)

with V denoting the speed of the sheet disappearance. If $V \gg v_c$, the field will be much greater than the initial field given by Equation (117). Alfvén and Carlqvist's electrostatic interpretation, however, also could be incorporated into Syrovatsky's theory.

Syrovatsky assumes that the whole acceleration process is accomplished during the phase of the sheet rupture. The accelerated electrons follow the magnetic field lines towards regions of higher density, producing hard X-rays through bremsstrahlung and exciting the EUV lines. Some of the accelerated particles may penetrate very deep and produce the white-light emission. Synchrotron radiation of the accelerated electrons gives rise to the microwave and type IV burst. A small fraction of the accelerated particles escape into interplanetary space.

D. *Deficiencies of Syrovatsky's Theory*

In spite of the fact that Syrovatsky's flare model is probably the best one available at present, at least from the point of view that it is able to build up a flare phenomenon very similar to the flare we actually observe, it can be attacked from several points of view.

As we mentioned in Section 6.5.1.E, Smith and Priest (1972) criticized the Alfvén-Carlqvist model on the ground that any strong current develops the ion-sound instability (due to electron drift with respect to the ions) at a rate faster than the rate of the instability that leads to current interruption. The ion-sound instability produces turbulence, thus heats the plasma, but cannot accelerate particles to suprathermal energies. The same objection, according to Smith and Priest, also applies to Syrovatsky's mechanism as described in Section C. Thus Syrovatsky's model can produce a thermal flare, but it needs some other additional mechanism for producing the impulsive phase through rapidly accelerated particles. However, as we said in Section 6.5.1.E, it is not yet clear whether Smith and Priest's objection has general validity, since some laboratory experiments do not seem to support it.

On the other hand, Anzer, (1973) argues that Syrovatsky needs too much compression of the magnetic field and an artificially low plasma density in order to make his acceleration mechanism work. Anzer finds that unless the restoring Lorentz force is completely neglected, the maximum admissible current-to-density ratio is smaller by a factor of 10^6 than Syrovatsky has found.

The preceding two objections concern only Syrovatsky's interpretation of the 'impulsive phase' of his model. Another question is whether the whole configuration of the Syrovatsky's model is realistic. As Piddington (1973) pointed out, Syrovatsky's overlying field H_0 must have strength of at least 100 G. Thus it hardly can be an 'external field' as Syrovatsky first assumed and we must place his bipolar configuration into an existing active region. The closed field is then a newly emerging field with magnetic polarities opposite to H_0. This, as we have seen, is a configuration intimately related to the impulsive acceleration processes for which Syrovatsky's explanation seems to be inadequate. Thermal flares, which Syrovatsky possibly can explain, probably do not need a configuration of this, magnetically complex, type (at least as far as the present resolution of our magnetographs allows to judge).

Syrovatsky's 'horizontal' configuration is also unfavourable for explaining any agents propagating upwards into the corona. It is true that the mechanism proposed by Syrovatsky can also be modified for other configurations of the neutral line, which might perhaps be more favourable for particle escape. Nevertheless, one is reminded of Sturrock's model (Section 6.5.2.B) which, while probably failing to explain the chromospheric flare, makes any coronal propagation very easy. Therefore, one wonders whether some combination of both these models might be eventually possible (cf. Figure 123 d).

6.6. The Particle Acceleration

As we have seen, several authors mentioned some acceleration process in their models, either by means of current interruption (Alfvén and Carlqvist, 1967; Syrovatsky, 1972) or through some kind of stochastic mechanism (Schatzman, 1967; Sturrock, 1968; Piddington, 1974). However it is questionable whether indeed one single process can accelerate particles to the very high energies observed in the most energetic flares.

In order to illustrate the difficulty, let us suppose that all the particles are accelerated during the short period when the impulsive phase is observed. This must have been the case, e.g., in the flare of 1972, August 4, when γ-rays requiring > 30 MeV protons for their production were observed practically simultaneously with the hard X-rays (Section 5.4.3). It also should be the case in the white-light flares if their emission is produced by > 20 MeV protons (Section 2.3.11). Even the ejection of > 500 MeV protons from the Sun, required for producing the GLE events (Section 5.1), seems to occur close to the impulsive phase (e.g. in 1972, August 7). Also type IV bursts, requiring relativistic electrons, set in immediately after the impulsive phase (Švestka, 1970a).

In such a case the whole acceleration must be accomplished within $\sim 10^2$ s. We have seen that the impulsive acceleration starts when the plasma is slightly preheated, let us say to 10^7 K. For a thermal distribution (Figure 3) the mean energy of the gas particles is then close to 1 keV, with only very few particles in excess of 10 keV. Thus within $\sim 10^2$ s particles must be accelerated from a few keV up to the GeV range with an integral power-law spectrum with the exponent $\gamma \simeq 2$, and in such a number that about 1 in 50 particles has energy exceeding 100 keV (Švestka, 1970a). It is very difficult to find a one-step acceleration process that can do it under reasonable physical assumptions.

Therefore, one is inclined to consider a two-step acceleration as a more likely interpretation. The first step would pre-accelerate the particles to energies of the order of 10^2 keV and the second step would accelerate them to the higher energies. The electrons accelerated in the first step give rise to the impulsive bursts, whilst the relativistic electrons and protons recorded in space are accelerated in the second step. De Jager (1969), who instigated this idea, has supposed that the first step, occurring during the flash phase, is due to current interruption, while the second step is accomplished through Fermi processes in shock waves, and this still seems to be the most acceptable interpretations of the particle acceleration in solar flares (Švestka, 1975).

However, the time difference between the first and second acceleration step must be very small (not tens of minutes as de Jager originally assumed). This follows from the observations of all the effects of high-energy protons and electrons mentioned above, from early onsets of radio type II bursts that seem to be visible manifestation of the second acceleration step (Švestka and Fritzová, 1974; Sections 3.2.5 and 5.1.1.A; Figure 78) and from X-ray observations in some particular flares (Frost and Dennis, 1971; Frost, 1974). The only fact so far that contradicts this conclusion is the observation of the late onsets of the relativistic electrons (Simnett, 1971; Section 5.3.2; Figure 114). This discrepancy might be perhaps explained through rigidity-dependent injection of the preaccelerated particles into the second-step acceleration process so that electrons cannot be accelerated as easily and as rapidly as protons (cf. Section 6.6.3).

Another problem is whether we always meet with the same acceleration mechanisms in all flares. When comparing the characteristics of electron and proton events, one is inclined to believe that we encounter at least two different types of the acceleration processes. One, which produces non-relativistic electrons in rather small flares, in association with type III bursts, and another, which gives rise to proton events, typically in large two-ribbon flares in association with type II and type IV bursts.

Thus, when combining this observation with the two-step acceleration, we actually may encounter three different acceleration processes in flares:

Process 1: An acceleration that gives rise to the impulsive phase. The accelerated electrons produce the different types of impulsive bursts, but no particles are observed in space.

Process 2: An acceleration that gives rise to collimated streams of electrons in the corona causing the type III bursts. When this acceleration is linked with (1), strong streams of electrons can travel very far from the Sun. We observe electron events in space.

Process 3: A powerful second-step acceleration that accelerates the particles preaccelerated in (1) to higher energies. Both the (1)+(3) and (1)+(2)+(3) combinations seem to be possible. We observe then proton events in space.

Another difficulty arises from the fact that observations still do not give us enough information about the effect of the acceleration on particles of different kinds. We know that all particles are not accelerated to the same velocity, since the presence of >10 MeV electrons in space without relativistic protons clearly shows that this cannot be the case. But we do not know whether all particles are accelerated to the same energy, since the acceleration process (implying the process of injection) can also be rigidity dependent. We have seen in Section 5.2 that rigidity-dependent injection has been proposed by Cartwright and Mogro-Campero (1973) as an explanation for the enhanced abundances of heavy nuclei in the particle streams. Also the low flux of relativistic electrons in proton events (Section 5.3.2) may indicate a rigidity-dependent injection into the second-step acceleration (process (3)). On the other hand, the absence of protons in the type III-associated electron events (process (2)) might indicate a rigidity-dependent acceleration. In the acceleration process (1) we have no information on the proton acceleration at all; only the fact that we 'need' protons to be preaccelerated in this phase in order to be able to enter into the acceleration process (3), speaks for an acceleration of all kinds of particles to about the same energy.

Let us now briefly summarize the different acceleration mechanisms that might be available in the flare region.

6.6.1. CURRENT INTERRUPTION

Charged elementary particles can be accelerated in strong electric fields. However, only the component of the electric intensity in the direction of the magnetic field can be effective in the acceleration process. Thus current interruptions in twisted force-free magnetic ropes or in current sheets are the likely mechanisms that might be considered.

In case we try to explain the whole acceleration to be accomplished in one step through current interruption, we need a voltage of at least 10^9 to 10^{11} V. If only the first-step acceleration is ascribed to this mechanism, some 10^5 to 10^6 V would satisfy.

In the Alfvén-Carlqvist theory (Carlqvist, 1969) the progressive rarefaction of the plasma in the region of current interruption gives rise to the creation of a space charge region of high impedance. With $n = 10^8$ cm^{-3} and $T = 10^6$ K Carlqvist finds that the interruption occurs if the current exceeds ~ 60 A m^{-2}. Taking the total current I equal to 3×10^{11} A (in agreement with Section 6.4.3.B) this critical current density is reached when the rope is confined to a diameter of about 80 km. The maximum potential drop then can be estimated at $\sim 5 \times 10^{10}$ V, and the characteristic time for release of energy W becomes $\lesssim 7 \times 10^2$ s for $W < 10^{32}$ erg. Hence, in principle, the mechanism is fast enough and able to explain the whole acceleration in one step.

In Syrovatsky's (1969) model one can estimate the maximum rigidity P_M which a particle can reach, putting it equal to the maximum value of the potential change given by Equation (116), i.e. $P_M \simeq h_0 \delta$. For our example in Table XVI we thus get $P_M \simeq 10^{11}$ V.

Thus the current interruption represents a mechanism which could accelerate particles to the energies required in one step. The only question is whether the instability assumed can actually develop. We mentioned earlier the objections Anzer (1973) raised against Syrovatsky's acceleration mechanism: the magnetic field needed is too high and the gas density too low to represent a realistic case (Section 6.5.3.D). Another objection was raised by Smith and Priest (1972): in both Carlqvist's and Syrovatsky's mode the current interruption will not develop, because other instabilities, and the ion-sound current instability in particular, develop faster (Section 6.5.1.E).

While laboratory experiments indicate that the objection by Smith and Priest might not be fully valid, one cannot escape the conclusion that turbulence certainly sets a limit to the energy gain. As soon as some turbulence is present – and according to Smith and Priest (1972) and even Syrovatsky himself (Section 6.5.3.B) this simply must be the case – the free run of particles along the rope or current sheet, and thus the acceleration as well, are greatly suppressed. Therefore, it appears indeed to be unrealistic to expect an energy gain responding to $\gtrsim 10^{10}$ V through this process. However, one may still suppose that particles can reach energies of $\sim 10^6$ eV through this mechanism. Therefore, current interruption can be considered a likely mechanism for the first step of the acceleration process.

In contrast to Syrovatsky, who uses density inhomogeneities to disrupt the current sheet, Sonnerup (1973) and Sturrock (1974) consider the quasi-stationary reconnection process in a current sheet as the source of the first-step acceleration. Since the reconnection process converts magnetic energy into plasma kinetic energy, this must give rise to some increase in particle energy. A simple estimate has been made by Sturrock from the energy balance equation

$$H^2/8\pi = 2n_e \times 1.6 \times 10^{-9} E, \tag{134}$$

where H is the magnetic field strength, n_e the electron density and E (in keV $= 1.6 \times 10^{-9}$ erg) the mean energy acquired by one particle. Hence

$$E = 1.2 \times 10^7 H^2 n_e^{-1}. \tag{135}$$

As an example, for $H = 30$ G and $n_e = 10^8$ cm^{-3} the mean acquired energy is ~ 100 keV, a value that is typical for the first-step acceleration process.

6.6.2. FERMI MECHANISM

Particles can also be accelerated through coupling with the magnetic field. One mechanism, proposed by Swann (1933), accelerates the particles on the Sun by the betatron principle. However, as Dungey (1958) and Parker (1958) have pointed out, the very high electrical conductivity of the solar plasma makes this mechanism inefficient. On the other hand, the mechanism proposed by Fermi (1949, 1954) appears to be applicable to the flare process in several different forms.

Fermi's (1949) original theory was based on the random reflection of charged particles at moving magnetic mirrors. It is well known that a magnetic mirror reflects a particle if its pitch-angle α in magnetic field H is larger than α_0, defined as

$$\sin^2 \alpha_0 = H/H_M , \tag{136}$$

where H_M is the maximum field the particle meets with. If the mirror moves in the direction opposite to the particle motion, the particle gains energy at the reflection. If there are two such mirrors approaching each other, the particle passes through a series of such reflections and is gradually accelerated.

This simple model encounters several difficulties. First, particles on the other side of the mirror are decelerated. That means that the number of head-on collisions must be larger than the number of overtaking collisions, in order to make the acceleration mechanism efficient. This can be achieved by means of some special configurations; examples of two of them are shown in Figure 128.

The case (a) illustrates two pairs of Parker's (1958) interpenetrating shocks crossing each other. Since particles are swept ahead of the shocks, the density of particles being accelerated exceeds the number of particles which are subject to deceleration. In the course of time the length of the accelerating trap shrinks to zero so that all particles in the trap must necessarily be accelerated up to their escape energy. In contrast, particles which have crossed a shock find themselves in a decelerating trap whose length is increasing so that decelerations become progressively less frequent. Moreover, when the shocks cross each other these particles again are in an acceleration trap.

The case (b) shows a plane shock crossing closed lines of force. The shock intersects each line in two points which defines a trap for particles ahead of the shock, and the trap becomes progressively shorter as the shock moves ahead. The particles escape before the points of intersection have moved completely together with a significant energy gain. A similar model was presented in Figure 73 to explain the acceleration of particles and radiopulsations.

However, there is another restricting effect in Fermi's mechanism which these models do not remove. An increase in energy means a decrease in the pitch-angle.

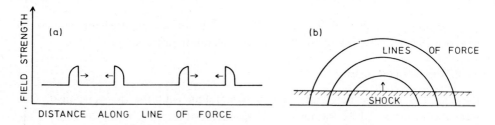

Fig. 128. A schematic drawing of Parker's interpenetrating shocks (left) and Wentzel's plane shock crossing closed lines of force. (After Wentzel, 1963.)

As soon as $\alpha < \alpha_0$ in Equation (136), the particle penetrates through the mirror. Therefore, the particles actually escape from the acceleration region after being only slightly accelerated and cannot be successively accelerated to the high energies we need. For example, the configuration in Figures 73 and 128 b can accelerate particles to a certain limit. But there will be a high-energy cutoff in the acceleration which the particles cannot exceed.

The only way how to make the Fermi mechanism effective is to randomise in some way the pitch-angle distribution. This randomisation can be achieved in a turbulent field, either in association with randomness of the turbulence itself (e.g. Hall and Sturrock, 1967) or through some independent scattering process, e.g. by resonant wave particle interactions (e.g. Kulsrud and Ferrara, 1971; Melrose, 1974a). Thus, as it seems, the only Fermi-type mechanism that can effectively accelerate particles to very high energies in flares is the stochastic acceleration in turbulent fields.

6.6.3. STOCHASTIC ACCELERATION

The turbulent field that accelerates particles may be either an enhanced turbulence in the active region normally responsible for the heating of the solar corona, as in the theories of Schatzman (1967) and Pneuman (1967), or a newly created turbulence due to flare-produced instabilities (e.g. Sturrock, 1966, 1968). As the observations indicate, the most plausible model is magnetohydrodynamic turbulence associated with a shock wave. We know one example of such a phenomemon: In the Earth's bow shock a collision-free shock leads to shocked gas in a state of MHD turbulence. Thus a shock giving rise to a solar type II radio burst, which is clearly associated with the acceleration process (cf. Figure 92), can lead to a turbulence and through it to a stochastic acceleration of particles in the flare region. The stochastic acceleration need not be necessarily of Fermi type. As Hall and Sturrock (1967) showed, it comprises a class of acceleration mechanisms of which Fermi's is a particular case.

Let us show, as an example, a mode of stochastic acceleration that Sturrock (1974) expects to occur in a gas after a shock has passed through it. Let the mean magnetic field strength in the shocked gas be H (G), mean plasma density n (cm^{-3}) and the thickness of the shocked region L (cm). The phase velocity of waves comprising the turbulence should be comparable with the Alfvén velocity given by Equation (112), which can be transcribed as

$$v_A \simeq 10^{11.3}\, H n^{-1/2}.$$ (137)

The maximum frequency in the turbulence spectrum will be close to the gyro-frequency of protons,

$$\omega_g \simeq 10^4 H.$$ (138)

Hence the maximum size of the turbulent cells can be estimated as

$$\lambda \simeq \frac{v_A}{\omega_g} \simeq 10^{7.3}\, n^{-1/2}.$$ (139)

Let us denote by δH the magnitude of the fluctuating magnetic field. The magnitude of the electric field E (esu) is then

$$E \simeq \frac{v_A}{c} \delta H \simeq 10^{0.8} \, \delta H^2 n^{-1/2}. \tag{140}$$

A particle that traverses the turbulent plasma samples the electric field in different cells; if it travels a total distance L, it samples (L/λ) cells and gains or loses an energy $eE\lambda$ (erg) per cell. If this is treated as a random walk process, the expectation value of the final energy $\bar{\varepsilon}$ is given by

$$\bar{\varepsilon} \simeq (L/\lambda)^{1/2} \, eE\lambda, \tag{141}$$

which can be rewritten for $\bar{\varepsilon}$ in eV by using Equations (139) and (140) as

$$\bar{\varepsilon} \simeq 10^{7.0} \, \delta L^{1/2} H^2 n^{-3/4}. \tag{142}$$

On the other hand, the maximum possible energy would be gained by a particle that moves along cells with the same sign of the electric field over the total path length L:

$$\varepsilon_{\text{Max}} = (L/\lambda) \, eE\lambda \tag{143}$$

or, in eV,

$$\varepsilon_{\text{Max}} = 10^{3.3} \, \delta L H^2 n^{-1/2}. \tag{144}$$

Sturrock assumes $\delta = 0.1$. Then an average energy of about 10 MeV can be obtained from (142) for reasonable values of L, H, and n, for example $L = 10^9$ cm, $n = 10^9$ cm^{-3} and $H \simeq 40$ G. The corresponding maximum energy is, according to (144), $E_M \simeq$ 10 GeV.

However, there is still one restriction in the Fermi mechanism and generally in stochastic acceleration processes and this is the fact that particles need a certain minimum energy to be able to enter into the acceleration. There are several reasons for it. The accelerated particles lose energy through collisions and they can be effectively accelerated only when the acceleration significantly exceeds the energy loss. Since the relative energy loss decreases with energy, obviously only particles above a certain minimum energy can be accelerated. In the Fermi acceleration generally, the particle speed must be significantly higher than the speed of the moving magnetic inhomogeneities to make the process effective (see, e.g., Wentzel, 1963). The same is true in some special models, e.g., that by Friedman and Hamberger (1969a). They suggest that particles moving against the plasma flow in Figure 124 are reflected by the fields MF back through the neutral sheet into the opposite flow region. They gain energy through repetitions of this process. However, in order to be able to pass through the turbulent sheet, their effective mean free path must exceed the sheet thickness; thus only protons with $E \gtrsim 50$ keV can participate in this process.

Also in stochastic acceleration mechanisms we encounter a similar restriction. For example, in the Sturrock's mechanism discussed above, a particle must have an energy high enough to transverse at least one turbulent cell of size λ given by Equation (139) in order to participate in the acceleration. This implies that only particles with gyro-radii greater than λ can experience stochastic acceleration. Thus, for the example considered above ($H = 40$ G, $n = 10^9$ cm^{-3}) only protons with energy in excess of 25 keV can enter into the acceleration process and for electrons the minimum energy is several MeV.

This very high limit for electrons may be lowered probably, e.g. by assuming that electrons with smaller gyro-radii are accelerated along the magnetic field in induced electric fields of which the less movable protons will make little use (Meyer, in discussion to Sturrock, 1974). Nevertheless, Sturrock's mechanism demonstrates how rigidity-dependent injection can greatly suppress the number of accelerated electrons as compared to protons. Such an effect actually is observed (Section 5.3.2). Sturrock's mechanism also fits the demands of Cartwright and Mogro-Campero (1973) interpretation of the enhanced abundance of heavy elements in the particle stream, which calls for rigidity-dependent injection as well (Section 5.2).

However, the most important conclusion is that even at temperatures of tens of million degrees there are very few thermal particles that might enter into the process of stochastic acceleration. Therefore, the Fermi acceleration process, and the stochastic acceleration in all its modifications, can work much more efficiently and faster if the particles which enter into it are pre-accelerated from the normal thermal Maxwellian energy distribution to somewhat higher energies. Thus we are brought back to de Jager's (1969) concept that the acceleration process in flares has two steps; pre-acceleration (possibly through current interruption in the upper chromosphere or low corona), and a second-step acceleration of the pre-accelerated particles through stochastic acceleration in shocks at higher altitudes.

REFERENCES

Abrami, A.: 1970, *Solar Phys.* **11**, 104.
Achong, A.: 1974, *Solar Phys.* **37**, 477.
Acton, L. W.: 1968, *Astrophys. J.* **152**, 305.
Acton, L. W., Catura, R.C., Meyerott, A. J., Wolfson, C. J., and Culhane, J. L.: 1972, *Solar Phys.* **26**, 183.
Akabane, K. and Cohen, M. H.: 1961, *Astrophys. J.* **133**, 258.
Akasofu, S. I.: 1967, *Space Sci. Rev.* **6**, 21.
Alexander, J. K., Malitson, H. H., and Stone, R. G.: 1969, *Solar Phys.* **8**, 388.
Alfvén, H. and Carlqvist, P.: 1967, *Solar Phys.* **1**, 220.
Allen, C. W.: 1940, *Monthly Notices Roy. Astron. Soc.* **100**, 635.
Altschuler, M. D.: 1973, Paper presented at the *IAFE Flare Conference*, Buenos Aires.
Altschuler, M. D.: 1974, *IAU Symp.* **57**, 3.
Altschuler, M. D. and Newkirk, G.: 1969, *Solar Phys.* **9**, 131.
Altschuler, M. D., Lillienquist, C. G., and Nakagawa, Y.: 1968, *Solar Phys.* **5**, 366.
Alvarez, H. and Haddock, F.T.: 1973, *Solar Phys.* **29**, 197.
Alvarez, H., Haddock, F. T., and Lin, R. P.: 1972, *Solar Phys.* **26**, 468.
Alvarez, H., Haddock, F. T., and Potter, W. H.: 1973, *Solar Phys.* **31**, 493.
Alvarez, H., Haddock, F. T., and Potter, W. H.: 1974, *Solar Phys.* **34**, 413.
Anastassiadis, M., Ilias, D., Caroubalos, C., Macris, C., and Elias, D. P.: 1964, *Nature* **201**, 357.
Anderson, G. F.: 1966, University of Colorado, Boulder (Thesis).
Anderson, K. A.: 1969, *Solar Phys.* **6**, 111.
Anderson, K. A.: 1970, Paper presented at the 11*th Int. Conf. on Cosmic Rays*, Budapest.
Anderson, K. A.: 1972a, Paper presented at the 15th Meeting of COSPAR, Madrid.
Anderson, K. A.: 1972b, *Solar Phys.* **27**, 442.
Anderson, K. A. and Mahoney, W. A.: 1974, *Solar Phys.* **35**, 419.
Anderson, K. A. and Winckler, J. R.: 1962, *J. Geophys. Res.* **67**, 4103.
Anderson, K. A., Kane, S. R., and Lin, R. P.: 1970, Paper presented at the Seminar on Cosmic Ray Generation, Leningrad.
Anglin, J. D., Dietrich, W. F., and Simpson, J. A.: 1973a, in *Symp. on High Energy Phenomena on the Sun*, NASA GSFC Preprint, p. 315.
Anglin, J. D., Dietrich, W. F., and Simpson, J. A.: 1973b, *Astrophys. J. Letters* **186**, L41.
Antalová, A. and Gnevyshev, M. N.: 1965, *Astron. Zh.* **42**, 253.
Anzer, U.: 1968, *Solar Phys.* **3**, 298.
Anzer, U.: 1973, *Solar Phys.* **30**, 459.
Arnoldy, R. L., Kane, S. R., and Winckler, J. R.: 1967, *Solar Phys.* **2**, 171.
Arnoldy, R. L., Kane, S. R., and Winckler, J. R.: 1968a, *Astrophys. J.* **151**, 711.
Arnoldy, R. L., Kane, S. R., and Winckler, J. R.: 1968b, *IAU Symp.* **35**, 490.
AS & E: 1975, Unpublished results from the S-054 experiment on Skylab, American Science & Engineering, Cambridge, USA.
Athay, R. G.: 1966, *Astrophys. J.* **145**, 784.
Athay, R. G.: 1972, *Radiation Transport in Spectral Lines*, D. Reidel Publ. Co., Dordrecht.
Athay, R. G. and Johnson, H. R.: 1960, *Astrophys. J.* **131**, 413.
Athay, R. G. and Moreton, G. E.: 1961, *Astrophys. J.* **133**, 935.
Athay, R. G. and Skumanich, A.: 1968, *Solar Phys.* **3**, 181.
Avignon, Y., Martres, M. J., and Pick, M.: 1963, *Compt. Rend. Acad. Sci. Paris* **256**, 2112.
Avignon, Y., Martres, M. J., and Pick, M.: 1964, *Ann. Astrophys.* **27**, 23.

Avignon, Y., Caroubalos, C., Martres, M. J., and Pick, M.: 1965, *IAU Symp.* **22**, 373.
Axford, W. I.: 1965, *Planetary Space Sci.* **13**, 1301.
Axisa, F.: 1972, private communication.
Axisa, F.: 1974, *Solar Phys.* **35**, 207.
Axisa, F., Martres, M. J., Pick, M., and Soru-Escaut, I.: 1973a, *Solar Phys.* **29**, 163.
Axisa, F., Martres, M. J., Pick. M., and Soru-Escaut, I.: 1973b, in *Symposium on High Energy Phenomena on the Sun*, NASA GSFC Preprint, p. 615.

Ballario, M. C.: 1963, *Mem. Soc. Astron. Ital.* **34**, Nos. 2 and 4.
Ballario, M. C.: 1968, *Nobel Symp.* **9**, 53.
Bame, S. J., Asbridge, J. R., Hundhausen, A. J., and Strong, I. B.: 1968, *J. Geophys. Res.* **73**, 5761.
Banin, V. G.: 1965a, *Izv. Krymsk. Astrofiz. Obs.* **33**, 118.
Banin, V. G.: 1965b, *Izv. Krymsk. Astrofiz. Obs.* **34**, 252.
Banin, V. G.: 1969, *Geodet. Geophys. Veröff. II*, **13**, 49.
Banin, V. G., Kopecký, M., and Rybina, A. A.: 1967, *Bull. Astron. Inst. Czech.* **18**, 319.
Banin, V. G., De Feiter, L. D., Fokker, A. D., Martres, M. J., and Pick, M.: 1969, *Ann. IQSY* **3**, 229.
Bappu, M. K. V. and Punetha, L. M.: 1962, *Observatory* **82**, 170.
Barnes, C. W. and Sturrock, P. A.: 1972, *Astrophys. J.* **174**, 659.
Barouch, E., Gros, M., and Masse, P.: 1971, *Solar Phys.* **19**, 483.
Bartley, W. C., Bukata, R. P., McCracken, K. G., and Rao, U. R.: 1966, *J. Geophys. Res.* **71**, 3297.
Baselyan, L. L., Boncharov, N. Y., Zaitsev, V. V., Zinichev, V. A., Rapoport, V. O., and Tsybko, Y. G.: 1974, *Solar Phys.* **39**, 223.
Batstone, R. M., Evans, K., Parkinson, J. H., and Pounds, K. A.: 1970, *Solar Phys.* **13**, 389.
Becker, U.: 1958a, *Z. Astrophys.* **44**, 243.
Becker, U.: 1958b, *Z. Astrophys.* **46**, 168.
Beigman, I. L.: 1974, P. N. Lebedev Physical Institute, Moscow, Preprint No. 68.
Beigman, I. L., Grineva, Y. I., Mandelstam, S. L., Vainstein, L. A., and Zhitnik, I. A.: 1969, *Solar Phys.* **9**, 160.
Beigman, I. L., Grineva, Y. I., Korneev, V. V., Krutov, V. V., Mandelstam, S. L., Vainstein, L. A., Vasilyev, B. N., and Zhitnik, I. A.: 1974, P. N. Lebedev Physical Institute, Moscow, Preprint No. 67.
Bell, B. and Glazer, H.: 1959, *Smithson. Contrib. Astrophys.* **3**, 4.
Bell, B. and Noci, G.: 1975, *J. Geophys. Res.* (in press).
Bertsch, D. L., Fichtel, C. E., and Reames, D. V.: 1969, *Astrophys. Letters* **157**, L53.
Bertsch, D. L., Fichtel, C. E., and Reames, D. V.: 1972, *Astrophys. J.* **171**, 169.
Bertsch, D. L., Giswas, S., Fichtel, C. E., Pellerin, C. J., and Reames, D. V.: 1973, *Solar Phys.* **31**, 247.
Bertsch, D. L., Biswas, S., and Reames, D. V.: 1974, *Solar Phys.* **39**, 479.
Bessey, R. J. and Kuperus, M.: 1970, *Solar Phys.* **12**, 216.
Bhatia, V. B. and Tandon, J. N.: 1970, *Astrophys. Letters* **6**, 113.
Bhonsle, R. V. and McNarry, L. R.: 1964, *Astrophys. J.* **139**, 1312.
Biermann, L., Haxel, O., and Schulter, A.: 1951, *Z. Naturforsch.* **6a**, 47.
Biswas, S. and Fichtel, C. E.: 1965, *Space Sci. Rev.* **4**, 705.
Biswas, S. and Radhakrishnan, B.: 1973, *Solar Phys.* **28**, 211.
Blaha, M.: 1971, *Solar Phys.* **17**, 99.
Blaha, M., Kopecký, M., and Švestka, Z.: 1960, *Nature* **187**, 224.
Blaha, M., Kopecký, M., and Švestka, Z.: 1962, *Bull. Astron. Inst. Czech.* **13**, 85.
Blocker, N. K., Chambers, W. H., Fehlau, P. E., Fuller, J. C., Kunz, W. E., and Milkey, R. W.: 1971, *Solar Phys.* **20**, 400.
Blumenthal, G. R., Drake, G. W. F., and Tucker, W. H.: 1972, *Astrophys. J.* **172**, 205.
Boischot, A.: 1957, *Compt. Rend. Acad. Sci. Paris* **244**, 1326.
Boischot, A.: 1958, *Ann. Astrophys.* **21**, 273.
Boischot, A.: 1972, in C. de Jager (ed.), *Solar-Terrestrial Physics*, Part I, p. 87.
Boischot, A. and Clavelier, B.: 1967a, *Astrophys. Letters* **1**, 7.
Boischot, A. and Clavelier, B.: 1967b, *Ann. Astrophys.* **31**, 445.
Boischot, A. and Denisse, J. F.: 1957, *Compt. Rend. Acad. Sci. Paris* **245**, 2199.
Boischot, A. and Pick, M.: 1962, *J. Phys. Soc. Japan* **17**, A II, 203.
Boischot, A., Clavelier, B., Mangeney, A., and Lacombe, C.: 1967, *Compt. Rend. Acad. Sci. Paris* **265**, 1151.

Bonnelle, C., Senemaud, C., Senemaud, G., Chambe, G., Guionnet, M., Henoux, J. C., and Michard, R.: 1973, *Solar Phys.* **29**, 341.

Boorman, J. A., McLean, D. J., Sheridan, K. V., and Wild, J. P.: 1961, *Monthly Notices Roy. Astron. Soc.* **123**, 87.

Bratenahl, A. and Baum, P. J.: 1974, Preprint, IGPP-UCR-74-22.

Brinkman, A. C. and Shaw, M. L.: 1972, *Solar Phys.* **23**, 120.

Brown, J. C.: 1971, *Solar Phys.* **18**, 489.

Brown, J. C.: 1972a, *Solar Phys.* **25**, 158.

Brown, J. C.: 1972b, *Solar Phys.* **26**, 441.

Brown, J. C.: 1973a, *Solar Phys.* **28**, 151.

Brown, J. C.: 1973b, *Solar Phys.* **31**, 143.

Brown, J. C.: 1973c, *Solar Phys.* **32**, 227.

Brown, J. C.: 1974, *Solar Phys.* **36**, 371.

Brueckner, G. E.: 1974a, *IAU Symp.* **57**, 333.

Brueckner, G. E.: 1974b, Paper presented at the *IAU Symp.* **68**, Buenos Aires. (Abstract in *IAU Symp.* **68**, 105.)

Brueckner, G. E.: 1974c, *IAU Symp.* **68**, 135.

Brueckner, G. E.: 1975, paper presented at the FBS Workshop at Falmouth, Mass.

Brueckner, G. E., Kjeldseth Moe, O., and Van Hoosier, M. E.: 1975, Paper presented at the AAS Meeting, Boulder, Colorado.

Bruns, A. V., Prokofiev, V. K., and Severny, A. B.: 1970, *IAU Symp.* **36**, 256.

Bruzek, A.: 1951, *Z. Astrophys.* **28**, 277.

Bruzek, A.: 1952a, *Z. Astrophys.* **31**, 99.

Bruzek, A.: 1952b, *Z. Astrophys.* **31**, 111.

Bruzek, A.: 1957, *Z. Astrophys.* **42**, 76.

Bruzek, A.: 1958, *Z. Astrophys.* **44**, 183.

Bruzek, A.: 1964a, in *AAS-NASA Symp. on Physics of Solar Flares*, p. 301.

Bruzek, A.: 1964b, *Astrophys. J.* **140**, 746.

Bruzek, A.: 1967, *Solar Phys.* **2**, 451.

Bruzek, A.: 1968, *Nobel Symp.* **9**, 67.

Bruzek, A.: 1969a, *Annals IQSY* **3**, 82.

Bruzek, A.: 1969b, in C. de Jager and Z. Švestka (eds.), *Solar Flares and Space Research*, p. 61.

Bruzek, A.: 1972, *Solar Phys.* **26**, 94.

Bruzek, A. and DeMastus, H.: 1970, *Solar Phys.* **12**, 447.

Bryant, D. A., Cline, T. L., Desai, U. D., and McDonald, F. B.: 1962, *J. Geophys. Res.* **67**, 4983.

Bryant, D. A., Cline, T. L., Desai, U. D., and McDonald, F. B.: 1965, *Astrophys. J.* **141**, 478.

Bukata, R. P., Rao, U. R., McCracken, K. G., and Keath, E. P.: 1972, *Solar Phys.* **26**, 229.

Bumba, V.: 1958, *Izv. Krymsk. Astrofiz. Obs.* **19**, 105.

Bumba, V. and Howard, R.: 1965, *Astrophys. J.* **141**, 1502.

Burger, M. and Dijkstra, J. H.: 1972, *Solar Phys.* **24**, 395.

Burlaga, L. F.: 1967, *J. Geophys. Res.* **72**, 4449.

Burlaga, L. F. and Ogilvie, K. W.: 1969, *J. Geophys. Res.* **74**, 2815.

Cameron, A. G. W.: 1968, *Origin and Distribution of Elements*, Pergamon Press, New York, p. 125.

Canfield, R. C.: 1974, *Solar Phys.* **34**, 339.

Canfield, R. C. and Athay, R. G.: 1974, *Solar Phys.* **34**, 193.

Carlqvist, P.: 1969, *Solar Phys.* **7**, 377.

Carmichael, H.: 1962, *Space Sci. Rev.* **1**, 28.

Carmichael, H.: 1964, in *AAS-NASA Symp. on Physics of Solar Flares*, p. 451.

Caroubalos, C.: 1964, *Ann. Astrophys.* **27**, 333.

Caroubalos, C., Couturier, P., and Prokakis, T.: 1973, *Astron. Astrophys.* **23**, 131.

Carrington, R. C.: 1859, *Monthly Notices Roy. Astron. Soc.* **20**, 13.

Cartwright, B. G. and Mogro-Campero, A.: 1972, *Astrophys. J. Letters* **177**, L43.

Cartwright, B. G. and Mogro-Campero, A.: 1973, in *Symp. on High Energy Phenomena on the Sun*, NASA GSFC Preprint, p. 393.

Castelli, J. P. and Guidice, D. A.: 1971, in *Proceedings of COSPAR Symposium on Solar Particle Event of November 1969*, p. 27.

Castelli, J. P. and Richards, D. W.: 1971, *J. Geophys. Res.* **76**, 8409.

Castelli, J. P., Aarons, J., and Michael, G. L.: *J. Geophys Res.* **72**, 5491.

Castelli, J. P., Barron, W. R., and Badillo, V. L.: 1973, WDC-A Report UAG – 28, p. 183.

Castelli, J. P., Guidice, D. A., Forrest, D. J., and Babcock, R. R.: 1974, *J. Geophys. Res.* **79**, 889.

Catalano, C. P. and Van Allen, J. A.: 1973, *Astrophys. J.* **185**, 335.

Chambe, G.: 1971, *Astron. Astrophys.* **12**, 210.

Chapman, R. D. and Neupert, W. M.: 1975, Paper presented at the AAS Meeting, Boulder, Colorado.

Chase, R. C., Krieger, A. S., Švestka, Z., and Vaiana, G. S.: 1975, *Space Res.* **16**, in press.

Cheng, C. C.: 1972a, *Solar Phys.* **22**, 178.

Cheng, C. C.: 1972b, *Space Sci. Rev.* **13**, 3.

Cheng, C. C. and Widing, K. G.: 1975, Paper presented at the AAS Meeting, Boulder, Colorado.

Cherki, G. Mercier, J. P., Raviart, A., Treguer, L., Maccagni, D., Perotti, F., and Villa, G.: 1974, *Solar Phys.* **34**, 223.

Chin, Y. C., Lusignan, B. B., and Fung, P. C. W.: 1971, *Solar Phys.* **16**, 135.

Chistyakov, V.: 1959, *Soln. Dann.* No. 9, 81.

Chubb, T. A.: 1972, in C. De Jager (ed.), *Solar-Terrestrial Physics*, Part I, p. 99.

Chubb, T. A., Friedman, H., and Kreplin, R. W.: 1960, *J. Geophys. Res.* **65**, 1831.

Chubb, T. A., Friedman, H., and Kreplin, R. W.: 1961, Les spectres des astres dans l'ultraviolet lointain, in *Liège Symp.*, p. 216.

Chubb, T. A., Friedman, H., and Kreplin, R. W.: 1966, *J. Geophys. Res.* **71**, 3611.

Chupp, E. L.: 1971, *Space Sci. Rev.* **12**, 486.

Chupp, E. L., Forrest, D. J., Higbie, P. R., Suri, A. N., Tsai, C., and Dunphy, P. P.: 1973, *Nature* **241**, 5388.

Chupp, E. L., Forrest, D. L., and Suri, A. N.: 1973, in *Symp. on High Energy Phenomena on the Sun*, NASA GSFC Preprint, p. 285.

Chupp, E. L., Forrest, D. J., and Suri, A. N.: 1974, in D. E. Page (ed.), *Correlated Interplanetary and Magnetospheric Observations*, D. Reidel Publ. Co., Dordrecht, p. 519.

Cline, T. L. and McDonald, F. G.: 1968, *Solar Phys.* **5**, 507.

Cline, T. L., Holt, S. S., and Hones, E. W.: 1968, *J. Geophys. Res.* **73**, 434.

Coffey, H. E. (ed.): 1973, WDC–A Report UAG – 28, NOAA, Boulder.

Cohen, M. H.: 1960, *Astrophys. J.* **131**, 664.

Coppi, B.: 1975, *Astrophys. J.* **195**, 545.

Coppi, B. and Friedland, A. B.: 1971, *Astrophys. J.* **169**, 379.

Cortelessa, P., Di Benedetto, P., and Paizis, C.: 1971, *Solar Phys.* **20**, 474.

Covington, A. E.: 1973, *Solar Phys.* **33**, 439.

Covington, A. E. and Harvey, G. A.: 1961, *Phys. Rev. Letters* **6**, 51.

Cowan, R. D. and Widing, K. G.: 1973, *Astrophys. J.* **180**, 285.

Cowley, C. and Marlborough, J. M.: 1969 *Astrophys. J.* **158**, 803.

Cox, D. P. and Tucker, W. H.: 1969, *Astrophys. J.* **157**, 1157.

Craig, I. J. D.: 1973, *Solar Phys.* **31**, 197.

Craig, I. J. D., Culhane, J. L., Phillips, K. J. H., and Vesecky, J. F.: 1973, in *Symp. on High Energy Phenomena on the Sun*, NASA GSFC Preprint, p. 276.

Crawford, H. J., Price, P. B., and Sullivan, J. D.: 1972, *Astrophys. J. Letters* **175**, L149.

Crawford, H. J., Price, P. B., Cartwright, B. G., and Sullivan, J. D.: 1975, *Astrophys. J.* **195**, 213.

Croom, D. L.: 1971, *Solar Phys.* **19**, 152.

Croom, D. L. and Powell, R. J.: 1971, *Solar Phys.* **20**, 136.

Culhane, J. L.: 1969, *Monthly Notices Roy. Astron. Soc.* **144**, 375.

Culhane, J. L. and Acton, L. W.: 1970, *Monthly Notices Roy. Astron. Soc.* **151**, 141.

Culhane, J. L., Sanford, P. W., Shaw, M. L., Phillips, K. J. H., Willmore, A. P., Bowen, P. J., Pounds, K. A., and Smith, D. G.: 1969, *Monthly Notices Roy. Astron. Soc.* **145**, 435.

Culhane, J. L., Vesecky, J. F., and Phillips, K. J. H.: 1970, *Solar Phys.* **15**, 394.

Datlowe, D.: 1971, *Solar Phys.* **17**, 436.

Datlowe, D. W. and Lin, R. P.: 1973, *Solar Phys.* **32**, 459.

Datlowe, D. W., L'Heureux, J., and Meyer, P.: 1969, Paper presented at the *11th Int. Conf. on Cosmic Rays*, Budapest.

Datlowe, D. W., Hudson, H. S., and Peterson, L. E.: 1974a, *Solar Phys.* **35**, 193.

Datlowe, D. W., Elcan, M. J., and Hudson, H. S.: 1974b, *Solar Phys.* **39**, 155.
D'Azambuja, L.: 1942, *L'Astronomie* **56**, 97.
D'Azambuja, L.: 1949, *L'Astronomie* **63**, 241.
D'Azambuja, M.: 1947, *L'Astronomie* **61**, 114.
D'Azambuja, M. and D'Azambuja, L.: 1948, *Ann. Obs. Paris Meudon* **6**, fasc. 7.
De, B. R.: 1973, *Solar Phys.* **31**, 437.
De, B. R.: 1975, Paper presented at the AAS Meeting, Boulder, Colorado.
De Feiter, L. D.: 1964, in *AAS-NASA Symp. on Physics of Solar Flares*, p. 81.
De Feiter, L. D.: 1966, *Rech. Astron. Obs. Utrecht* **18**, 2.
De Feiter, L. D.: 1971, *Solar Phys.* **19**, 207.
De Feiter, L. D.: 1973, Paper presented at the *IAFE Flare Conference*, Buenos Aires.
De Feiter, L. D.: 1974, *Space Sci. Rev.* **16**, 3.
De Feiter, L. D. and de Jager, C.: 1973, *Solar Phys.* **28**, 183.
De Feiter, L. D. and Švestka, Z.: 1964, *Bull. Astron. Inst. Czech.* **15**, 117.
De Feiter, L. D. and Švestka, Z.: 1972, *Space Res.* **12**, 1547.
De Feiter, L. D. and Švestka, Z.: 1975, *Solar Phys.* **41**, 415.
De Groot, T.: 1970, *Solar Phys.* **14**, 176.
De Jager, C.: 1960, *Space Res.* **1**, 628.
De Jager, C.: 1962, *Space Sci. Rev.* **1**, 487.
De Jager, C.: 1965, *Ann. Astrophys.* **28**, 263.
De Jager, C.: 1967, *Solar Phys.* **2**, 327.
De Jager, C.: 1969, in C. de Jager and Z. Švestka (eds.), *Solar Flares and Space Research*, p. 1.
De Jager, C.: 1972, in C. de Jager (ed.), *Solar-Terrestrial Physics*, Part I, p. 1.
De Jager, C.: 1975, *Solar Phys.* **40**, 133.
DeMastus, H. L. and Stover, R. R.: 1967, *Publ. Astron. Soc. Pacific* **79**, 615.
Denisse, J. F.: 1960, in *Proc. XIIIth URSI General Assembly*, London.
Dere, K. P., Horan, D. M., and Kreplin, R. W.: 1974, *Solar Phys.* **36**, 459.
Deshpande, S. D. and Tandon, J. N.: 1972, *Astrophys. J.* **175**, 253.
Dietrich, W. F.: 1973, *Astrophys. J.* **180**, 955.
Dilworth, C., Maccagni, D., Perotti, F., Tanzi, E. G., Mercier, J. P., Raviart, A., Treguer, L., and Gros, M.: 1972, *Solar Phys.* **23**, 487.
Dizer, M.: 1969, *Solar Phys.* **10**, 416.
Dodge, J. C.: 1972, University of Colorado, Boulder (Thesis).
Dodson, H. W.: 1949, *Astrophys. J.* **110**, 382.
Dodson, H. W. and Cornwall, E.: 1939, *Publ. Am. Astron. Soc.* **9**, 258.
Dodson, H. W. and Hedeman, E. R.: 1949, *Astrophys. J.* **110**, 242.
Dodson, H. W. and Hedeman, E. R.: 1960, *Astron. J.* **65**, 51.
Dodson, H. W. and Hedeman, E. R.: 1964a, in *AAS-NASA Symp. on Physics of Solar Flares*, p. 15.
Dodson, H. W. and Hedeman, E. R.: 1964b, *J. Geophys. Res.* **69**, 3965.
Dodson, H. W. and Hedeman, E. R.: 1968a, *Solar Phys.* **4**, 229.
Dodson, H. W. and Hedeman, E. R.: 1968b, *Nobel Symp.* **9**, 37.
Dodson, H. W. and Hedeman, E. R.: 1969a, *Annals IQSY* **4**, 3.
Dodson, H. W. and Hedeman, E. R.: 1969b, *Solar Phys.* **9**, 278.
Dodson, H. W. and Hedeman, E. R.: 1970, *Solar Phys.* **13**, 401.
Dodson, H. W. and Hedeman, E. R.: 1971, WDC–A Report UAG – 14.
Dodson, H. W. and Hedeman, E. R.: 1973, WDC–A Report UAG – 28, Part I, p. 16.
Dodson, H. W. and McMath, R. R.: 1952, *Astrophys. J.* **115**, 78.
Dodson, H. W., Hedeman, E. R., and Owren, L.: 1953, *Astrophys. J.* **118**, 169.
Dodson, H. W., Hedeman, E. R., Kahler, S. W., and Lin, R. P.: 1969, *Solar Phys.* **6**, 294.
Dodson, H. W., Hedeman, E. R., and Miceli, M. R.: 1972, *Solar Phys.* **23**, 360.
Dolginova, Y. N. and Korchak, A. A.: 1968, *Soln. Dann.* No. 6, 81.
Dolginova, Y. N. and Odincova, I. N.: 1967, *Geomagnetizm i Aeronomiya* **7**, 958.
Donnelly, R. F.: 1969, *Astrophys. J. Letters* **158**, L165.
Donnelly, R. F.: 1970, ESSA Techn. Rep. ERL 169 – SDL 14.
Donnelly, R. F.: 1971, *Solar Phys.* **20**, 188.
Donnelly, R. F.: 1973, in *Symp. on High Energy Phenomena on the Sun*, NASA GSFC Preprint, p. 242.
Donnelly, R. F. and Hall, L. A.: 1973, *Solar Phys.* **31**, 411.

Donnelly, R. F., Wood, A. T., and Noyes, R. W.: 1973, *Solar Phys.* **29**, 107.

Doschek, G. A. and Meekins, J. F.: 1970, *Solar Phys.* **13**, 220.

Doschek, G. A. and Meekins, J. F.: 1973, *Solar Phys.* **28**, 517.

Doschek, G. A., Meekins, J. F., Kreplin, R. W., Chubb, T. A., and Friedman, H.: 1971, *Astrophys. J.* **164**, 165.

Doschek, G. A., Meekins, J. F., Kreplin, R.W., and Chubb, T. A.: 1972, Paper presented at the 14th COSPAR Meeting in Seattle, Washington, U.S.A.

Doschek, G. A., Meekins, J. F., and Cowan, R. D.: 1973, *Solar Phys.* **29**, 125.

Doschek, G. A., Feldman, U., Dere, K. P., Sandlin, G. D., Van Hoosier, M. E., Brueckner, G. E., Purcell, J. D., and Tousey, R.: 1975, *Astrophys. J.* **196**, L83.

Drake, J. F.: 1971, *Solar Phys.* **16**, 152.

Dryer, M.: 1974, *Space Sci. Rev.* **15**, 403.

Duggal, S. P. and Pomerantz, M. A.: 1973, *J. Geophys. Res.* **78**, 7205.

Dulk, G. A.: 1973, *Solar Phys.* **32**, 491.

Dulk, G. A. and Altschuler, M. D.: 1971, *Solar Phys.* **20**, 438.

Dulk, G. A., Altschuler, M. D., and Smerd, S. F.: 1971, *Astrophys. Letters* **8**, 235.

Dunckel, N., Helliwell, R. A., and Vesecky, J.: 1972, *Solar Phys.* **25**, 197.

Dungey, J. W.: 1958, *Cosmic Electrodynamics*, Cambridge Univ. Press, London.

Durgaprasad, N., Fichtel, C. E., Guss, D. E., and Reames, D. B.: 1968, *Astrophys. J.* **154**, 307.

Elgaroy, O. and Lyngstad, E.: 1972, *Astron. Astrophys.* **16**, 1.

Eliseev, G. F. and Moiseev, I. G.: 1965, *Izv. Krymsk. Astrofiz. Obs.* **34**, 3.

Eliseeva, L. A.: 1967, *Izv. Krymsk. Astrofiz. Obs.* **37**, 153.

Elliot, H.: 1964, *Planet. Space Sci.* **12**, 657.

Elliot, H.: 1969, in C. de Jager and Z. Švestka (eds.), *Solar Flares and Space Research*, p. 356.

Elliot, H.: 1972, in C. de Jager (ed.), *Solar-Terrestrial Physics*, Part I, p. 134.

Elliot, J., Ellison, M. A., and Reid, J. H.: 1960, *Monthly Notices Roy. Astron. Soc.* **121**, 463.

Ellison, M. A.: 1946, *Monthly Notices Roy. Astron. Soc.* **106**, 500.

Ellison, M. A.: 1949, *Monthly Notices Roy. Astron. Soc.* **109**, 3.

Ellison, M. A. and Hoyle, F.: 1947, *Observatory* **67**, 181.

Ellison, M. A., McKenna, S. M. P., and Reid, J. H.: 1960, *Dunsink Obs. Publ.* **1**, 1.

Ellison, M. A., McKenna, S. M. P., and Reid, J. H.: 1961, *Dunsink Obs. Publ.* **1**, 53.

Elwert, G.: 1964, in *AAS-NASA Symp. on Physics of Solar Flares*, p. 365.

Elwert, G. and Haug, E.: 1971, *Solar Phys.* **20**, 413.

Enomé, S.: 1964, *Publ. Astron. Soc. Japan* **16**, 135.

Enomé, S.: 1969, *Annals IQSY* **3**, 186.

Enomé, S.: 1971, WDC–A Report UAG – 12, Part I, p. 88.

Enomé, S., Kakinuma, T., and Tanaka, H.: 1969, *Solar Phys.* **6**, 428.

Epstein, S. and Taylor, H. P.: 1970, *Geochim. Cosmochim. Acta* **2**, Suppl. 1, 1085.

Erjushev, N. N. and Tsvetkov, L. I.: 1973, *Izv. Krymsk. Astrofiz. Obs.* **48**, 85.

Evans, J. W.: 1957, *Publ. Astron. Soc. Pacific* **69**, 421.

Evans, L. G., Fainberg, J., and Stone, R. G.: 1971, *Solar Phys.* **21**, 198.

Evans, L. G., Fainberg, J., and Stone, R. G.: 1973, *Solar Phys.* **31**, 501.

Eyles, C. J., Linney, A. D., and Rochester, G. K.: 1972, *Solar Phys.* **24**, 483.

Fainberg, J.: 1974, *IAU Symp.* **57**, 183.

Fainberg, J. and Stone, R. G.: 1970, *Solar Phys.* **15**, 433.

Fainberg, J. and Stone, R. G.: 1971, *Bull. Amer. Astron. Soc.* **3**, 482.

Fainberg, J. and Stone, R. G.: 1974, *Space Sci. Rev.* **16**, 145.

Fainberg, J., Evans, L. G., and Stone, R. G.: 1972, *Science* **178**, 743.

Fan, C. Y., Pick, M., Pyle, R., Simpson, J. A., and Smith, D. R.: 1968, *J. Geophys. Res.* **73**, 1555.

Feibelman, W. A.: 1974, *Solar Phys.* **39**, 409.

Feit, J.: 1973, *Solar Phys.* **29**, 211.

Fermi, E.: 1949, *Phys. Rev.* **75**, 1169.

Fermi, E.: 1954, *Astrophys. J.* **119**, 1.

Fichtel, C. E. and McDonald, F. B.: 1967, *Ann. Rev. Astron. Astrophys.* **5**, 351.

Fisher, R. R.: 1974, *Solar Phys.* **35**, 401.

Fisk, L. A. and Schatten, K. H.: 1972, *Solar Phys.* **23**, 204.

Fokker, A. D.: 1963a, *Bull. Astron Inst. Neth.* **17**, 84.

Fokker, A. D.: 1963b, *Space Sci. Rev.* **2**, 70.

Fokker, A. D.: 1967, *Solar Phys.* **2**, 316.

Fokker, A. D.: 1969, *Solar Phys.* **8**, 376.

Fokker, A. D.: 1970, *Solar Phys.* **11**, 92.

Fokker, A. D.: 1971a, *Solar Phys.* **19**, 472.

Fokker, A. D.: 1971b, *Proc. 2nd Meeting of CESRA*, Trieste, p. 110.

Fokker, A. D., Gon, T., Landre, E., and Roosen, J.: 1967, *Utrechtse Sterrekundige Overdrukken*, No. 23.

Fomichev, V. V. and Chertok, I. M.: 1965, *Astron. Zh.* **42**, 1256.

Fomichev, V. V. and Chertok, I. M.: 1967, *Astron. Zh.* **44**, 493.

Fontenla, J. M. and Seibold. J. R.: 1973, WDC–A Report UAG – 28, Part I, p. 150.

Fortini, T. and Torelli, M.: 1970, *Solar Phys.* **11**, 425.

Frank, L. A. and Gurnett, D. A.: 1972, *Solar Phys.* **27**, 446.

Freeman, F. F., Gabriel, A. H., Jones, B. B., and Jordan, C.: 1971, *Phil. Trans. Roy. Soc. London*, **A270**, 127.

Freier, P. S. and Webber, W. R.: 1963, *J. Geophys. Res.* **68**, 1605.

Friedman, H.: 1969, in C. de Jager and Z. Švestka (eds.), *Solar Flares and Space Research*, p. 87.

Friedman, H. and Kreplin, R. W.: 1969, *Annals IQSY* **3**, 78.

Friedman, M. and Hamberger, S. M.: 1968, *Astrophys. J.* **152**, 667.

Friedman, M. and Hamberger, S. M.: 1969a, *Solar Phys.* **8**, 104.

Friedman, M. and Hamberger, S. M.: 1969b, *Solar Phys.* **8**, 398.

Fritzová, L.: 1959, *Bull. Astron. Inst. Czech.* **10**, 145.

Fritzová, L.: 1960, *Bull. Astron. Inst. Czech.* **11**, 177.

Fritzová, L.: 1961, *Bull. Astron. Inst. Czech.* **12**, 254.

Fritzová, L.: 1964, *Bull. Astron. Inst. Czech.* **15**, 34.

Fritzová, L. and Chase, R. C.: 1975, paper presented at the AAS Meeting, San Diego, California.

Fritzová, L. and Švestka, Z.: 1966, *Bull. Astron. Inst. Czech.* **17**, 249.

Fritzová, L. and Švestka, Z.: 1967, *Solar Phys.* **2**, 87.

Fritzová, L. and Švestka, Z.: 1971, *Solar Phys.* **17**, 212.

Fritzová, L. and Švestka, Z.: 1973, *Solar Phys.* **29**, 417.

Frost, K. J.: 1969, *Astrophys. J. Letters* **158**, L159.

Frost, K. J.: 1974, *IAU Symp.* **57**, 421.

Frost, K. J. and Dennis, B. R.: 1971, *Astrophys. J.* **165**, 655.

Furth, H. P., Killeen, J. and Rosenbluth, M. N.: 1963, *Phys. Fluids* **6**, 459.

Gabriel, A. H. and Jordan, C.: 1969a, *Nature* **221**, 947.

Gabriel, A. H. and Jordan, C.: 1969b, *Monthly Notices Roy. Astron. Soc.* **145**, 241.

Gabriel, A. H. and Jordan, C.: 1970, *Phys. Letters* **32A**, 166.

Garrard, T. L., Stone, E. C., and Vogt, R. E.: 1973, in *Symp. on High Energy Phenomena on the Sun*, NASA GSFC Preprint, p. 341.

Geiss, J. Eberhardt, P., Buehler, F., Meister, J., and Signor, P.: 1970, *J. Geophys. Res.* **75**, 5972.

Gergely, T. E. and Erickson, W. D.: 1975, *Solar Phys.* **42**, 467.

Gergely, T. E. and Kundu, M. R.: 1974, *Solar Phys.* **34**, 433.

Gingerich, O. and De Jager, C.: 1968, *Solar Phys.* **3**, 5.

Gingerich, O., Noyes, R. W., Kalkofen, W., and Cuny, Y.: 1971, *Solar Phys.* **18**, 347.

Ginzburg, V. L. and Syrovatsky, S. I.: 1965, *Ann. Rev. Astron. Astrophys.* **3**, 297.

Ginzburg, V. L. and Zheleznyakov, V. V.: 1958, *Astron. Zh.* **35**, 694.

Ginzburg, V. L. and Zheleznyakov, V. V.: 1959, *Astron. Zh.* **36**, 233.

Giovanelli, R. G.: 1939, *Astrophys. J.* **89**, 555.

Giovanelli, R. G.: 1947, *Monthly Notices Roy. Astron. Soc.* **107**, 338.

Giovanelli, R. G.: 1948, *Monthly Notices Roy. Astron. Soc.* **108**, 163.

Giovanelli, R. G.: 1959, in R. M. Bracewell (ed.), *Paris Symp. on Radio Astronomy*, p. 214.

Giovanelli, R. and McCabe, M.: 1958, *Australian J. Phys.* **11**, 130.

Giovanelli, R. G. and Roberts, J. A.: 1958, *Australian J. Phys.* **11**, 353.

Gnevyshev, M. N.: 1963, *Astron. Zh.* **40**, 401.

Gnevyshev, M. N.: 1967, *Solar Phys.* **1**, 107.

Gold, T. and Hoyle, F.: 1960, *Monthly Notices Roy. Astron. Soc.* **120**, 89.

Goldberg-Rogozinskaya, N. M.: 1952, *Izv. Astron. Obs. Pulkovo* **22**, No. 170, 52.

Golub, L., Krieger, A. S., Silk, J. K., Timothy, A. F., and Vaiana, G. S.: 1974, *Astrophys. J. Letters* **189**, L93.

Golub, L., Krieger, A., Simon, R., Vaiana, G., and Timothy, A. F.: 1975, Paper presented at the AAS meeting, Boulder, Colorado.

Gopasyuk, S. I.: 1958, *Izv. Krymsk. Astrofiz. Obs.* **19**, 100.

Gopasyuk, S. I.: 1960, *Izv. Krymsk. Astrofiz. Obs.* **23**, 331.

Gopasyuk, S. I.: 1961, *Izv. Krymsk. Astrofiz. Obs.* **25**, 114.

Gopasyuk, S. I.: 1962, *Izv. Krymsk. Astrofiz. Obs.* **27**, 110.

Gopasyuk, S. I. and Křivský, L.: 1967, *Bull. Astron. Inst. Czech.* **18**, 135.

Gopasyuk, S. I. and Ogir, M. B.: 1963, *Izv. Krymsk. Astrofiz. Obs.* **30**, 185.

Gopasyuk, S. I., Ogir, M. B., Severny, A. B., and Shaposhnikova, E. F.: 1962, *Izv. Krymsk. Astrofiz. Obs.* **29**, 15.

Gordon, I. M.: 1954, *Doklady A. N. SSSR* **94**, 813.

Gosling, J. T. and Dulk, G. A.: 1975, Paper presented at the AAS Meeting, Boulder, Colorado.

Gosling, J. T., Asbridge, J. R., Bame, S. L., Hundhausen, A. J., and Strong, I. B.: 1967, *J. Geophys. Res.* **72**, 3357.

Gosling, J. T., Hildner, E., Mac Queen, R. M., Munro, R. H., Poland, A. I., and Ross, C. L.: 1974, *J. Geophys. Res.* **79**, 4581.

Gosling, J. T., Hildner, E., MacQueen, R. M., Munro, R. H., Poland, A. I., and Ross, C. L.: 1975, *Solar Phys.* **40** 439.

Gotwols, B. L.: 1972, *Solar Phys.* **25**, 232.

Green, R. M. and Sweet, P.: 1967, *Astrophys. J.* **147**, 1153.

Grevesse, N.: 1969, *Mem. Soc. Roy. Sci. Liège* **19**, 249.

Griem, H. R.: 1960, *Astrophys. J.* **132**, 883.

Griem, H. R.: 1962, *Astrophys. J.* **136**, 422.

Griem, H. R.: 1967, *Astrophys. J.* **147**, 1092.

Grineva, Y. I., Karev, V. I., Korneev, V. V., Krutov, V. V., Mandelstam, S. L., Vainstein, L. A., Vasilyev, B. N., and Zhitnik, I. A.: 1973a, *Solar Phys.* **29**, 441.

Grineva, Y. I., Karev, V. I., Korneev, V. V., Krutov, V. V., Mandelstam, S. L., Safronova, U. I., Urnov, A. M., Vainstein, L. A., Vasilyev, B. N., and Zhitnik, I.A.: 1973b, Lebedev Physical Institute, Laboratory of Spectroscopy, Moscow, Preprint No. 80.

Gros, M., Masse, P., Engelmann, J., and Barouch, E.: 1971, in Proceedings of *COSPAR Symp. on November 1969 Solar Particle Event*, p. 115.

Grossi Gallegos, H., Molnar, M., and Seibold, J. R.: 1971, *Solar Phys.* **16**, 120.

Gruber, D. E., Peterson, L. E., and Vette, J. I.: 1973, in *Symp. on High Energy Phenomena on the Sun*, NASA GSFC Preprint, p. 147.

Gulbrandsen, A.: 1974, *Planetary Space Sci.* **22**, 841.

Gurtovenko, E. A.: 1967, *Voprosy Astrofiziky*, Kiev, p. 74.

Gurtovenko, E. A. and Didychenko, E. I.: 1960, *Izv. Glav. Astron. Obs. Kiev* **3**, No. 2.

Guseynov, P. E.: 1967, *Astron. Zh.* **43**, 1159.

Guseynov, P. E. and Gasanalizade, A. G.: 1960, *Soln. Dann.* No. 9, 75.

Hachenberg, O.: 1959, *Rendiconti S. I. F.* **12**, 245.

Hachenberg, O. and Wallis, Z.: 1961, *Z. Astrophys.* **52**, 42.

Haddock, F.: 1958, *Proc. I.R.E.* **46**, 1.

Haddock, F. T. and Alvarez, H.: 1973, *Solar Phys.* **29**, 183.

Hagen, J. P. and Neidig, D. F.: 1971, *Solar Phys.* **18**, 305.

Hakura, Y.: 1974, *Solar Phys.* **39**, 493.

Hakura, Y. and Goh, T.: 1959, *J. Radio Res. Lab. Japan* **6**, 635.

Hall, D. E. and Sturrock, P. A.: 1967, *Phys. Fluids* **10**, 2620.

Hall, L. A.: 1971, *Solar Phys.* **21**, 167.

Hall, L. A. and Hinteregger, H. E.: 1969, in C. de Jager and Z. Švestka (eds.), *Solar Flares and Space Research*, p. 81.

Hansen, S. F., Hansen, R. T., and Garcia, C. J.: 1972, *Solar Phys.* **26**, 202.

Hartz, T. R.: 1969, *Planetary Space Sci.* **17**, 267.

Harvey, J. W.: 1969, NCAR Cooperative Thesis No. 17, Univ. of Colorado, Boulder.

Harvey, K. L. and Martin, S. F.: 1973, *Solar Phys.* **32**, 389.

Harvey, K. L., Martin, S. F., and Riddle, A. C.: 1974, *Solar Phys.* **36**, 151.

Harvey, K. L., Harvey, J. W., and Martin, S. F.: 1975, *Solar Phys.* **40**, 87.

Haug, E.: 1972, *Solar Phys.* **25**, 425.

Hauge, O. and Engvold, O.: 1970, Inst. Theor. Astrophys. Blindern Publ. No. 31.

Haurwitz, M. W.: 1968, *Astrophys. J.* **151**, 351.

Haurwitz, M. W., Yoshida, S., and Akasofu, S. I.: 1965, *J. Geophys. Res.* **70**, 2977.

Heitler, W.: 1954, *The Quantum Theory of Radiation*, Oxford Univ. Press, Cambridge, England.

Heristchi, D. and Trottet, G.: 1971, *Phys. Rev. Letters* **26**, 197.

Heristchi, D., Kangas, J., Kremser, G., Legrand, J. P., Masse, P., Palous, M., Pfotzer, G., Riedler, W., and Wilhelm, K.: 1969, *Annals IQSY* **3**, 267.

Herring, J. R. H.: 1974, *Solar Phys.* **39**, 1975.

Herring, J. R. H. and Craig, I. J. D.: 1973, *Solar Phys.* **28**, 169.

Heyvaerts, J.: 1974, *Solar Phys.* **38**, 419.

Hinteregger, H. E. and Hall, L. A.: 1969, *Solar Phys.* **6**, 175.

Hirayama, T.: 1961, *Publ. Astron. Soc. Japan* **13**, 152.

Hirayama, T.: 1974, *Solar Phys.* **34**, 323.

Hirayama, T. and Endler, F.: 1975, Paper presented at the AAS Meeting, Boulder, Colorado.

Hirshberg, J., Alksne, A., Colburn, D. S., Bame, S. J., and Hundhausen, A. J.: 1970, *J. Geophys. Res.* **75**, 1.

Hirshberg, J., Asbridge, J. R., and Robbins, D. E.: 1971, *Solar Phys.* **18**, 313.

Hirshberg, J., Bame, S. J., and Robbins, D. E.: 1972, *Solar Phys.* **23**, 467.

Holt, S. S. and Cline, T. L.: 1968, *Astrophys. J.* **154**, 1027.

Holt, S. S. and Ramaty, R.: 1969, *Solar Phys.* **8**, 119.

Horan, D. M.: 1971, *Solar Phys.* **21**, 188.

Horan, D. M. and Kreplin, R. W.: 1969, Paper presented at XVth Electromagnetic Wave Propagation Committee Technical Symp. on Ionospheric Forecasting.

Howard, R.: 1963, *Astrophys. J.* **138**, 1312.

Howard, R. and Babcock, H. W.: 1960, *Astrophys. J.* **132**, 218.

Howard, R. and Severny, A. B.: 1963, *Astrophys. J.* **137**, 1242.

Howard, R., Cragg, T., and Babcock, H. W.: 1959, *Nature* **184**, 351.

Hsieh, K. C. and Simpson, J. A.: 1970, *Astrophys. J. Letters* **162**, L191.

Hudson, H. S.: 1972, *Solar Phys.* **24**, 414.

Hudson, H. S.: 1973, in *Symp. on High Energy Phenomena on the Sun*, NASA GSFC Preprint, p. 207.

Hudson, H. S. and Ohki, K.: 1972, *Solar Phys.* **23**, 155.

Hudson, H. S., Peterson, L. E., and Schwartz, D. A.: 1969, *Astrophys, J.* **157**, 389.

Hultqvist, B.: 1969, in C. de Jager and Z. Švestka (eds.), *Solar Flares and Space Research*, p. 215.

Hundhausen, A. J.: 1970, in V. Manno and D. F. Page (eds.), *Intercorrelated Satellite Observations Related to Solar Events*, p. 111.

Hundhausen, A. J.: 1972a, in E. R. Dyer *et al.* (eds.), *Solar-Terrestrial Physics, Part II*, p. 1.

Hundhausen, A. J.: 1972b, *Coronal Expansion and Solar Wind*, Springer Verlag, New York.

Hyder, C. L.: 1964, *Astrophys. J.* **140**, 817.

Hyder, C. L.: 1967a, *Solar Phys.* **2**, 49.

Hyder, C. L.: 1967b, *Solar Phys.* **2**, 267.

Hyder, C. L.: 1973, in *Symp. on High Energy Phenomena on the Sun*, Preprint, p. 19.

Hyder, C. L., Epstein, G. L., and Hobbs, R. W.: 1973, *Astrophys. J.* **185**, 985.

Ifedili, S. O.: 1974, *Solar Phys.* **39**, 233.

Iospha, B. A.: 1962, *Geomagn. Aeron.* **2**, 149.

Ioshpa, B. A.: 1963, *Geomagn. Aeron.* **3**, 903.

Jaggi, R. K.: 1964, in *AAS-NASA Symp. on Physics of Solar Flares*, p. 419.

Janssens, T. J.: 1972, *Solar Phys.* **27**, 149.

Janssens, T. J. and White, K. P., III: 1970, *Solar Phys.* **11**, 299.

Janssens, T. J., White, K. P., III, and Broussard, R. M.: 1973, *Solar Phys.* **31**, 207.

Jefferies, J. T.: 1957, *Monthly Notices Roy. Astron. Soc.* **117**, 493.

Jefferies, J. T. and Orrall, F. Q.: 1961a, *Astrophys. J.* **133**, 946.
Jefferies, J. T. and Orrall, F. Q.: 1961b, *Astrophys. J.* **133**, 963.
Jefferies, J. T. and Orrall, F. Q.: 1963, in C. de Jager (ed.), *The Solar Spectrum*, p. 254.
Jefferies, J. T. and Orrall, F. Q.: 1964, in *AAS-NASA Symp. on Physics of Solar Flares*, p. 71.
Jefferies, J. T. and Orrall, F. Q.: 1965a, *Astrophys. J.* **141**, 505.
Jefferies, J. T. and Orrall, F. Q.: 1965b, *Astrophys. J.* **141**, 519.
Jefferies, J. T., Smith, E. v. P., and Smith, H. J.: 1959, *Astrophys. J.* **129**, 146.
Jokipii, J. R. and Parker, E. N.: 1969, *Astrophys. J.* **155**, 777.

Kahler, S. W.: 1969, *Solar Phys.* **8**, 166.
Kahler, S. W.: 1971, *Astrophys. J.* **164**, 365.
Kahler, S. W.: 1972, *Solar Phys.* **25**, 435.
Kahler, S. W.: 1973a, *Solar Phys.* **32**, 477.
Kahler, S. W.: 1973b, *Solar Phys.* **23**, 239.
Kahler, S. W. and Kreplin, R. W.: 1970, *Solar Phys.* **14**, 372.
Kahler, S. W. and Kreplin, R. W.: 1971, *Astrophys. J.* **168**, 531.
Kahler, S. W. and Lin, R. P.: 1969, *IQSY Annals* **3**, 299.
Kahler, S. W., Primbsch, J. H., and Anderson, K. A.: *Solar Phys.* **2**, 179.
Kahler, S. W., Meekins, J. F., Kreplin, R. W., and Bowyer, C. S.: 1970, *Astrophys. J.* **162**, 293.
Kahler, S. W., Krieger, A. S., and Vaiana, G. S.: 1975, *Astrophys. J. Letters* **199**, L57.
Kai, K.: 1965a, *Publ. Astron. Soc. Japan* **17**, 294.
Kai, K.: 1965b, *Publ. Astron. Soc. Japan* **17**, 309.
Kai, K.: 1967, *Proc. Astron. Soc. Australia* **1**, 49.
Kai, K.: 1969, *Proc. Astron. Soc. Australia* **1**, 189.
Kai, K.: 1970, *Solar Phys.* **11**, 310.
Kai, K. and McLean, D. J.: 1968, *Proc. Astron. Soc. Australia*, No. 4.
Kai, K. and Sheridan, K. V.: 1974, *IAU Symp.* **57**, 97.
Kakinuma, T.: 1958, *Proc. Res. Inst. Atmospherics* **5**, 71.
Kakinuma, T., Yamashita, T., and Enomé, S.: 1969, *Proc. Res. Inst. Atmospherics* **16**, 127.
Kandel, R. S., Papagiannis, M. D., and Strauss, F. M.: 1971, *Solar Phys.* **21**, 176.
Kane, S. R.: 1969, *Astrophys. J. Letters* **157**, L139.
Kane, S. R.: 1971, *Astrophys. J.* **170**, 587.
Kane, S. R.: 1972, *Solar Phys.* **27**, 174.
Kane, S. R.: 1973, in *Symp. on High Energy Phenomena on the Sun*, NASA GSFC Preprint, p. 55.
Kane, S. R.: 1974, *Space Sci. Lab. Series* 14, Issue 76.
Kane, S. R.: 1975, Paper presented at AAS Meeting, Boulder, Colorado.
Kane, S. R. and Anderson, K. A.: 1970, *Astrophys. J.* **162**, 1003.
Kane, S. R. and Donnelly, R. F.: 1971, *Astrophys. J.* **164**, 151.
Kane, S. R. and Hudson, H. S.: 1970, *Solar Phys.* **14**, 414.
Kane, S. R. and Lin, R. P.: 1972, *Solar Phys.* **23**, 457.
Kane, S. R. and Winckler, J. R.: 1969, *Solar Phys.* **6**, 304.
Kane, S. R., Kreplin, R. W., Martres, M. J., Pick, M., and Soru-Escaut, I.: 1974, *Solar Phys.* **38**, 483
Kaplan, S. A. and Tsytowitch, V. N.: 1967, *Astron. Zh.* **44**, 1194.
Karzas, W. J. and Latter, R.: 1961, *Astrophys J. Suppl.* **6**, 167.
Kassinsky, V. V. and Krat, V. A.: 1973, *Solar Phys.* **31**, 219.
Kastner, S. O., Neupert, W. M., and Swartz, M.: 1974, *Astrophys. J.* **191**, 261.
Kawabata, K.: 1960, *Rep. Iono. Space Res. Japan* **14**, 403.
Kawabata, K.: 1966, *Rep. Iono. Space Res. Japan* **20**, 118.
Kazachevskaya, T. V.: 1958, *Izv. Krymsk. Astrofiz. Obs.* **20**, 80.
Kazachevskaya, T. V. and Severny, A. B.: 1958, *Izv. Krymsk. Astrofiz. Obs.* **19**, 46.
Keath, E. P., Bukata, R. P., McCracken, K. G., and Rao, U. R.: 1971, *Solar Phys.* **18**, 503.
Kellogg, P. J., Lai, J. C., and Cartwright, D. G.: 1973, WDC–A Report UAG-28, p. 288.
Kelly, P. T. and Rense, W. A.: 1972, *Solar Phys.* **26**, 431.
Khokhlova, V. L.: 1958, *Izv. Krymsk. Astrofiz. Obs.* **21**, 190.
Kiepenheuer, K. O.: 1960, *Proc. Scuola Internaz. Fisica Enrico Fermi*, p. 39.
Kippenhahn, R. and Schlüter, A.: 1957, *Z. Astrophys.* **34**, 36.
Kirsch, E.: 1973, *Solar Phys.* **28**, 233.

Kirsch, E. and Münch, J. W.: 1974, *Solar Phys.* **39**, 459.

Kleczek, J.: 1953, *Bull. Astron. Inst. Czech.* **4**, 9.

Kleczek, J.: 1957, *Bull. Astron. Inst. Czech.* **8**, 120.

Kleczek, J.: 1958, *Bull. Astron. Inst. Czech.* **9**, 115.

Kleczek, J.: 1964, in *AAS-NASA Symp. on Physics of Solar Flares*, p. 77.

Koch, H. W. and Motz, J. W.: 1959, *Rev. Mod. Phys.* **31**, 920.

Koechlin, Y., Raviart, A., Treguer, L., Bland, C. J., Degliantoni, G., Dilworth, C., Maccagni, D., and Tanzi, E. G.: 1969, Paper presented at the *11th Int. Conf. on Cosmic Rays*, Budapest.

Komesaroff, M.: 1958, *Australian J. Phys.* **11**, 201.

Kopecký, M., Letfus, V., Blaha, M. and Švestka, Z.: 1963, *Bull. Astron. Inst. Czech.* **14**, 146.

Kopp, R. A.: 1972, *Proc. Asilomar Conf. on Solar Wind*, p. 252.

Korchak, A. A.: 1965a, *Geomagn. Aeron.* **5**, 32 and 601.

Korchak, A. A.: 1967, *Astron. Zh.* **44**, 328.

Korchak, A. A.: 1971, *Solar Phys.* **18**, 284.

Korchak, A. A.: 1972, *Space Res.* **13**, 731.

Korchak, A. A.: 1974, *Comments Astrophys. Space Phys.* **6**, 57.

Kostjuk, N. D. and Pikelner, S. B.: 1974, *Astron. Zh.* **51**, 1002.

Koval, A. N.: 1967, *Izv. Krymsk. Astrofiz. Obs.* **37**, 62.

Koval, A. N.: 1972, *Izv. Krymsk. Astrofiz. Obs.* **44**, 94.

Koval, A. N. and Steshenko, N. V.: 1963, *Izv. Krymsk. Astrofiz. Obs.* **30**, 200.

Krat, V. A. and Sobolev, V. M.: 1960, *Izv. Astron. Obs. Pulkovo* **21**, No. 163, 2.

Kreplin, R. W.: 1961, *Ann. Geophys.* **17**, 151.

Krieger, A. S., Vaiana, G. S., and van Speybroeck, L. P.: 1971, *IAU Symp.* **43**, 397.

Krieger, A. S., Paolini, F., Vaiana, G. S., and Webb, D.: 1972, *Solar Phys.* **22**, 150.

Krieger, A. S., Timothy, A. F., and Roelof, E. C.: 1973, *Solar Phys.* **29**, 505.

Krieger, A. S., Timothy, A. F., Vaiana, G. S., Lazarus, A. J., and Sullivan, J. D.: 1974, Paper presented at the *Third Solar Wind Conference*.

Krimigis, S. M.: 1973, in *Symp. on High Energy Phenomena on the Sun*, NASA GSFC Preprint, p. 479.

Křivský, L.: 1963, *Bull. Astron. Inst. Czech.* **14**, 77.

Křivský, L.: 1968, *IAU Symp.* **35**, 465.

Křivský, L.: 1969, *Solar Phys.* **9**, 194.

Křivský, L. and Krüger, A.: 1966, *Bull. Astron. Inst. Czech.* **17**, 243.

Křivský, L. and Nestorov, G.: 1968, *Bull. Astron. Inst. Czech.* **19**, 197.

Křivský, L., Valníček, B., Böhme, A., Fürstenberg, F., and Krüger, A.: 1973, Paper presented at the 7th Regional Consultation on Solar Physics, Starý Smokovec, Czechoslovakia.

Krüger, A.: 1969a, *Annals IQSY* **3**, 70.

Krüger, A.: 1969b, *Geodät. Geophys. Veröff. II* **10**, 41.

Krüger, A.: 1972a, *Physics of Solar Continuum Radio Bursts*, Akademie Verlag Berlin.

Krüger, A.: 1972b, *Solar Phys.* **27**, 217.

Kubeš, P.: 1964, *Publ. Astron. Inst. Prague* **51**, 89.

Kubota, J., Tamenaga, T., Kawaguchi, I., and Kitai, R.: 1974, *Solar Phys.* **38**, 389.

Kuiper, T. B.: 1973, *Solar Phys.* **33**, 461.

Kuiper, T. B. and Pasachoff, J. M.: 1973, *Solar Phys.* **28**, 187.

Kulsrud, R. M. and Ferrara, A.: 1971, *Astrophys. Space Sci.* **12**, 302.

Kundu, M. R.: 1961, *J. Geophys. Res.* **66**, 4308.

Kundu, M. R.: 1963, *Space Sci. Rev.* **2**, 438.

Kundu, M. R.: 1964, in *AAS-NASA Symp. on Physics of Solar Flares*, p. 335.

Kundu, M. R.: 1965, *Solar Radio Astronomy*, Interscience Publishers, New York.

Kundu, M. R. and Erickson, W. C.: 1974, *Solar Phys.* **36**, 179.

Kundu, M. R. and Firor, J. W.: 1961, *Astrophys. J.* **134**, 389.

Kundu, M. R. and McCullough, T. P.: 1972, *Solar Phys.* **27**, 182.

Kundu, M. R. and Sou-Yang, L.: 1973, *Solar Phys.* **29**, 409.

Kundu, M. R., Velusamy, T., and Becker, R. H.: 1974, *Solar Phys.* **34**, 217.

Künzel, H.: 1960, *Astron. Nachr.* **285**, 271.

Kurochka, L. N.: 1969, *Astron. Zh.* **46**, 85.

Kurochka, L. N.: 1970, *Astron. Zh.* **47**, 111.

Kurochka, L. N. and Maslennikova, L. B.: 1970, *Solar Phys.* **11**, 33.

Labrum, N. R. and Smerd, S. F.: 1968, *Proc. Astron. Soc. Australia* **1**, 140.
Labrum, N. R. and Stewart, R. T.: 1970, *Proc. Astron. Soc. Australia* **1**, 316.
Lambert, D.: 1967, *Nature* **215**, 43.
Landini, M. and Monsignori-Fossi, B. C.: 1970a, *Astron. Astrophys.* **6**, 468.
Landini, M. and Monsignori-Fossi, B. C.: 1970b, *Mem. Soc. Astron. Italiana* **41**, 467.
Landini, M., Monsignori-Fossi, B. C., and Pallavicini, R.: 1972, *Solar Phys.* **27**, 164.
Landini, M., Monsignori-Fossi, B. C., and Pallavicini, R.: 1973, *Solar Phys.* **29**, 93.
Lanzerotti, L. J.: 1969, WDC–A Report UAG-5, p. 56.
Lanzerotti, L. J.: 1973, in *Symp. on High Energy Phenomena on the Sun*, NASA GSFC Preprint, p. 427.
Lanzerotti, L.J. and Maclennan, C. G.: 1971, in Proceedings of *COSPAR Symp. on November 1969 Solar Particle Event*, p. 85.
Lanzerotti, L. J., Maclennan, C. G., and Graedel, T. E.: 1972, *Astrophys. J.* **173**, 39.
Lazarus, A. J. and Binsack, J. H.: 1969, *Annals IQSY* **3**, 378.
LeBlanc, Y., Kuiper, T. B. H., and Hansen, S. F.: 1974, *Solar Phys.* **37**, 215.
Leroy, J. L.: 1962, *Ann. Astrophys.* **25**, 127.
Leroy, J. L.: 1969, *Solar Phys.* **7**, 221.
Letfus, V.: 1964, *Bull. Astron. Inst. Czech.* **15**, 211.
Levine, R. H. and Altschuler, M. D.: 1974, *Solar Phys.* **36**, 345.
Levitzky, L. S.: 1972, *Izv. Krymsk. Astrofiz. Obs.* **41–42**, 203.
Levy, G. S., Sato, T., Seidel, B. L., Stelzried, C. T., Ohlson, J. E., and Rusch, W. V. T.: 1969, *Science* **166**, 596.
Lewis, M.: 1961, *Phys. Rev.* **121**, 501.
L'Heureux, J., Fan, G. Y., and Meyer, P.: 1972, *Astrophys. J.* **171**, 363.
Lin, R. P.: 1970a, *Solar Phys.* **12**, 266.
Lin, R. P.: 1970b, *Solar Phys.* **15**, 453.
Lin, R. P.: 1971, Paper presented at the Seminar on the Acceleration of Particles, Leningrad.
Lin, R. P.: 1973, in *Symp. on High Energy Phenomena on the Sun*, NASA GSFC Preprint, p. 439.
Lin, R. P.: 1974, *Space Sci. Rev.* **16**, 189.
Lin, R. P. and Anderson, K. A.: 1967, *Solar Phys.* **1**, 446.
Lin, R. P. and Hudson, H. S.: 1971, *Solar Phys.* **17**, 412.
Lin, R. P., Kahler, S. W., and Roelof, E. C.: 1968, *Solar Phys.* **4**, 338.
Lin, R. P., Evans, L. G., and Fainberg, J.: 1973, *Space Sci. Lab. Series* 14, Issue 26.
Lincoln, J. V. and Leighton, H. I.: 1972, WDC–A Report UAG-21, NOAA, Boulder.
Lingenfelter, R. E.: 1969, *Solar Phys.* **8**, 341.
Lingenfelter, R. E. and Ramaty, R.: 1967, *High Energy Nuclear Reactions in Astrophysics*, W. A. Benjamin, New York, p. 99.
Lingenfelter, R. E., Flamm, E. J., Canfield, E. H., and Kellman, S.: 1965, *J. Geophys. Res.* **70**, 4087.
Linsky, J. L., Glackin, D. L., Chapman, R. D., Neupert, W. M., and Thomas, R. J.: 1975, Paper presented at the AAS Meeting, Boulder, Colorado.
Lites, B. L.: 1973, University of Colorado (Thesis).
Lockwood, J. A.: 1968, *J. Geophys. Res.* **73**, 4247.
Lodén, K.: 1958, *Arkiv Astron.* **2**, 153.
Loughhead, R. E., Roberts, J. A., and McCabe, M. K.: 1957, *Australian J. Phys.* **10**, 483.
Low, B. C.: 1973, *Astrophys. J.* **181**, 209.
Lüst, R. and Zirin, H.: 1960, *Z. Astrophys.* **49**, 8.

Machado, M. E.: 1971, *Solar Phys.* **17**, 389.
Machado, M. E. and Linsky, J. L.: 1975, *Solar Phys.* **42**, 395.
Machado, M. E. and Rust, D. M.: 1974, *Solar Phys.* **38**, 499.
Machado, M. E. and Seibold, J. R.: 1973, *Solar Phys.* **29**, 75.
Machado, M. E., Grossi-Gallegos, H., and Silva, A. F.: 1972, *Solar Phys.* **25**, 402.
Malinovsky, M. and Heroux, L.: 1973, *Astrophys. J.* **181**, 1009.
Malitson, H. H. and Erickson, W. C.: 1966, *Astrophys. J.* **144**, 337.
Malitson, H. H., Fainberg, J., and Stone, R. G.: 1974, *IAU Symp.* **57**, 349.
Malville, J. M.: 1961, University of Colorado (Doctoral Thesis).
Malville, J. M.: 1967, *Solar Phys.* **2**, 484.
Malville, J. M. and Moreton, G. E.: 1963, *Publ. Astron. Soc. Pacific* **75**, 176.

Malville, J. M. and Smith, S. F.: 1963, *J. Geophys. Res.* **68**, 3181.

Mandelstam, S. L.: 1974, Report to COSPAR, Sao Paulo.

Martres, M. J.: 1968, *IAU Symp.* **35**, 25.

Martres, M. J. and Pick, M.: 1962, *Ann. Astrophys.* **25**, 4.

Martres, M. J., Michard, R., and Soru-Iscovici, I.: 1966, *Ann. Astrophys.* **29**, 249.

Martres, M. J., Michard, R., Soru-Iscovici, I., and Tsap, T.: 1968a, *IAU Symp.* **35**, 318.

Martres, M. J., Michard, R., Soru-Iscovici, I., and Tsap, T.: 1968b, *Solar Phys.* **5**, 187.

Martres, M. J., Soru-Escaut, I., and Rayrole, J.: 1971, *IAU Symp.* **43**, 435.

Martyn, D. F.: 1947, *Nature* **159**, 26.

Matsuura, O. T. and Nave, M. F. F.: 1971, *Solar Phys.* **16**, 417.

Maxwell, A. and Rinehart, R.: 1974, *Solar Phys.* **37**, 437.

Maxwell, A. and Swarup, G.: 1958, *Nature* **181**, 36.

Maxwell, A. and Thompson, A. R.: 1962, *Astrophys. J.* **135**, 138.

Maxwell, A., Defouw, R. J., and Cummings, P.: 1964, *Planetary Space Sci.* **12**, 435.

Mayfield, E. B.: 1971, *IAU Symp.* **43**, 376.

McCabe, M.: 1970, *Solar Phys.* **12**, 115.

McCabe, M.: 1973, *Solar Phys.* **30**, 439.

McClinton, A. T.: 1968, *NRL Space Research Seminar*, p. 63.

McCracken, K. G.: 1959, *Nuovo Cimento* **13**, 1081.

McCracken, K. G.: 1969, in C. de Jager and Z. Švestka (eds.), *Solar Flares and Space Research*, p. 202.

McCracken, K. G. and Palmeira, R. A. R.: 1960, *J. Geophys. Res.* **65**, 2673.

McCracken, K. G. and Rao, U. R.: 1970, *Space Sci. Rev.* **11**, 155.

McCracken, K. G., Rao, U. R., and Bukata, R. P.: 1967, *J. Geophys. Res.* **73**, 4293.

McDonald, F. B.: 1970, in V. Manno and D. E. Page (eds.), *Intercorrelated Satellite Observations Related to Solar Events*, p. 34.

McDonald, F. B. and Dessai, U. D.: 1971, *J. Geophys. Res.* **76**, 808.

McDonald, F. B., Cline, T. L., and Simnett, G. M.: 1972, *J. Geophys. Res.* **77**, 2213.

McIntosh, P. S.: 1969, WDC–A Report UAG-5, p. 14.

McIntosh, P. S.: 1970, WDC–A Report UAG-8.

McIntosh, P. S. and Donnelly, R. F.: 1972, *Solar Phys.* **23**, 444.

McKenna, S. M. P.: 1965, *Monthly Notices Roy. Astron. Soc.* **129**, 437.

McKenna-Lawler, S. M. P.: 1968, *Astrophys. J.* **153**, 367.

McKenzie, D. L.: 1972, *Astrophys. J.* **175**, 481.

McKenzie, D. L.: 1975, *Solar Phys.* **40**, 183.

McKenzie, D. L., Datlowe D. W., and Peterson, L. E.: 1972, *Solar Phys.* **28**, 175.

McLean, D. J.: 1967, *Proc. Astron. Soc. Australia* **1**, 47.

McLean, D. J.: 1970, *Proc. Astron. Soc. Australia* **1**, 315.

McLean, D. J., Sheridan, K. V., Stewart, R. T., and Wild, J. P.: 1971, *Nature* **234**, 140.

McMath, R. R., Mohler, O. C., and Dodson, H. W.: 1960, *Proc. Natl. Acad. Sci.* **46**, 165.

Meekins, J. F. and Doschek, G. A.: 1970, *Solar Phys.* **13**, 213.

Meekins, J. F., Kreplin, R. W., Chubb, T. A., and Friedman, H.: 1968, *Science* **162**, 891.

Meekins, J. F., Doschek, G. A., Friedman, H., Chubb, T. A., and Kreplin, R. W.: 1970, *Solar Phys.* **13**, 198.

Melrose, D. B.: 1970a, *Australian J. Phys.* **23**, 871.

Melrose, D. B.: 1970b, *Australian J. Phys.* **23**, 885.

Melrose, D. B.: 1974a, *Solar Phys.* **37**, 353.

Melrose, D. B.: 1974b, *Solar Phys.* **38**, 205.

Mercier, C.: 1973, *Solar Phys.* **33**, 177.

Mercier, C.: 1974, *Solar Phys.* **39**, 193.

Mewe, R.: 1972a, *Solar Phys.* **22**, 114.

Mewe, R.: 1972b, *Solar Phys.* **22**, 459.

Meyer, F.: 1968, *IAU Symp.* **35**, 485.

Meyer, P. and Vogt, R. E.: 1962, *Phys. Rev. Letters* **8**, 387.

Michalitsanos, A. G. and Kupferman, P.: 1974, *Solar Phys.* **36**, 403.

Michard, R.: 1959, *Ann. Astrophys.* **22**, 887.

Michard, R.: 1971, *IAU Symp.* **43**, 359.

Michard, R., Servajean, R., and Laborde, G.: 1959, *Ann. Astrophys.* **22**, 877.
Michard, R., Mouradian, Z., and Semel, M.: 1961, *Ann. Astrophys.* **24**, 54.
Milkey, R. W.: 1971, *Solar Phys.* **16**, 465.
Minaeva, L. A.: 1968, *Astron. Zh.* **45**, 578.
Minaeva, L. A. and Sobelman, I. I.: 1968, *J. Quant. Spectr. Radiative Transfer* **8**, 783.
Minaeva, L. A., Sorochenko, P. L., and Sobelman, I. I.: 1967, *Astron. Zh.* **44**, 995.
Mogro-Campero, A. and Simpson, J. A.: 1972a, *Astrophys. J. Letters* **171**, L5.
Mogro-Campero, A. and Simpson, J. A.: 1972b, *Astrophys. J. Letters* **177**, L37.
Moreton, G. E.: 1964, *Astrophys. J.* **69**, 145.
Moreton, G. E. and Ramsey, H. E.: 1960, *Publ. Astron. Soc. Pacific* **72**, 357.
Moreton, G. E. and Severny, A. B.: 1968, *Solar Phys.* **3**, 282.
Mullan, D. J.: 1973, *Astrophys. J.* **185**, 353.
Munro, R. H.: 1975, Paper presented at the AAS Meeting, Boulder, Colorado.
Munro, R. H., Dupree, A. K., and Withbroe, G. L.: 1971, *Solar Phys.* **19**, 347.
Mustel, E. R.: 1955, *Izv. Krymsk. Astrofiz. Obs.* **15**, 54.

Nagasawa, S., Takakura, T., Tanaka, H., and Koyama, H.: 1961, *Publ. Astron. Soc. Japan* **13**, 129.
Najita, K. and Orrall, F. Q.: 1970, *Solar Phys.* **15**, 176.
Nakada, M. P., Neupert, W. M., and Thomas, R. J.: 1974, *Solar Phys.* **37**, 429.
Nakagawa, Y. and Raadu, M. A.: 1972, *Solar Phys.* **25**, 127.
Nakagawa, Y., Wu, S. T., and Han, S. M.: 1973, *Solar Phys.* **30**, 111.
Ness, N. F.: 1965, *J. Geophys. Res.* **70**, 2989.
Neupert, W. M.: 1968, *Astrophys. J. Letters* **153**, L59.
Neupert, W. M.: 1969, *Ann. Rev. Astron. Astrophys.* **7**, 121.
Neupert, W. M.: 1971, *Solar Phys.* **18**, 474.
Neupert, W. M. and Swartz, M.: 1970, *Astrophys. J. Letters* **160**, L189.
Neupert, W. M., Gates, W., Swartz, M., and Young, R.: 1967, *Astrophys. J.* **149**, L79.
Neupert, W. M., Swartz, M., and White, W. A.: 1968, *Astron. J.* **73**, S73.
Neupert, W. M., White, W. A., Gates, W. J., Swartz, M., and Young, R. M.: 1969, *Solar Phys.* **6**, 183.
Neupert, W. M., Swartz, M., and Kastner, S. O.: 1973, *Solar Phys.* **31**, 171.
Neupert, W. M., Thomas, R. J., and Chapman, R. D.: 1974, *Solar Phys.* **34**, 349.
Newkirk, G.: 1957, *Ann. Astrophys.* **20**, 127.
Newkirk, G.: 1973, in *Symp. on High Energy Phenomena on the Sun*, NASA GSFC Preprint, p. 453.
Newkirk, G., Hansen, R. T., and Hansen, S.: 1969, *Annals IQSY* **3**, 49.
Newkirk, G., Trotter, D. E., Altschuler, M. D., and Howard, R.: 1972, A Microfilm Atlas of Magnetic Fields in the Solar Corona, NCAR-TN/STR-85.
Nishi, K. and Nakagomi, T.: 1963, *Publ. Astron. Soc. Japan* **15**, 56.
Noë, J. de la and Boischot, A.: 1972, *Astron. Astrophys.* **20**, 55.
Noyes, R. W.: 1973, in *Symp. on High Energy Phenomena on the Sun*, NASA GSFC Preprint, p. 231
Noyes, R. W., Withbroe, G. L., and Kirshner, R. P.: 1970, *Solar Phys.* **11**, 388.

Obayashi, T.: 1964, *Space Sci. Rev.* **3**, 79.
Obayashi, T.: 1975, *Solar Phys.* **40**, 217.
Ogawa, H. and Kawabata, K. A.: 1975, *Solar Phys.* **40**, 159.
Ogilvie, K. W. and Wilkinson, T. D.: 1969, *Solar Phys.* **8**, 435.
Ogilvie, K. W., Burlaga, L. F., and Wilkerson, T. D.: 1968, *J. Geophys. Res.* **73**, 6809.
Ohki, K.: 1969, *Solar Phys.* **7**, 260.
Öhman, Y.: 1968a, *Highlights of Astronomy*, p. 533.
Öhman, Y.: 1968b, *Solar Phys.* **3**, 354.
Öhman, Y.: 1972, *Solar Phys.* **23**, 134.
Öhman, Y. and Öhman, N.: 1953, *Observatory* **73**, 203.
Öhman, Y., Lindgren, A., and Lindgren, U.: 1962, *Arkiv Astron.* **3**, 121.
Öhman, Y., Hosinsky, G., and Kusoffsky, U.: 1968, *Nobel Symp.* **9**, 95.
Olson, C. A. and Lykoudis, P. S.: 1967, *Astrophys. J.* **150**, 303.
Oster, L. and Sofia, S.: 1966, *Astrophys. J.* **143**, 944.
Osterbrock, D. E.: 1961, *Astrophys. J.* **134**, 347.

Pallavicini, R., Vaiana, G. S., Kahler, S. W., and Krieger, A. S.: 1975, *Solar Phys.*, in press.
Palmeira, R. A. R. and Allum, F. R.: 1973, *Solar Phys.* **30**, 243.
Palmeira, R. A. R., Allum, F. R., and Rao, U. R.: 1971, *Solar Phys.* **21**, 204.
Palmer, I. D.: 1974, *Solar Phys.* **37**, 443.
Palmer, I. D. and Smerd, S. F.: 1972, *Solar Phys.* **26**, 460.
Papadopoulos, K., Goldstein, M. L., and Smith, R. A.: 1974, *Astrophys. J.* **190**, 175.
Papagiannis, M. D., Zerefos, C. S., and Repapis, C. C.: 1972, *Solar Phys.* **27**, 208.
Parker, E. N.: 1958, *Phys. Rev.* **109**, 1328.
Parker, E. N.: 1963, *Astrophys. J. Suppl.* **8**, 177.
Parker, E. N.: 1965, *Planetary Space Sci.* **13**, 9.
Parker, E. N.: 1973a, *J. Plasma Phys.* **9**, 49.
Parker, E. N.: 1973b, *Astrophys. J.* **180**, 247.
Parkinson, J. H.: 1973a, *Solar Phys.* **28**, 137.
Parkinson, J. H.: 1973b, *Solar Phys.* **28**, 487.
Parks, G. K. and Winckler, J. R.: 1971, *Solar Phys.* **16**, 186.
Payne-Scott, R., Yabsley, D. E., and Bolton, J. G.: 1947, *Nature* **64**, 697.
Peterson, L. E. and Winckler, J. R.: 1959, *J. Geophys. Res.* **64**, 697.
Peterson, L. E., Datlowe, D. W., and McKenzie, D. L.: 1973, in *Symp. on High Energy Phenomena on the Sun*, NASA GSFC Preprint, p. 132.
Petrasso, R. D., Kahler, S. W., Krieger, A. S., Silk, J. K., and Vaiana, G. S.: 1975, *Astrophys. J. Letters* **199**, L127.
Petrosian, V.: 1973, *Astrophys. J.* **186**, 291.
Petschek, H. E.: 1964, in *AAS-NASA Symp. on Physics of Solar Flares*, p. 425.
Petschek, H. E. and Thorne, K. S.: 1968, *Astrophys. J.* **147**, 1157.
Philip, K. W.: 1964, *Astrophys. J.* **139**, 723.
Phillips, K. J. H. and Neupert, W. M.: 1973, *Solar Phys.* **32**, 209.
Phillips, K. J. H., Neupert, W. M., and Thomas, R. J.: 1974, *Solar Phys.* **36**, 383.
Pick, M.: 1961, *Ann. Astrophys.* **24**, 183.
Pick, M.: 1966, *Ann. Geophys.* **22**, 310.
Piddington, J. H.: 1970, *Cosmic Electrodynamics*, J. Wiley and Sons, New York.
Piddington, J. H.: 1973, *Solar Phys.* **31**, 229.
Piddington, J. H.: 1974, *Solar Phys.* **38**, 465.
Pikelner, S. B.: 1971, *Solar Phys.* **17**, 44.
Pikelner, S. B. and Ginzburg, M. A.: 1963, *Astron. Zh.* **40**, 842.
Pintér, S.: 1969, *Solar Phys.* **8**, 149.
Pizzichini, G., Spizzichino, A., and Vespignani, G. R.: 1974, *Solar Phys.* **35**, 431.
Pneuman, G. W.: 1967, *Solar Phys.* **2**, 462.
Polupan, P. N.: 1968, *Solnechnaya Aktivnost* **3**, 125.
Polupan, P. N. and Yakovkin, N. A.: 1966, *Astron. Zh.* **42**, 764.
Pomeranz, M. A. and Duggal, S. P.: 1973, *Nature* **241**, 331.
Posener, D. W.: 1959, *Australian J. Phys.* **12**, 184.
Price, P. B.: 1973, in *Symp. on High Energy Phenomena on the Sun*, NASA GSFC Preprint, p. 377.
Price, P. B., Hutcheon, I. D., Cowsik, R., and Barber, D. J.: 1971, *Phys. Rev. Letters* **26**, 916.
Priest, E. R.: 1973, *Astrophys. J.* **181**, 227.
Priest, E. R. and Heyvaerts, J.: 1974, *Solar Phys.* **36**, 433.
Priest, E. R. and Smith, D. F.: 1972, *Astrophys. Letters* **12**, 25.
Purcel, J. D. and Widing, K. G.: 1972, *Astrophys. J.* **176**, 239.
Pustilnik, L. A.: 1973, *Astron. Zh.* **50**, 1211.

Raadu, M. A. and Kuperus, M.: 1973, *Solar Phys.* **28**, 77.
Ramaty, R.: 1968, *J. Geophys. Res.* **72**, 879.
Ramaty, R.: 1973, in *Symp. on High Energy Phenomena on the Sun*, NASA GSFC Preprint, p. 188.
Ramaty, R. and Holt, S. S.: 1970, *Nature* **226**, 68.
Ramaty, R. and Lingenfelter, R. E.: 1967, *J. Geophys. Res.* **72**, 879.
Ramaty, R. and Lingenfelter, R. E.: 1973, in *Symp. on High Energy Phenomena on the Sun*, NASA GSFC Preprint, p. 301.
Ramaty, R., Kozlovsky, B., and Lingenfelter, R. E.: 1974, X-660-74-368 GSFC Preprint.

Ramsey, H. E. and Smith, S. F.: 1965, Lockheed Rep. LR 19038.
Ramsey, H. E. and Smith, S. F.: 1966a, *Astron. J.* **71**, 197.
Ramsey, H. E. and Smith, S. F.: 1966b, Lockheed Rep. LR 20444, p. 21.
Rao, U. R., Mc Cracken, K. G., and Bukata, R. P.: 1967, *J. Geophys. Res.* **72**, 4325.
Rao, U. V. G.: 1965, *Australian J. Phys.* **18**, 283.
Razin, V. A.: 1960, *News of Higher Educational Institutions*, Radio Physics Series **3**, 73.
Redman, R. O. and Suemoto, Z.: 1954, *Monthly Notices Roy. Astron. Soc.* **114**, 524.
Reeves, H., Andouze, J., Fowler, W. A., and Schramm, D. N.: 1973, *Astrophys. J.* **179**, 909.
Reid, G. C.: 1964, *J. Geophys. Res.* **69**, 2659.
Reid, J. H.: 1959, *Observatory* **79**, 96.
Reinhard, R. and Wibberenz, G.: 1974, *Solar Phys.* **36**, 473.
Ribes, E.: 1969, *Astron. Astrophys.* **2**, 316.
Richardson, R. S.: 1936, *Annual Rep. Director Mt. Wilson Obs.* **35**, 871.
Richardson, R. S.: 1951, *Astrophys. J.* **114**, 356.
Richardson, R. S. and Minkowski, R. 1939, *Astrophys. J.* **89**, 347.
Riddle, A. C.: 1970, *Solar Phys.* **13**, 448.
Riddle, A. C. and Sheridan, K. V.: 1971, *Proc. Astron. Soc. Australia* **2**, 62.
Riddle, A. C., Tandberg-Hanssen, E., and Hansen, R. T.: 1974, *Solar Phys.* **35**, 171.
Roberts, J. A.: 1959, *Australian J. Phys.* **12**, 327.
Robinson, R. and MacQueen, R. M.: 1975, Paper presented at the AAS Meeting, Boulder, Colorado.
Roelof, E. C.: 1973, in *Symp. on High Energy Phenomena on the Sun*, NASA GSFC Preprint, p. 487.
Roelof, E. C. and Krimigis, S. M.: 1973, *J. Geophys. Res.* **78**, 5375.
Rosenberg, J.: 1970, *Astron. Astrophys.* **9**, 159.
Rosseland, S. and Tandberg-Hanssen, E. A.: 1957, *Astrophys. Norwegica* **5**, 279.
Roy, J. R.: 1972, *Solar Phys.* **26**, 418.
Roy, J. R.: 1973, *Solar Phys.* **28**, 95.
Roy, J. R. and Datlowe, D. W.: 1975, *Solar Phys.* **40**, 165.
Roy, J. R. and Tang, F.: 1975, *Solar Phys.* **42**, 425.
Rugge, H. R. and Walker, A. B. C.: 1968, *Space Res.* **8**, 439.
Rugge, H. R. and Walker, A. B. C.: 1971, *Solar Phys.* **18**, 244.
Rust, D. M.: 1968a, *Astron. J.* **73**, 75.
Rust, D. M.: 1968b, *IAU Symp.* **35**, 77.
Rust, D. M.: 1972, *Solar Phys.* **25**, 141.
Rust, D. M.: 1973a, *Solar Phys.* **33**, 205.
Rust, D. M.: 1973b, Sacramento Peak Obs. Contr. No. 209.
Rust, D. M.: 1975a, 'Optical and Magnetic Measurements of the Photosphere and Low Chromo-
    sphere', presented at the Meeting on the Physics of the Solar Atmosphere, Royal Soc. London.
Rust, D. M.: 1975b, 'Observations of Flare-Associated Magnetic Field Changes', presented at the
    Meeting on the Physics of the Solar Atmosphere, Royal Soc. London.
Rust, D. M. and Bar, V.: 1973, *Solar Phys.* **33**, 445.
Rust, D. M. and Bridges, C. A.: 1975, *Solar Phys.* (in press).
Rust, D. M. and Hegwer, F.: 1975, *Solar Phys.* **40**, 141.
Rust, D. M. and Roy, J. R.: 1971, *IAU Symp.* **43**, 569.
Rust, D. M. and Roy, J. R.: 1974, Sacramento Peak Obs. Contr. No. 221.
Rust, D. M.,Nakagawa, Y., and Neupert, W. M.: 1975, *Solar Phys.* **41**, 397.

Sagdeev, R. Z.: 1962, *Proc. Symp. on Electromagnetic and Fluid Dynamics of Gaseous Plasma*,
    Polytechnic Press, Brooklyn, New York, p. 443.
Saito, K. and Billings, D. E.: 1964, *Astrophys. J.* **140**, 760.
Sakurai, K.: 1966, *Publ. Astron. Soc. Japan* **19**, 350.
Sakurai, K.: 1967, *Rep. Iono. Space Res. Japan* **21**, 29.
Sakurai, K.: 1971, *Solar Phys.* **20**, 147.
Sakurai, K.: 1972, *Solar Phys.* **23**, 142.
Sakurai, K.: 1974, *Solar Phys.* **36**, 171.
Santangelo, N., Horstman, H., and Horstman-Moretti, E.: 1973, *Solar Phys.* **29**, 143.
Sarris, E. T. and Shawhan, S. D.: 1973, *Solar Phys.* **28**, 519.
Sawyer, C.: 1968, *IAU Symp.* **35**, 543.

Sawyer, C. and Smith, S. F.: 1970, WDC–A Report UAG – 9, p. 9.

Scalise, E.: 1970, *Publ. Astron. Soc. Japan* **22**, 483.

Schatten, K. H.: 1970, *Solar Phys.* **12**, 484.

Schatzman, E.: 1965, in C. de Jager (ed.), *The Solar Spectrum*, p. 313.

Schatzman, E.: 1967, *Solar Phys.* **1**, 411.

Schlüter, A.: 1957, *IAU Symp.* **4**, 356.

Schmidt, H. U.: 1964, in *AAS-NASA Symp. on Physics of Solar Flares*, p. 107.

Schmidt, H. U.: 1969, in C. de Jager and Z. Švestka (eds.), *Solar Flares and Space Research*, p. 331.

Schoolman, S. A.: 1973, *Solar Phys.* **32**, 379.

Schröter, E. H.: 1973, private communication.

Semel, M. and Soru-Escaut, I.: 1971, *Astron. Astrophys.* **12**, 340.

Sen, H. K. and White, M. L.: 1972, *Solar Phys.* **23**, 146.

Servajean, R. and Olivieri, G.: 1946, *L'Astronomie* **60**, 215.

Severny, A. B.: 1957, *Astron. Zh.* **34**, 684.

Severny, A. B.: 1958a, *Izv. Krymsk. Astrofiz. Obs.* **19**, 72.

Severny, A. B.: 1958b, *Izv. Krymsk. Astrofiz. Obs.* **20**, 22.

Severny, A. B.: 1959, *Izv. Krymsk. Astrofiz. Obs.* **21**, 131.

Severny, A. B.: 1960, *Izv. Krymsk. Astrofiz. Obs.* **22**, 12.

Severny, A. B.: 1962, *Astron. Zh.* **39**, 961.

Severny, A. B.: 1963, *IAU Symp.* **22**, 238.

Severny, A. B.: 1964a, *Ann. Rev. Astron. Astrophys.* **2**, 363.

Severny, A. B.: 1964b, in *AAS-NASA Symp. on Physics of Solar Flares*, p. 95.

Severny, A. B.: 1964c, *Izv. Krymsk. Astrofiz. Obs.* **31**, 159.

Severny, A. B.: 1965, *Izv. Krymsk. Astrofiz. Obs.* **33**, 34.

Severny, A. B.: 1968, *Nobel Symp.* **9**, 71.

Severny, A. B.: 1969a, *Annals IQSY* **3**, 11.

Severny, A. B.: 1969b, in C. de Jager and Z. Švestka (eds.), *Solar Flares and Space Research*, p. 38.

Severny, A. B.: 1971, *IAU Symp.* **43**, 359.

Severny, A. B. and Khokhlova, V. L.: 1959, *Izv. Krymsk. Astrofiz. Obs.* **21**, 190.

Severny, A. B. and Shaposhnikova, E. F.: 1960, *Izv. Krymsk. Astrofiz. Obs.* **24**, 235.

Severny, A. B. and Steshenko, N. V.: 1969, *Solnechno-zemnaya fizika* **1**, 3.

Severny, A. B. and Steshenko, N. V.: 1972, in C. de Jager (ed.), *Solar-Terrestrial Physics*, Part I, p. 173.

Severny, A. B., Steshenko, N. V., and Khokhlova, V. L.: 1960, *Astron. Zh.* **37**, 23.

Shaw, M. L.: 1972, *Solar Phys.* **27**, 436.

Sheridan, K. V.: 1970, *Proc. Astron. Soc. Australia* **1**, 376.

Sheridan, K. V., Trent, G. H., and Wild, J. P.: 1959, *Observatory* **79**, 51.

Shimabukuro, F. I.: 1968, *Solar Phys.* **5**, 498.

Shimabukuro, F. I.: 1972, *Solar Phys.* **23**, 169.

Shine, R. A. and Linsky, J. L.: 1974a, *Solar Phys.* **37**, 145.

Shine, R. A. and Linsky, J. L.: 1974b, *Solar Phys.* **39**, 49.

Shirk, E. K.: 1974, *Astrophys. J.* **190**, 695.

Shmeleva, O. P. and Syrovatsky, S. I.: 1973, *Solar Phys.* **33**, 341.

Silk, J. K., Kahler, S. W., Krieger, A. S., and Vaiana, G. S.: 1975, *Space Res.* **16**, in press.

Simnett, G. M.: 1971, *Solar Phys.* **20**, 448.

Simnett, G. M.: 1972a, *Solar Phys.* **22**, 189.

Simnett, G. M.: 1972b, *Space Res.* **13**, 745.

Simnett, G. M.: 1974a, *Space Sci. Rev.* **16**, 257.

Simnett, G. M.: 1974b, *Solar Phys.* **34**, 377.

Simnett, G. M. and Holt, S. S.: 1971, *Solar Phys.* **16**, 208.

Simnett, G. M., Cline, T. L., Holt, S. S., and McDonald, F. B.: 1970, *Acta Phys.* **29**, Suppl. 2, 649.

Simon, G. W. and Noyes, R. W.: 1972, *Solar Phys.* **22**, 450.

Simon, M.: 1969, *Astrophys. Letters* **3**, 23.

Simon, P. and Švestka, Z.: 1969, *Annals IQSY* **3**, 469.

Simon, P., Martres, M. J., and Legrand, J. P.: 1969, in C. de Jager and Z. Švestka (eds.), *Solar Flares and Space Research*, p. 405.

Sivaraman, K. R.: 1969, *Solar Phys.* **6**, 152.

Slonim, Y. M.: 1962, *Astron. Zh.* **39**, 798.

Slonim, Y. M.: 1969, *Astron. Zh.* **46**, 570.
Slonim, Y. M. and Korobova, Z. B.: 1975, *Solar Phys.* **40**, 397.
Slottje, C.: 1972, *Solar Phys.* **25**, 210.
Slysh, V. I.: 1967, *Astron. Zh.* **44**, 487.
Smart, D. F. and Shea, M. A.: 1971, *Solar Phys.* **16**, 484.
Smerd, S. F.: 1970, *Proc. Astron. Soc. Australia* **1**, 305.
Smerd, S. F. and Dulk, G. A.: 1971, *IAU Symp.* **43**, 616.
Smerd, S. F., Sheridan, K. V., and Stewart, R. T.: 1974, *IAU Symp.* **57**, 389.
Smith, D. F.: 1970a, *Adv. Astron. Astrophys.* **7**, 147.
Smith, D. F.: 1970b, *Solar Phys.* **15**, 202.
Smith, D. F.: 1972a, *Solar Phys.* **23**, 191.
Smith, D. F.: 1972b, *Astrophys. J.* **174**, 643.
Smith, D. F.: 1974, *IAU Symp.* **57**, 253.
Smith, D. F. and Fung, P. C. W.: 1971, *J. Plasma Phys.* **5**, 1.
Smith, D. F. and Pneuman, G. W.: 1972, *Solar Phys.* **25**, 461.
Smith, D. F. and Priest, E. R.: 1972, *Astrophys. J.* **176**, 487.
Smith, E. v. P.: 1963, *Astrophys. J.* **137**, 580.
Smith, E. v. P.: 1968, *Nobel Symp.* **9**, 137.
Smith, H. J.: 1957, Techn. Memorandum GRD-TM-57-6.
Smith, H. J.: 1962a, *GRD Research Note* AFCRL-62-827.
Smith, H. J.: 1962b, *GRD Solar Research Note* No. 58, AFCRL-472 (III).
Smith, H. J. and Smith, E. v. P.: 1963, *Solar Flares*, The Macmillan Co., New York.
Smith, S. F. and Harvey, K. L.: 1971, in C. J. Macris (ed.), *Physics of the Solar Corona*, p. 156.
Smith, S. F. and Ramsey, H. E.: 1964, *Z. Astrophys.* **60**, 1.
Smith, S. F. and Ramsey, H. E.: 1967, *Solar Phys.* **2**, 158.
Snijders, R.: 1968, *Solar Phys.* **4**, 432.
Sobolev, V. M.: 1962, *Soln. Dann.* No. 3, 69.
Somov, B. V.: 1975, *Solar Phys.* **42**, 235.
Somov, B. V. and Syrovatsky, S. I.: 1974, *Solar Phys.* **39**, 415.
Sonnerup, B. U. O.: 1973, in *Symp. on High Energy Phenomena on the Sun*, NASA GSFC Preprint, p. 357.
Spangler, Z. R. and Shawhan, S. D.: 1974, *Solar Phys.* **37**, 189.
Spicer, D. S.: 1975, Paper presented at the AAS Meeting, Boulder, Colorado.
Spicer, D. S. and Davis, J.: 1975, *Solar Phys.* **43**, 107.
Spitzer, 1962, *Physics of Fully Ionized Gases*, Interscience, New York.
Stein, W. A. and Ney, E. P.: 1963, *J. Geophys. Res.* **68**, 1605.
Stenflo, J. O.: 1973, *Solar Phys.* **32**, 41.
Stepanyan, N. N.: 1963, *Izv. Krymsk. Astrofiz. Obs.* **29**, 68.
Stepanyan, N. N.: 1969, *Astron. Zh.* **46**, 580.
Steshenko, N. V.: 1971a, *Izv. Krymsk. Astrofiz. Obs.* **44**, 130.
Steshenko, N. V.: 1971b, *Izv. Krymsk. Astrofiz. Obs.* **44**, 152.
Steshenko, N. V. and Khokhlova, V. L.: 1960a, *Izv. Krymsk. Astrofiz. Obs.* **23**, 322, and **24**, 258.
Steshenko, N. V. and Khokhlova, V. L.: 1962, *Izv. Krymsk. Astrofiz. Obs.* **27**, 120.
Stewart, R. T.: 1966, *Australian J. Phys.* **19**, 209.
Stewart, R. T. and Sheridan, K. V.: 1970, *Solar Phys.* **12**, 299.
Stewart, R. T., McCabe, M. K., Koomen, M. J., Hansen, J. T., and Dulk, G. A.: 1947a, *Solar Phys.* **36**, 203.
Stewart, R. T., Howard, R. A., Hansen, F., Gergely, T., and Kundu, M.: 1974b, *Solar Phys.* **36**, 219.
Stone, R. G. and Fainberg, J.: 1971, *Solar Phys.* **20**, 106.
Stone, R. G. and Fainberg, J.: 1973, in *Symp. on High Energy Phenomena on the Sun*, NASA GSFC Preprint, p. 519.
Strauss, F. M. and Papagiannis, M. D.: 1971, *Astrophys. J.* **164**, 369.
Sturrock, P. A.: 1961, *Nature* **192**, 58.
Sturrock, P. A.: 1964, in *AAS-NASA Symp. on Physics of Solar Flares*, p. 357.
Sturrock, P. A.: 1965, *Phys. Fluids* **8**, 281.
Sturrock, P. A.: 1966, *Phys. Rev.* **141**, 186.
Sturrock, P. A.: 1968, *IAU Symp.* **35**, 471.

Sturrock, P. A.: 1972a, *Solar Phys.* **23**, 438.

Sturrock, P. A.: 1972b, in P. S. McIntosh and M. Dryer (eds.), *Solar Activity – Observations and Predictions*, MIT Press, Cambridge, U.S.A., p. 173.

Sturrock, P. A.: 1973, in *Symp. on High Energy Phenomena on the Sun*, NASA GSFC Preprint, p. 3.

Sturrock, P. A.: 1974, *IAU Symp.* **57**, 437.

Suemoto, Z. and Hiei, E.: 1959, *Publ. Astron. Soc. Japan* **11**, 185.

Sullivan, J. D.: 1970, University of Chicago (Thesis).

Švestka, Z.: 1951, *Bull. Astron. Inst. Czech.* **2**, 100 and 120.

Švestka, Z.: 1957, *Publ. Astron. Inst. Prague*, No. 32.

Švestka, Z.: 1959, *Bull. Astron. Inst. Czech.* **10**, 10.

Švestka, Z.: 1961, *Bull. Astron. Inst. Czech.* **12**, 73.

Švestka, Z.: 1962, *Bull. Astron. Inst. Czech.* **13**, 190.

Švestka, Z.: 1963, *Bull. Astron. Inst. Czech.* **14**, 234.

Švestka, Z.: 1964a, *Bull. Astron. Inst. Czech.* **15**, 38.

Švestka, Z.: 1964b, *Bull. Astron. Inst. Czech*, **15**, 162.

Švestka, Z.: 1965, *Adv. Astron. Astrophys.* **3**, 119.

Švestka, Z.: 1966a, *Bull. Astron. Inst. Czech.* **17**, 137.

Švestka, Z.: 1966b, *Bull. Astron. Inst. Czech.* **17**, 262.

Švestka, Z.: 1966c, *Space Sci. Rev.* **5**, 388.

Švestka, Z.: 1967, *Bull. Astron. Inst. Czech.* **18**, 55.

Švestka, Z.: 1968a, *IAU Symp.* **35**, 287.

Švestka, Z.: 1968b, *Nobel Symp.* **9**, 17.

Švestka, Z.: 1968c, *Solar Phys.* **4**, 18.

Švestka, Z.: 1968d, *Solar Phys.* **4**, 361.

Švestka, Z.: 1969, *Solar Phys.* **8**, 400.

Švestka, Z.: 1970a, *Solar Phys.* **13**, 471.

Švestka, Z.: 1970b, *Space Res.* **10**, 797.

Švestka, Z.: 1971a, *Report of the Meeting of CESRA*, 15.

Švestka, Z.: 1971b, *Solar Phys.* **19**, 202.

Švestka, Z.: 1971c, *Phil. Trans. Roy. Soc. London* **A270**, 157.

Švestka, Z.: 1972a, *Ann. Rev. Astron. Astrophys.* **10**, 1.

Švestka, Z.: 1972b, in C. de Jager (ed.), *Solar-Terrestrial Physics*, Part I, p. 72.

Švestka, Z.: 1972c, *Solar Phys.* **24**, 154.

Švestka, Z.: 1973, *Solar Phys.* **31**, 389.

Švestka, Z.: 1975, *IAU Symp.* **68**, 427.

Švestka, Z. and Fritzová, L. 1967, *Solar Phys.* **2**, 75.

Švestka, Z. and Fritzová-Švestková, L.: 1974, *Solar Phys.* **36**, 417.

Švestka, Z. and Olmr, J.: 1966, *Bull. Astron. Inst. Czech.* **17**, 4.

Švestka, Z. and Simon, P.: 1969, *Solar Phys.* **10**, 3.

Švestka, Z., Kopecký, M., and Blaha, M.: 1961, *Bull. Astron. Inst. Czech.* **12**, 229.

Švestka, Z., Kopecký, M., and Blaha, M.: 1962, *Bull. Astron. Inst. Czech.* **13**, 37.

Švestka, Z., Castelli, J. P., Dizer, M., Dodson, H. W., McIntosh, P. S., and Urbarz, H.: 1974, Paper presented at the CINOF Meeting in Sao Paulo, Brazil.

Swann, W. F. G.: 1933, *Phys. Rev.* **43**, 217.

Swarup, G., Stone, P. H., and Maxwell, A.: 1960, *Astrophys. J.* **131**, 725.

Sweet, P.: 1958, *IAU Symp.* **6**, 123.

Sweet, P.: 1969, *Ann. Rev. Astron. Astrophys.* **7**, 149.

Sweet, P.: 1971, *IAU Symp.* **43**, 457.

Syrovatsky, S. I.: 1966, *Astron. Zh.* **43**, 340.

Syrovatsky, S. I.: 1969a, in C. de Jager and Z. Švestka (eds.), *Solar Flares and Space Research*, p. 346.

Syrovatsky, S. I.: 1969b, *Trudy Mezhdunar. Seminara*, Leningrad, p. 7.

Syrovatsky, S. I.: 1972, in C. de Jager (ed.), *Solar-Terrestrial Physics*, Part I, p. 119.

Syrovatsky, S. I. and Shmeleva, O. P.: 1972, *Astron. Zh.* **49**, 334.

Takakura, T.: 1959, in R. N. Bracewell (ed.), *Paris Symp. on Radio Astronomy*, p. 562.

Takakura, T.: 1960, *Publ. Astron. Soc. Japan* **12**, 55.

Takakura, T.: 1961, *Publ. Astron. Soc. Japan* **13**, 312.

Takakura, T.: 1962, *J. Phys. Soc. Japan* **17**, Suppl. A II, 243.

Takakura, T.: 1963, *Publ. Astron. Soc. Japan* **15**, 327.

Takakura, T.: 1964, in *AAS-NASA Symp. on Physics of Solar Flares*, p. 383.

Takakura, T.: 1966, *Space Sci. Rev.* **5**, 80.

Takakura, T.: 1967, *Solar Phys.* **1**, 304.

Takakura, T.: 1969a, *Solar Phys.* **6**, 133.

Takakura, T.: 1969b, in C. de Jager and Z. Švestka (eds.), *Solar Flares and Space Research*, p. 165.

Takakura, T.: 1972, *Solar Phys.* **26**, 151.

Takakura, T.: 1973, in *Symp. on High Energy Phenomena on the Sun*, NASA GSFC Preprint, p. 179.

Takakura, T. and Kai, K.: 1966, *Publ. Astron. Soc. Japan* **18**, 57.

Takakura, T. and Ono, M.: 1962, *J. Phys. Soc. Japan* **17**, Suppl. A II, 207.

Takakura, T. and Scalise, E.: 1970, *Solar Phys.* **11**, 434.

Takakura, T., Uchida, Y., and Kai, K.: 1968, *Solar Phys.* **4**, 45.

Takakura, T., Ohki, K., Shibuya, N., Fujii, M., Matsuoka, M., Miyamoto, S., Nishimura, J., Oda, M., Ogawara, Y., and Ota, S.: 1971, *Solar Phys.* **16**, 454.

Talon, R., Vendrenne, G., Melioransky, A. S., Pissarenko, N. F., Shamolin, V. M., and Likin, O.B.: 1975, *IAU Symp.* **68**, 315.

Tanaka, H.: 1969, *Solar Terrestrial Activity Chart, Jan.–June, 1969*.

Tanaka, H. and Enomé, S.: 1970, *Nature* **225**, 435.

Tanaka, H. and Enomé, S.: 1975, *Solar Phys.* **40**, 123.

Tanaka, H. and Kakinuma, T.: 1959, in R. N. Bracewell (ed.), *Paris Symp. on Radio Astronomy*, p. 215.

Tanaka, H. and Kakinuma, T.: 1961, *Proc. Res. Inst. Atmospherics* **8**, 39.

Tanaka, H. and Kakinuma, T.: 1964, *Rep. Iono. Space Res. Japan* **18**, 132.

Tanaka, H., Kakinuma, T., and Enomé, S.: 1967, *Proc. Res. Inst. Atmospherics* **14**, 23.

Tanaka, H., Kakinuma, T., and Enomé, S.: 1969, *Annals IQSY* **3**, 63.

Tanaka, K. and Nakagawa, Y.: 1973, *Solar Phys.* **33**, 187.

Tandberg-Hanssen, E.: 1963, *Astrophys. J.* **137**, 26.

Tandberg-Hanssen, E.: 1967a, *Solar Phys.* **2**, 98.

Tandberg-Hanssen, E.: 1967b, *Solar Activity*, Blaisdell Publ. Co., Waltham Mass., U.S.A.

Tandberg-Hanssen, E. and Malville, J. M.: 1974, *Solar Phys.* **39**, 107.

Teegarden, B. J., von Rosenvinge, T. T., and McDonald, F. B.: 1973, *Astrophys. J.* **180**, 571.

Teske, R. G.: 1967, *Astron. J.* **72**, 832.

Teske, R. G.: 1971a, *Solar Phys.* **17**, 76.

Teske, R. G.: 1971b, *Solar Phys.* **17**, 181.

Teske, R. G.: 1971c, *Solar Phys.* **21**, 146.

Thomas, R. J. and Neupert, W. M.: 1975, Paper presented at the AAS Meeting, Boulder, Colorado.

Thomas, R. J. and Teske, R. G.: 1971, *Solar Phys.* **16**, 431.

Thompson, A. R.: 1962, *J. Phys. Soc. Japan* **17**, Suppl. A II, 198.

Thompson, A. R. and Maxwell, A.: 1962, *Astrophys. J.* **136**, 546.

Tidman, D. A.: 1965, *Planetary Space Sci.* **13**, 781.

Tidman, D. A. and Dupree, T. H.: 1965, *Phys. Fluids* **8**, 1860.

Tidman, D. A., Birmingham, T. J., and Stainer, H. M.: 1966, *Astrophys, J.* **146**, 207.

Tindo, I. P., Ivanov, V. D., Mandelstam, S. L., and Skurygin, A. I.: 1970, *Solar Phys.* **14**, 204.

Tindo, I. P., Ivanov, V. D., Mandelstam, S. L., and Skurygin, A. I.: 1972, *Solar Phys.* **24**, 429.

Tindo, I. P., Ivanov, V. D., Valníček, B., and Livshits, M. A.: 1972, *Solar Phys.* **27**, 26.

Tindo, I. P., Mandelstam, S. L., and Shurygin, A. I.: 1973, *Solar Phys.* **32**, 469.

Tomblin, F. F.: 1972, *Astrophys. J.* **171**, 377.

Tousey, R., Bartoe, J. D. F., Bohlin, J. D., Brueckner, G. E., Purcell, J. D., Scherrer, V. E., Sheeley, N. R., Schumacher, R. J., and Vanhoosier, M. E.: 1973, *Solar Phys.* **33**, 265.

Tsap, T. T.: 1965, *Izv. Krymsk. Astrofiz. Obs.* **33**, 92.

Tsytovitch, V. N.: 1966, *Astron. Zh.* **43**, 528.

Tsytovitch, V. N.: 1967, *Non-Linear Effects in Plasma*, Nauka, Moscow.

Tucker, W. H.: 1973, *Astrophys. J.* **186**, 285.

Tucker, W. H. and Koren, M.: 1971, *Astrophys. J.* **168**, 283.

Uchida, Y.: 1960, *Publ. Astron. Soc. Japan* **12**, 376.

Uchida, Y.: 1968, *Solar Phys.* **4**, 30.
Uchida, Y.: 1974, *Solar Phys.* **39**, 431.
Uchida, Y. and Kaburaki, O.: 1974, *Solar Phys.* **35**, 451.
Uchida, Y., Altschuler, M. D., and Newkirk, G.: 1973, *Solar Phys.* **28**, 495.
Unsöld, A.: 1968, *Roy. Astron. Soc. Quart. J.* **9**, 294.

Vaiana, G. S. and Giacconi, R.: 1969, in D. G. Wentzel and D. A. Tidman (eds.), *Plasma Instabilities in Astrophysics*, p. 91.
Vaiana, G. S., Reidy, W. P., Zehnpfennig, T., Van Speybroeck, L., and Giacconi, R.: 1968, *Science* **161**, 564.
Vaiana, G. S., Krieger, A. S., and Timothy, A. F.: 1973, *Solar Phys.* **32**, 81.
Valdes, J. and Altschuler, M. D.: 1970, *Solar Phys.* **15**, 446.
Valníček, B.: 1961, *Bull. Astron. Inst. Czech.* **12**, 237.
Valníček, B.: 1962, *Bull. Astron. Inst. Czech.* **13**, 91.
Valníček, B., Letfus, V., Blaha, M., Švestka, Z., and Seidl, Z.: 1959, *Bull. Astron. Inst. Czech.* **10**, 149.
Van Allen, J. A. and Krimigis, S. M.: 1965, *J. Geophys. Res.* **70**, 5737.
Van Allen, J. A. and Ness, N. F.: 1967, *J. Geophys. Res.* **72**, 935.
Van Beek, H. F., De Feiter, L. D., and de Jager, C.: 1974, *Space Res.* **15**, in press.
Van de Hulst, H. C.: 1950, *Bull. Astron. Inst. Neth.* **11**, 135.
Van Hollebeke, M. A. I., Ma Sung, L. S., and McDonald, F. B.: 1975, *Solar Phys.* **41**, 189.
Vernov, S. N. and Lyubimov, G. P.: 1972, in E. R. Dyer *et al.* (eds.), *Solar-Terrestrial Physics*, Part II, p. 92.
Vernov, S. N., Chudakov, A. E., Vakulov, P. V., Gorchakov, E. V., Kontor, N. N., Logachev, Y. I., Lyubimov, G. P., Peresligina, N. V., and Timofeev, G. A.: 1970, in V. Manno and D. E. Page (eds.), *Intercorrelated Satellite Observations Related to Solar Events*, p. 53.
Verzariu, P. and Krimigis, S. M.: 1972, *J. Geophys. Res.* **77**, 3985.
Vitinskij, J. I.: 1969a, *Soln. Dann.* No. 4, 88.
Vitinskij, J. I.: 1969b, *Solar Phys.* **7**, 210.
Vorpahl, J. A.: 1972, *Solar Phys.* **26**, 397.
Vorpahl, J. A.: 1973a, *Solar Phys.* **28**, 115.
Vorpahl, J. A.: 1973b, *Solar Phys.* **29**, 447.
Vorpahl, J. A.: 1975, *Astrophys. J.*, in press.
Vorpahl, J. A. and Pope, T.: 1972, *Solar Phys.* **25**, 347.
Vorpahl, J. A. and Takakura, T.: 1974, *Astrophys. J.* **191**, 563.
Vorpahl, J. A. and Zirin, H.: 1970, *Solar Phys.* **11**, 285.
Vorpahl, J. A., Gibson, E. G., Landecker, P. B., McKenzie, D. L., and Underwood, J. H.: 1975, *Solar Phys.*, in press.
Vrabec, C.: 1973, *IAU Symp.* **56**, 201.

Wagner, W. J., Hansen, R. T., and Hansen, S. F.: 1974, *Solar Phys.* **34** 453.
Waldmeier, M.: 1938, *Z. Astrophys.* **15**, 299.
Waldmeier, M.: 1945, *Astron. Mitt. Zürich*, No. 146.
Waldmeier, M.: 1947, *Astron. Mitt. Zürich*, No. 151.
Waldmeier, M.: 1955, *Ergebnisse und Probleme der Sonnenforschung*, Akad. Verlagsgesellschaft, Leipzig, p. 246.
Waldmeier, M.: 1956–1970, Definite Sunspot Numbers, *Bull. Eidg. Sternw. Zürich*.
Waldmeier, M.: 1960, *Z. Astrophys.* **51**, 1.
Waldmeier, M.: 1963, *Z. Astrophys.* **56**, 291.
Waldmeier, M.: 1973, *Solar Phys.* **30**, 129.
Walker, A. B. C.: 1972, *Space Sci. Rev.* **13**, 672.
Walker, A. B. C. and Rugge, H. R.: 1969, in C. de Jager and Z. Švestka (eds.), *Solar Flares and Space Research*, p. 102.
Walker, A. B. C. and Rugge, H. R.: 1970, *Astron. Astrophys.* **5**, 4.
Walker, A. B. C. and Rugge, H. R.: 1971, *Astrophys. J.* **164**, 181.
Walker, A. B. C., Rugge, H. R., and Weiss, K.: 1974, *Astrophys. J.* **188**, 423, and **192**, 169.
Wang, H. T. and Ramaty, R.: 1974, *Solar Phys.* **36**, 129.
Warwick, C. S.: 1965, *Astrophys. J.* **141**, 500.

Warwick, C. S.: 1966, *Astrophys. J.* **145**, 215.
Warwick, C. S. and Haurwitz, M. W.: 1962, *J. Geophys. Res.* **67**, 1317.
Warwick, C. S. and Wood, M.: 1959, *Astrophys. J.* **129**, 801.
Warwick, J. W.: 1957, *Astrophys. J.* **125**, 811.
Warwick, J. W.: 1962, *Publ. Astron. Soc. Pacific* **74**, 302.
Warwick, J. W.: 1965, in J. Aarons (ed.), *Solar System Radio Astronomy*, p. 131.
Warwick, J. W.: 1973, in *Symp. on High Energy Phenomena on the Sun*, NASA GSFC Preprint, p. 625.
Webber, W. R.: 1964, in *AAS-NASA Symp. on Physics of Solar Flares*, p. 215.
Weiss, A. A.: 1963a, *Australian J. Phys.* **16**, 240.
Weiss, A. A.: 1963b, *Australian J. Phys.* **16**, 526.
Weiss, A. A.: 1965, *Australian J. Phys.* **18**, 167.
Weiss A. A. and Sheridan K. V.: 1962, *J. Phys. Soc. Japan* **17**, Suppl. A II, 223.
Weiss, A. A. and Stewart, R. T.: 1965, *Australian J. Phys.* **18**, 143.
Welly, J. D.: 1963, *Z. Naturforsch.* **18A**, 1157.
Wentzel, D. G.: 1963, *Astrophys. J.* **137**, 135.
Wentzel, D. G.: 1974, *Solar Phys.* **39**, 129.
Westfold, K. C.: 1957, *Phil. Mag.* **2**, 1287.
Westin, H.: 1969, *Solar Phys.* **7**, 393.
White, K. P. III and Janssens, T. J.: 1970, *Solar Phys.* **11**, 291.
Wibberentz, G. and Witte, M.: 1970, in V. Manno and D. E. Page (eds.), *Intercorrelated Satellite Observations Related to Solar Events*, p. 499.
Widing, K. G. and Cheng, C. C.: 1974, *Astrophys. J.* **194**, L111.
Wiehr, E.: 1972, *Solar Phys.* **24**, 129.
Wild, J. P.: 1950, *Australian J. Sci. Res.* **A3**, 399.
Wild, J. P.: 1962, *J. Phys. Soc. Japan* **17**, Suppl. A II, 249.
Wild, J. P.: 1964, in *AAS-NASA Symp. on Physics of Solar Flares*, p. 161.
Wild, J. P.: 1968, in D. G. Wentzel and D. A. Tidman (eds.), *Plasma Instabilities in Astrophysics*, p. 119.
Wild, J. P.: 1969, *Solar Phys.* **9**, 260.
Wild, J. P.: 1973, in *Symp. on High Energy Phenomena on the Sun*, NASA GSFC Preprint, p. 589.
Wild, J. P. and McCready, L. L.: 1950, *Australian J. Sci. Res.* **A3**, 387.
Wild, J. P. and Smerd, S. F.: 1972, *Ann. Rev. Astron. Astrophys.* **10**, 159.
Wild, J. P., Murray, J. D., and Rowe, W. C.: 1953, *Nature* **172**, 533.
Wild, J. P., Roberts, J. A., and Murray, J. D.: 1954, *Nature* **173**, 532.
Wild, J. P., Sheridan, K. V., and Trent, G. H.: 1959a, in R. N. Bracewell (ed.), *Paris Symp. on Radio Astronomy*, p. 176.
Wild, J. P., Sheridan, K. V., and Noylan, A. A.: 1959b, *Australian J. Phys.* **12**, 369.
Wild, J. P., Sheridan, K. V., and Kai, K.: 1968, *Nature* **218**, 536.
Winckler, J. R.: 1963, in D. P. LeGalley (ed.), *Space Science*, Chapter 11.
Withbroe, G. L.: 1971, in the *Menzel Symposium on Solar Physics, Atomic Spectra, and Gaseous Nebulae*, NBS Spec. Publ. No. 353, p. 127.
Wood, A. T. and Noyes, R. W.: 1972, *Solar Phys.* **24**, 180.
Wood, A. T., Noyes, R. W., Dupree, A. K., Huber, M. C. E., Parkinson, W. H., Reeves, E. M., and Withbroe, G. L.: 1972, *Solar Phys.* **24**, 169.

Yajima, S.: 1971, *Tokyo Astr. Bull.* No. 207.
Yeh, T. and Axford, W. I.: 1970, *J. Plasma Phys.* **4**, 207.
Yoshimura, H., Tanaka, K., Shimizu, M., and Hiei, E.: 1971, *Publ. Astron. Soc. Japan* **23**, 443.
Young, C. W., Spencer, C. L., Moreton, G. E., and Roberts, J. A.: 1961, *Astrophys. J.* **133**, 243.
Yurovsky, Y. F. and Babin, A. N.: 1970, *Izv. Krymsk. Astrofiz. Obs.* **41**, 53.

Zaitsev, V. V.: 1965, *Astron. Zh.* **42**, 740.
Zaitsev, V. V.: 1968, *Astron. Zh.* **45**, 766.
Zaitsev, V. V., Mityakov, N. A., and Rapoport, V. O.: 1972, *Solar Phys.* **24**, 444.
Zanstra, H.: 1950, *Astron. Inst. Univ. Amsterdam Circ.* No. 1.
Zaumen, W. T. and Acton, L. W.: 1974, *Solar Phys.* **36**, 139.
Zheleznyakov, V. V.: 1965, *Astron. Zh.* **42**, 244.

Zheleznyakov, V. V. and Zaitsev, V. V.: 1970a, *Astron. Zh.* **47**, 60.
Zheleznyakov, V. V. and Zaitsev, V. V.: 1970b, *Astron. Zh.* **47**, 308.
Zirin, H.: 1964, *Astrophys. J.* **140**, 1216.
Zirin, H.: 1966, *The Solar Atmosphere*, Blaisdell Publ. Co., Waltham, Massachusetts, U.S.A.
Zirin, H.: 1972, *Solar Phys.* **26**, 393.
Zirin, H. and Acton, L. W.: 1967, *Astrophys. J.* **148**, 501.
Zirin, H. and Lazareff, B.: 1975, *Solar Phys.* **41**, 425.
Zirin, H. and Russo Lackner, D.: 1969, *Solar Phys.* **6**, 86.
Zirin, H. and Severny, A. B.: 1961, *Observatory* **81**, 155.
Zirin, H. and Tanaka, K.: 1972, private communication to D. M. Rust.
Zirin, H. and Tanaka, K.: 1973, *Solar Phys.* **32**, 173.
Zirin, H. and Werner, S.: 1967, *Solar Phys.* **1**, 66.
Zirin, H., Pruss, G., and Vorpahl, J.: 1971, *Solar Phys.* **19**, 463.
Zvereva, A. M. and Severny, A. B.: 1970, *Izv. Krymsk. Astrofiz. Obs.* **41**, 97.

INDEX OF SUBJECTS

GEOPHYSICS AND ASTROPHYSICS MONOGRAPHS

AN INTERNATIONAL SERIES OF FUNDAMENTAL TEXTBOOKS